AF389100

Vibration Energy Harvesting for Wireless Sensors

Vibration Energy Harvesting for Wireless Sensors

Editors

Zdenek Hadas
Saša Zelenika
Vikram Pakrashi

MDPI • Basel • Beijing • Wuhan • Barcelona • Belgrade • Manchester • Tokyo • Cluj • Tianjin

Editors
Zdenek Hadas
Brno University of
Technology
Czech Republic

Saša Zelenika
University of Rijeka
Croatia

Vikram Pakrashi
University College Dublin
Ireland

Editorial Office
MDPI
St. Alban-Anlage 66
4052 Basel, Switzerland

This is a reprint of articles from the Special Issue published online in the open access journal *Sensors* (ISSN 1424-8220) (available at: https://www.mdpi.com/journal/sensors/special_issues/veh_wsensors).

For citation purposes, cite each article independently as indicated on the article page online and as indicated below:

LastName, A.A.; LastName, B.B.; LastName, C.C. Article Title. *Journal Name* **Year**, *Volume Number*, Page Range.

ISBN 978-3-0365-4463-2 (Hbk)
ISBN 978-3-0365-4464-9 (PDF)

Contents

About the Editors

Zdenek Hadas

Zdenek Hadas attained his M.Sc. degree at the Brno University of Technology in 2003. He defended his PhD thesis in vibration energy harvesting in 2007. Since 2015, he has been an Associate Professor in the Faculty of Mechanical Engineering, Brno University of Technology, Czech Republic. He is active in the field of dynamic mechanical and electro-mechanical systems, multidisciplinary simulation, metamaterials and development of kinetic-energy-harvesting devices and smart sensing systems. He has authored more than 120 scientific publications and 4 patents and actively participated in 6 EU FP and many national research projects.

Saša Zelenika

Saša Zelenika attained his M.Sc. degree at the University of Rijeka, Croatia, and the D.Sc. degree at the Polytechnic University of Turin, Italy. He was Head of Mechanical Engineering at the Paul Scherrer Institute in Switzerland. Since 2004, he has been a faculty member of the University of Rijeka, Faculty of Engineering (since 2015, full professor with tenure), where he was Dean's Assistant, Department Head and is Laboratory Head. From 2012 to 2014, he was Assistant and then Deputy Minister at the Croatian Ministry of Science, Education and Sports. Currently, he is Vice-rector for Strategic Projects and Deputy Head of the Centre for Micro- and Nanoscience and Technologies at the University of Rijeka. Since 2016, he has been a Member of the Croatian Academy of Technical Sciences. His research interests encompass precision engineering and microsystems technologies. He has authored more than 170 scientific and professional publications and actively participated in 8 EU FP and several international and national scientific projects.

Vikram Pakrashi

Vikram Pakrashi (BEng, PhD, C.ENG MIEI) is an Associate Professor in the School of Mechanical and Materials Engineering, University College Dublin, Ireland. He is the Director of the UCD Centre for Mechanics. Vikram is active in the field of dynamical systems, where he assimilates fundamental theory, sensors and analysis for applications in a range of engineering sectors, typically focusing on the topic of infrastructure systems. Vikram is the President of Eurostruct (European Association on Quality Control of Bridges and Structures) and a member of the Engineering and Computer Sciences Multidisciplinary Committee of the Royal Irish Academy. Vikram runs a happy and inquisitive research group and has published more than 120 journal papers around his topic of research. He works closely with industries and has been one of the champions of using energy harvesting techniques for Structural Health Monitoring.

Preface to "Vibration Energy Harvesting for Wireless Sensors"

Mechanical vibrations occur in most technical systems in operation. High level vibrations could indicate an overloaded or damaged technical system; these states and behaviours can be monitored or reported. The ambient vibrations may, in turn, be used as a source of energy. The vibrational energy harvesters used in this frame could be an alternative for the supply of low-power autonomous electronic systems for remote sensing of operations. However, a level of energy harvested by using such an approach is usually very low and the whole concept of vibration energy harvesting system operations (including power management electronics and wireless sensors) must be adapted for factual target applications.

This special issue, through twelve diverse contributions, intends to capture some of the contemporary challenges, solutions and insights around the outlined issues and provides an overview on this rapidly evolving topic. The papers collected in the SI represent the variety of energy harvesting sources, as well as the need to create numerical and experimental evidence bases around them. Qualifying and quantifying their performances in terms of energy harvesting levels, as well as their consistency and potential applications, are all reflected in the papers. The importance of fundamental understanding of the ensuing sensors, along with their possible integration within the respective application areas, including those related to the effective communication of measurement data, are also emphasized through this special issue.

In the SI Mech et. al present magnetomechanical harvesting and related possi-bility of data transfer and subsequently investigate a novel hybrid solution focusing on rapid demagnetization [2]. Hadas et al. [3] investigate an electromagnetic harvester for railway applications, where the energy is transferred from the vibration of the rails during their operational conditions, thereby leading to applications that can encompass monitoring. Litak et al. [4] on the other hand focus on the fundamentals of energy harvesting and investigate bifurcation aspects due to nonlinearities in them, whereby different domains of harvesting exist. In particular, the impact of hysteresis is ana-lysed, which should be considered in more detail by the energy harvesting community. Koszewnik et al. [5] demonstrate a smart beam with Macro Fiber Composites (MFC) and demonstrate numerically and experimentally how energy harvesting can be used for damage detection in this field. Machu et al. [6] further investigate the design of en-ergy harvesting-powered sensors through various configurations, creating experi-mental verifications of analytical models, which creates the possibility of developing robust models with minimized computational effort, in-keeping with fundamental physics. Kunz et al [7] focus their efforts towards novel methods to assess the performance of these harvesters in terms of power flow. Bae and Kim [8], on the other hand, approach sensor performance issues in terms of load resistance optimisation for a bi-stable system. Okosun et al. [9] address one of the core sustainable development goals of availability of drinking water and experimentally demonstrate how energy harvesting patches can be used for pipeline leak detection, creating a respective benchmark. The topic of experimental validation is continued in Chen et al. [10] for impact driven harvesters in a magnetic field; for such harvesters, the source can be important, and a comparison between piezoelectric and triboelectric harvesting of en-ergy is investigated by Thainiramit et al. [11]. Finally, Phan et al. [12] provide a short and impactful investigation of electromagnetic harvesters with linear and nonlinear springs.

The dynamism and breadth of the sector is clearly observed in the contributions to this special issue, as is the variety of approaches that are available. We expect the considered sector to move further in an interdisciplinary manner in the near future, giving rise to new sensors, methods of measurement and impactful applications around a range of sectors, established through scientific rigour, along with numerical and experimental benchmarks.

Zdenek Hadas, Saša Zelenika, and Vikram Pakrashi
Editors

Article

Use of Magnetomechanical Effect for Energy Harvesting and Data Transfer

Rafał Mech *, Przemysław Wiewiórski and Karol Wachtarczyk

Faculty of Mechanical Engineering, Wroclaw University of Science and Technology, 50-370 Wroclaw, Poland; przemyslaw.wiewiorski@pwr.edu.pl (P.W.); karol.wachtarczyk@pwr.edu.pl (K.W.)
* Correspondence: rafal.mech@pwr.edu.pl; Tel.: +48-(71)-3202899

Abstract: The presented paper describes a method where, with the use of a dedicated SMART Ultrasonic Resonant Power System (SURPS) developed by the authors, a power and data transfer between two devices can be performed at the same time. The proposed solution allows power to be supplied to the sensor, located in a hardly accessible place, with simultaneous data transfer in a half-duplex way (e.g., "question–response"). The power transmission mechanism is based on the excitation of a construction with a sinusoidal wave, with an actuator transforming this wave into useful, electrical power through a harvester device. Data transfer is achieved with the use of the F2F (Frequency Double Frequency) procedure, which is a kind of frequency modulation. To receive optimized parameters for each construction, an original software is developed, which allows the selection of the proper type of actuator, modulation, and frequency.

Keywords: smart materials; magnetostriction; Terfenol-D; wireless sensors; ultrasonic system

Citation: Mech, R.; Wiewiórski, P.; Wachtarczyk, K. Use of Magnetomechanical Effect for Energy Harvesting and Data Transfer. *Sensors* **2022**, *22*, 3304. https://doi.org/10.3390/s22093304

Academic Editors: Vikram Pakrashi, Zdenek Hadas and Saša Zelenika

Received: 16 March 2022
Accepted: 20 April 2022
Published: 26 April 2022

Publisher's Note: MDPI stays neutral with regard to jurisdictional claims in published maps and institutional affiliations.

1. Introduction

In the past few decades, the development of wearable and wireless devices has been growing significantly. It became possible to reduce the power which is needed to supply these devices to only tens of milliwatts [1]. At those power levels, traditional batteries are limited to only short-term operation, mainly due to dimension limitations. Additionally, in the case of long-term operation, batteries need to be replaced or recharged while, at the same time, undergoing degradation. Meanwhile, other components behind wearable and portable devices improved rapidly following Moore's law [2]. To overcome the problem of traditional batteries, researchers started to work on energy harvesting. This is a technique that can extract electrical power from ambient sources and might supplement and even replace batteries.

Energy Harvesting (EH), originally known as power harvesting or energy scavenging, is a set of techniques that provide electrical energy through energy conversion from different sources, such as mechanical, thermal, solar, and electromagnetic energy and salinity gradients, etc., e.g., [3]. Generally, the main goal is to use sources that are commonly available in the environment, which, in most cases, are undesirable and suppressed (e.g., noise, impact, and mechanical vibration from equipment and constructions and different sources of heat from friction or combustion or as a result of electric current flow and engine cooling, etc.). Energy harvesting is also based on commonly available energy sources (solar light, wave energy, salinity differences, and biochemical processes, e.g., plants), as well as on energy connected with human biology (motion, body heat, etc.). Nowadays, it is said that EH might be a useful source of "low-cost or cost-free" (excluding installation costs) power supply to low-power electric devices [4–8]. Currently, many types of research are being carried out in relation to vast energy harvester networks which provide a relatively large amount of energy in a short time.

One of the sources of wasted energy is structural vibrations. In many cases, it can be a consistent source of energy, even though its amplitude and frequency can vary significantly, depending on the location. Vibrations in civil engineering constructions, such as buildings or bridges, have low amplitudes and frequencies (0.1 g and 0.1 Hz); at the same time, various small electrical devices, such as ovens, microwaves, and others, have higher amplitudes and frequencies (0.5 g and about 150 Hz, respectively) [9]. In other constructions, such as cars, planes, or helicopters, vibration amplitudes are relatively high, while amplitudes are varied dependent on operation conditions [10,11]. The above conditions have inspired multiple types of harvester to be developed and described in the literature.

Two types of harvester device can be distinguished, i.e., passive materials and active materials. Those which are passive-type harvesters can be divided into electromagnetic and electrostatic. The electromagnetic devices use Faraday's law, and they are built from a coil and permanent magnet. The relative motion of these two elements generates AC voltage [12,13]. The electrostatic devices are variable capacitors with movable electrodes and a dielectric layer between them. The motion of the layers caused by the vibrations induces AC currents [14–17]. In the case of active harvesters, most devices are based on magnetostrictive or piezoelectric (PZT) materials [18,19]. At this point, it should be noted that piezoelectric harvesters are capacitive sources of energy; therefore, they have a high output impedance. This implies that appropriate energy management circuits must be used to be able to supply electrical devices. On the other hand, there are magnetostrictive harvesters, which are inductive. Thus, they can provide low impedance at frequencies characteristic of most common sources of vibrations (described above).

Of the passive vibration energy harvesters, magnetostrictive harvesters supply higher energy density. What is more, a comparison of magnetostrictive devices with those based on piezoelectric material showed that both can generate similar levels of output energy; however, there is no need for additional special power management circuits in the case of solutions based on magnetostrictive material. Among the magnetostrictive devices, the two most common types are the axial type and bending type, based on the stress state in the material. The axial-type devices are usually mounted in places where a large excitation force is provided [20–28]. Because of these high loads, they can generate relatively high power density levels up to even 10 W/cm^{-3} [23]. However, to protect core material from damage, proper mechanisms of protection are needed, especially in the case of brittle Terfenol-D [23,24]. Contrary to axial-type harvesters, the bending type of vibration energy harvesting device can be mounted on any source of vibrations [29–33]. However, the output power density is much lower, and the typical level is 10 mW/cm^{-3} [30,32]. Additionally, bending harvesters can be divided into three different types: a single-layer magnetostrictive beam, the output of which is relatively small [29,30]; a double layer, the output of which is greater than for single layer [34] but is still relatively small, mainly because of dominant shear stresses; and a composite magnetostrictive beam [31,32], which provides the highest amount of energy of these three types but requires further investigations. It should be noted that, in the case of magnetostrictive harvesters, their efficiency varies and depends on many factors, such as load, frequency of operation, method of mounting, or the material on which the given device is based. In study [32], it was seen that the efficiency of the energy conversion of the proposed device was 16% at 395 Hz; however, in [31], the maximum conversion efficiency was 35%, and it was achieved at an input frequency of 202 Hz.

In the literature on the subject, it can be seen that the research focused mainly on piezoelectric transducers [5–7,35,36]. However, it turns out that, in some cases, a better solution is the use of magnetostrictive harvesters [37]. Taking into account the previous research conducted in this area, the main goals of this research are:

- development of a system that allows the transmission of energy and information through a solid in the case of ultrasonic frequencies (the system should operate at frequencies above 20 kHz, i.e., inaudible to most people);
- use of smart materials (piezoelectric and magnetostrictive);
- development of the test stands which allow the determination of the operating parameters of the device.

In this work, the axial-type harvester is presented. Such a harvester was chosen after analysis of already developed harvesters, which can be found in the literature. Additionally, such a solution was also connected with the predicted amount of energy which might be generated with the use of this type of harvester. It can be seen, based on the literature research, that higher amounts of energy can be obtained from axial-type harvesters.

2. Materials and Methods

The results presented in the paper are related to the magnetomechanical effect. The main material used in the research was Terfenol-D (a material with so-called giant magnetostriction). Additionally, the research used material prepared by the authors, which was produced by the suction casting method.

The prepared material was $Fe_{57}Co_{10}B_{20}Si_5Nb_4V_4$, which is given using atomic notation. The elements used for the preparation of the alloy were melted into uniform material with the use of a laboratory furnace. High accuracy of chemical composition was obtained by using elements of very high purity and using a scale with high accuracy. The weighted elements were mixed and placed in the furnace to create alloy material. After melting the elements three times in the argon atmosphere, the obtained alloy was again heated to the temperature of the liquid and then rapidly sucked into the copper mold, which was constantly water-cooled. The materials prepared with this method were in form of rods.

To achieve the main goal set by the authors, the design of a multiphase magnetostrictive actuator was developed. This device served as an actuator of a multiphase mechanical vibration regulator, which allowed the positioning of objects that are freely placed on the beam. Positioning was performed using mechanical vibrations. It was assumed that the proposed solution will make it possible to supply energy in a controlled manner and remotely perform mechanical work (e.g., displacement in a given direction, rotation of a mechanism, unscrewing an element by pressing a switch). The described process should be possible to carry out with the use of vibrations up to 30 kHz. To accomplish this task with the use of magnetostrictive actuators, the idea of generating phase-shifted vibrations that have the same frequency was implemented. The implementation of this idea consisted of switching on the actuators, which were placed one after the other, and the signals sent by them were shifted relative to each other by an appropriate angle, which was controlled by the developed algorithm. The whole system used feedback obtained from a vibration sensor, thanks to which the presented method allowed the use of the so-called Structural Stiffness Code (CSS).

2.1. Magnetostrictive Actuators

To be able to generate mechanical vibrations, first, it is necessary to understand the operation of the different types of actuator. The selection of an actuator for a given task depends on the vibration level (PSD—Power Spectral Density), the operating frequency, and the level of amplitude-phase distortions. When we consider devices without moving elements, i.e., solid-state devices, we limit the choice between two types of transducer, namely, piezoelectric (PZT) and magnetostrictive (usually based on Terfenol-D). In the case of piezoelectric devices, it should be noted that they only work in a strictly defined range of resonance frequencies. This range is related to the design of the transducer. In the case of the operation of such devices outside their scope, the actuator overheats, and the ceramic material is usually destroyed. In the case of actuators based on a magnetostrictive material, the frequency range in which they can operate is much wider compared to PZT actuators. The operating frequency of the magnetostrictive actuator can reach 100 kHz [38] and also includes the resonant frequencies of the actuator itself [39]. Additionally, such actuators generate large forces, which, depending on the design and size of the device, can reach up to several hundred kN. Despite the above advantages, these actuators also have some drawbacks, the most important of which are the non-linear operating characteristics, low vibration amplitude, and the limited operating temperature associated with the use of an induction coil. Additionally, the price of such actuators is relatively high when compared to devices based on PZT. However, due to the abovementioned advantages, it was decided to use magnetostrictive actuators in the solution presented in this paper.

2.2. Magnetostrictive Harvesters

In the case of magnetostrictive harvester actuators, the most important element is the magnetostrictive core. Such a core may consist of one or more elements, depending on the size, length, and purpose of the device. The material from which the core is made is also an important issue. Such material must be characterized by gigantic magnetostriction (GMM— Giant Magnetostrictive Materials, e.g., Terfenol-D, nano-ferrite cobalt). An additional element is a system that allows you to adjust the pre-magnetization of the material, which are usually properly selected neodymium magnets. The number of elements that make up the core of the device has a significant impact on the frequency of operation. The smaller the number of elements, the higher the operating frequency can be. To optimize the construction of the actuator, the magnetostrictive material and neodymium magnets are arranged alternately.

In the case of magnetostrictive materials, the parameters of the physical fields that affect the material from the outside, i.e., the magnetic field and also the mechanical field, are of great importance. Therefore, to obtain the best working parameters, permanent magnets and appropriate spring systems are used. The springs are used to create initial stress in the magnetostrictive material, the appropriate level of which makes it possible to increase the amplitude of work. In addition, the main task of the permanent magnets is to shift the starting point of the material's operation, thanks to which such devices can work in the linear range of their characteristics.

The actuators/harvesters presented in this paper have relatively large dimensions: the diameter of the device was 44 mm and its height was 47 mm. The geometry was forced mainly by the necessity of applying the appropriate actuator pressure to the tested structure. Inside the casing was a coil with a resistance $R_{coil} = 5.5\ \Omega$. The devices worked in a wide frequency range from 10 Hz to 30 kHz to find the resonant frequency of the system within which the system obtained the highest voltage values. Additionally, both harvesters and actuators were pre-stressed with a force of 400 N. This value was determined based on experimental tests, during which the magnetomechanical response of the system depending on the applied load was determined.

The above information is based on the authors' extensive experience in the construction and modernization of magnetostrictive actuators [38]. Subsequent constructions and modifications allow more and more power to be obtained; therefore, these devices began to be used as a source of electricity (Figure 1). The amount of electricity, i.e., the values of current and voltage, must be appropriate for the power supply of the sensor and the built-in processor (with a matched converter) and the communication system. In addition to these parameters, the generator voltage/current conditioning system is equally important. Additionally, to properly design the harvester's electrical circuits, it is necessary to know the characteristics of the receiving device.

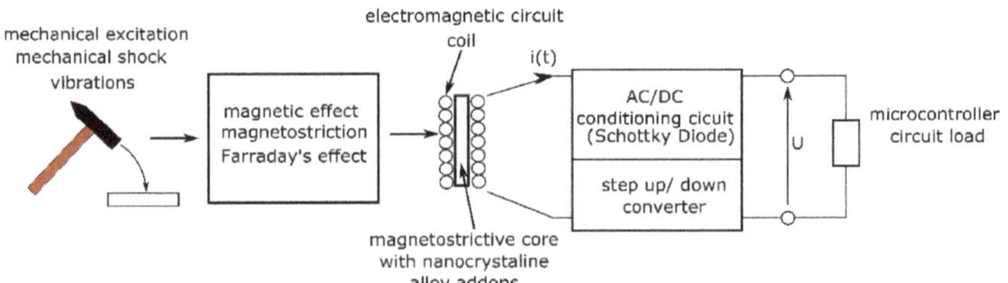

Figure 1. Harvester scheme.

2.3. Electric Transducer

Harvesters can be divided into direct current and alternating current. DC harvesters include devices that use the thermoelectric or photovoltaic effect. On the other hand, AC harvesters are devices that generate energy from vibrations (magnetostrictive, piezoelectric, and the Faraday effect). Harvesters that use a mechanical impulse (shock) to generate energy are a special case [38]. In the event of an impact, electricity is produced and available for a very short time. Impact harvesters are characterized by a strong current pulse that appears in the form of an alternating current. At the same time, in the generated signal, one can observe the frequencies related to the magnetic resonance between the core and the coil of the device. Harvesters of this type differ in the amount of magnetostrictive material and their form (solid, composite, powdered) in the magnetic circuit. The power that can be obtained from this type of harvester depends on the type of material, layout, and dimensions of the core.

3. Results

3.1. Multiphase Actuator

Mechanical vibrations in a wide frequency range were generated with the use of a multiphase magnetostrictive actuator. The system also had an integrated sensory part. The control of the system was carried out using the HERON Advanced Multiphase software. This software allowed us to supply the actuator with control signals, which kept the whole structure in resonance. The vibration controller used an electronic system containing DSP (Digital Signal Processor) and measurement modules, i.e., input–output modules, DDS (Direct Digital Synthesis) generators, and an ICP (Integrated Electronics Piezoelectric) sensor from PCB Piezotronics. A dedicated CDM-P1 device was responsible for conditioning the signal and changing the vibration amplitude.

To generate the so-called multiphase vibrations, it is necessary to use a head that contains many magnetostrictive actuators. The research presented in the paper was carried out with the use of a head consisting of four actuators equipped with Terfenol-D cores and a nanocrystalline alloy prepared by the authors. Such an arrangement of actuators allowed for their analog control similar to that of a typical stepper motor. In addition, a PCB-type vibration sensor, which was located at the central point of the system, was used to measure the vibration values in the tested object. A diagram of the arrangement of actuators with the PCB sensor is presented in Figure 2.

The prototype head was manufactured according to the design shown in Figure 2. The actuators were arranged symmetrically (Figure 2b). In addition, Figure 2c schematically shows the positioning system that used the generated vibrations for a straight rail. This circuit worked with a feedback loop, which is described below.

Figure 3 shows the vibration control system with the designed head. Importantly, in the case of this system, it was possible to control each actuator separately, thanks to which it was possible to generate even very complex mechanical vibrations. In addition to the designed head, the vibration controller consisted of many advanced electronic systems based on the HERON card with a floating-point DSP and the Texas Instruments C6000 card with expansion modules. Moreover, the system was supplemented with dedicated software using API.

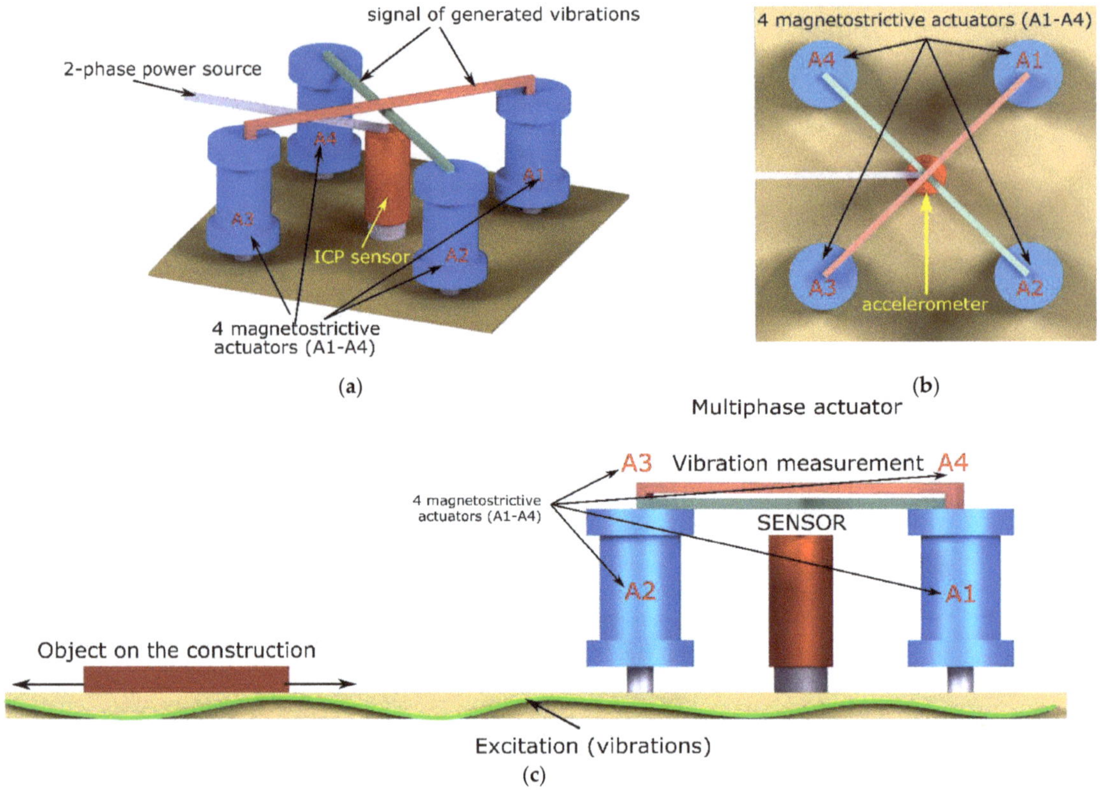

Figure 2. The idea of the four magnetostrictive actuators (A1–A4) system: (**a**) actuators operating simultaneously with vibration measurement path; (**b**) symmetrically distributed actuators; (**c**) linear positioning system using vibration.

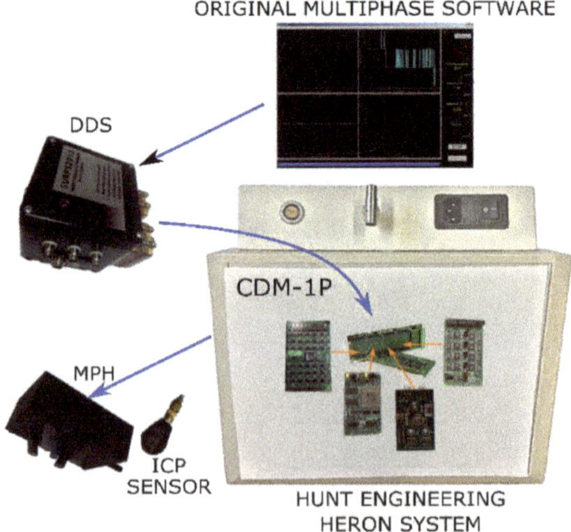

Figure 3. Elements of the control system based on Hunt Engineering DSP.

The Hunt Engineering HERON system was responsible for signal processing in the system. Thanks to the use of this system, it was possible to comprehensively design the experiment, because the software allowed for the acquisition and conditioning of the sensor signal and the generation of phase-shifted control signals. For this purpose, both a DAC (Digital-to-Analog Converter) and a DDS were used.

The PCB sensor received vibrations in the form of electrical signals and then, after being appropriately supplied, they were collected by a module of 16-bit ADCs-HEGD12. They were used as the basis for the determination of the setpoints of the digital vibration controller which released subsequent DDS values in the feedback loop through HEGD4. The role of CDM-1P was only the conditioning of sensor signals and power amplification for the four magnetostrictive actuators.

The HERON Advanced Multiphase software used in the system was applied to maintain the structure in resonance, although the resonant frequencies were changing. To achieve such an effect, the system generated a control signal for actuators with the appropriate frequency of resonant vibration. Determining the appropriate resonant frequency was possible based on the analysis of the signal with a given moment of frequency. In the case of a decrease in the vibration amplitude at a given excitation, a deviation from the resonance state was found. In the next step, the system checked how the system would behave in the case of excitation with a higher and lower frequency. If, in any of the cases, the amplitude of the vibrations increased, then the system defined this frequency as the new resonant frequency. In the next step, the operation of checking the amplitude value was repeated.

3.2. Remote Object Positioning

One of the tasks that were assigned to the designed system was the positioning of the object on the structure with the use of vibrations. To carry out this task, in the first stage, it was necessary to generate vibrations with a resonance frequency in the structure. Then, a non-magnetic object weighing about 30 g was placed on the vibrating structure (a steel beam 6 m long). The vibrations caused by the beam set the mass in motion—the element was jumping on the beam and hitting it. These impacts were used to determine its position on the object through changes in the phase shifts in the generated signals.

Due to the application of the acceleration sensor in the developed multiphase head, it was possible to register the acoustic events that occurred when the mass separated from the beam. Due to the small mass of the object (20 g), the frequency of the system only slightly changed (0.04%). In addition, the previously described acoustic event related to the separation of the mass from the beam had a much higher frequency than the resonant frequency of the beam itself. The changes in vibrations were recorded as a sinusoidal signal with a frequency of 667 Hz.

3.3. Code of the Structure Stiffness—CSS

The above-described acoustic event that occurred in the system during the impact of the mass against the beam was characteristic of each structure. In the event of a change of mass or construction change, the response changed. It was related to the different amounts of energy accumulated over time. This response served as a control signal for the vibration controller. In addition, the ability to control the amplitude of vibrations and measure the response of the structure made it possible to detect changes in it, including the appearance of defects or damage.

Based on these tests, it can be concluded that the number and nature of recorded acoustic events are mainly influenced by the energy (in the form of mechanical vibrations) supplied to the tested structure and the temporary state of the structure in which they are located. Hence, the sequence of acoustic events as a modulated binary waveform (via F2F-frequency/dual-frequency modulation or modified MFM frequency) depends on the changing parameters of the medium in which the vibrations propagate, including propagation obstructions. The obtained results were mainly influenced by: the type of mechanical structure (its stiffness), the frequency of induction characteristic for any of the structural elements, and material defects such as pores or cracks. The resulting sequence of acoustic events was characteristic of each object and depended on the energy supplied to the object. This relationship was referred to as the Structural Stiffness Code (CSS). The use of this relationship and the ability to correctly interpret the received binary signal enabled the operation of the vibration controller in a wide frequency range, thanks to which it was possible to move the object with a specific algorithm.

Figure 4 shows the application of the CSS method. As a result of moving the object along with the structure, it was possible to isolate several acoustic events in the signal of the ICP sensor. The essence of the CSS method is the determination of a binary series based on the duration of acoustic events and the appropriate energy released into the system by a moving object. The main influence on the amount of released energy is the period in which a high level is maintained from the initiation of a given event. In this way, a unique code connected with the state of the structure can be obtained. Work on this code and its use are the subjects of further work.

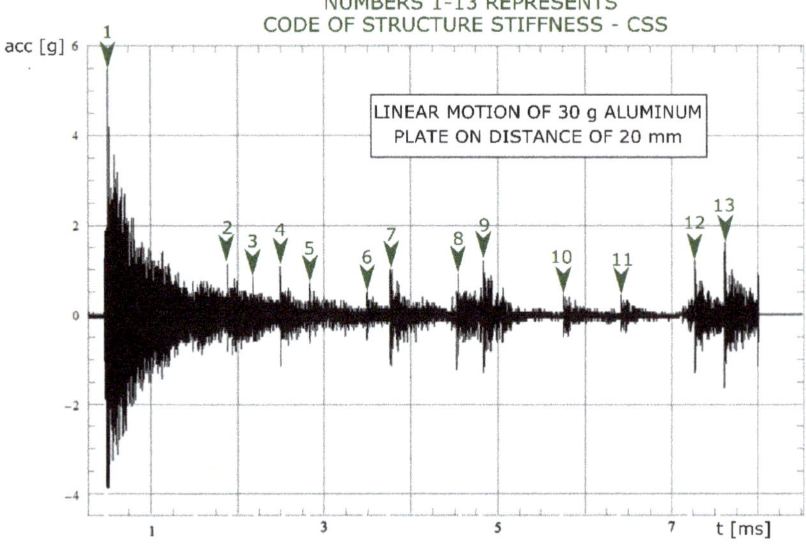

Figure 4. Decoded signals from ICP sensor, where the numbers from 1 to 13 represent code of the structure stiffness (CCS).

The number of acoustic events is characteristic of a specific state of the structure. Thanks to the use of a vibration regulator in a wide frequency range, it is possible to determine a specific structure "code" and to determine the dynamic characteristics of a mechanical structure. To accomplish this task, it is necessary to use a programmable pure sinusoidal excitation.

This issue requires further research, but, even at this stage, according to the authors, it can be stated that it was possible to develop a unique method that allows for the positioning of objects and the assessment of their structure/condition. The measurement technique based on the proposed solution can be an alternative to vibroacoustics and can potentially be used in SHM (Structural Health Monitoring).

3.4. Power and Data Transmission

In previous research by the authors, it turned out that SMART magnetic materials can be effectively used for wireless transmission of power and information [38]. The achieved results also indicated the high effectiveness of this method. The system that allowed for simultaneous data and power transmission was developed by the authors and is called

SURPS (SMART Ultrasonic Resonant Power System). This system provides transmission through various solids, as well as through liquids. An additional advantage of the system is the possibility of using various transmitter–receiver configurations [38].

The results presented in this paper are a continuation of work on the use of the magnetomechanical effect in the case of energy harvesting which was more widely described in [38]. As was shown earlier (Figure 2), four actuators were used instead of one. This was due to the desire to check whether it was possible to transmit energy continuously during data transmission. For this purpose, the Phase Shift Vibration Algorithm (PSVA) was used, which is schematically presented in Figure 5. In the case of this algorithm, the vibrations caused by the actuators were out of phase with each other, while each of the harvesters received a signal with a specific phase.

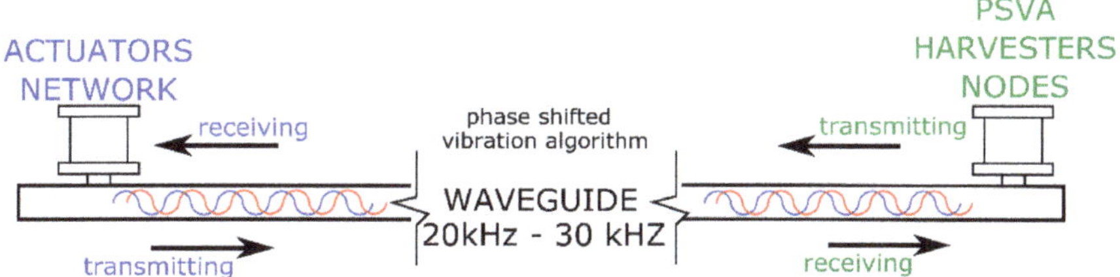

Figure 5. Power transmission through ultrasonic vibration—scheme.

The transmission was carried out by an actuator that transmits mechanical energy in the form of a pure sinusoidal ultrasonic wave and then this wave was picked up by the harvester, which converted this wave into an electric current using the magneto- or electrostrictive material contained therein. This way of transmitting energy also allowed information to be sent. What is more, it was possible to send energy through different materials, and the choice of material depended mainly on the distance the energy needed to be sent.

To transmit the information, the F2F procedure was used, which is a type of frequency modulation. The modulation worked in such a way that the data transmission frequency was one order lower than the resonant frequency of the structure. Figure 6 shows, schematically, how the data were transmitted via a magnetostrictive (AT) actuator and how the signal was received on the harvester with a magnetostrictive core. In the case of using a single harvester, it is visible that there was a break in the collection of energy while receiving the signal, which means that the harvester was not powered at the moment. In the case of using more harvesters, this was transferred to each of them. The use of PSVA allowed the creation of a dedicated circuit in which each harvester received the signal in the appropriate phase so that, when data were sent to one of the harvesters, others were still powered continuously.

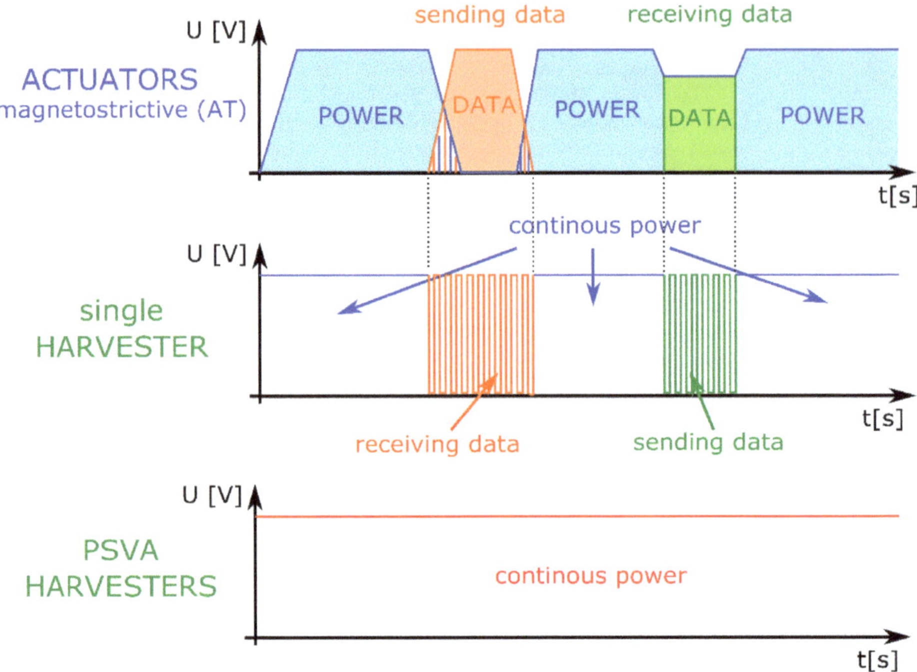

Figure 6. Data transmission and receiving information for PSVA.

3.5. System Structure

The SURPS was designed to work with various actuator–harvester configurations. One such configuration is a system where harvesters are connected in series and are located between two parallel beams. Such a system is characterized by a resonance frequency above 20 kHz. The test stand, consisting of two steel rails with magnetostrictive transducers between them, is shown in Figure 7. This system made it possible to simultaneously power the microprocessor on the side of the energy harvesters and transmit data in both directions.

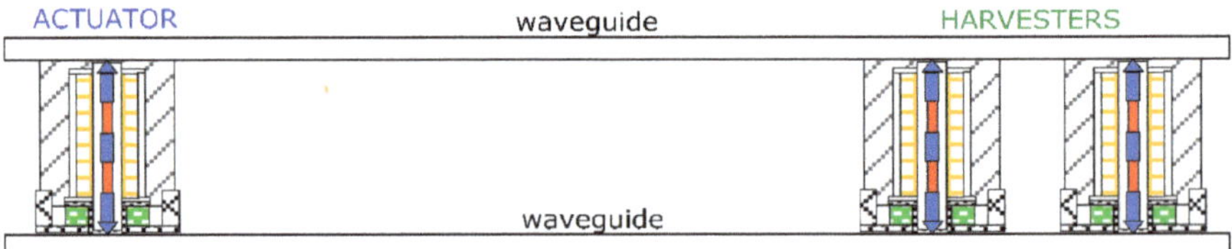

Figure 7. Actuator–harvester magnetostrictive system based on two beams.

Based on the above-described solution, supplemented with the current state of knowledge in the field of ultrasonic techniques and wireless power transmission, an original and innovative transmitting–receiving system was developed, equipped with a microcontroller, frequency modulators, and dedicated proprietary software. Figure 8 shows a schematic overview of how the system works. The developed system has several variants, depending on what medium is used for data and energy transmission. More information on the tested system can be found in [3].

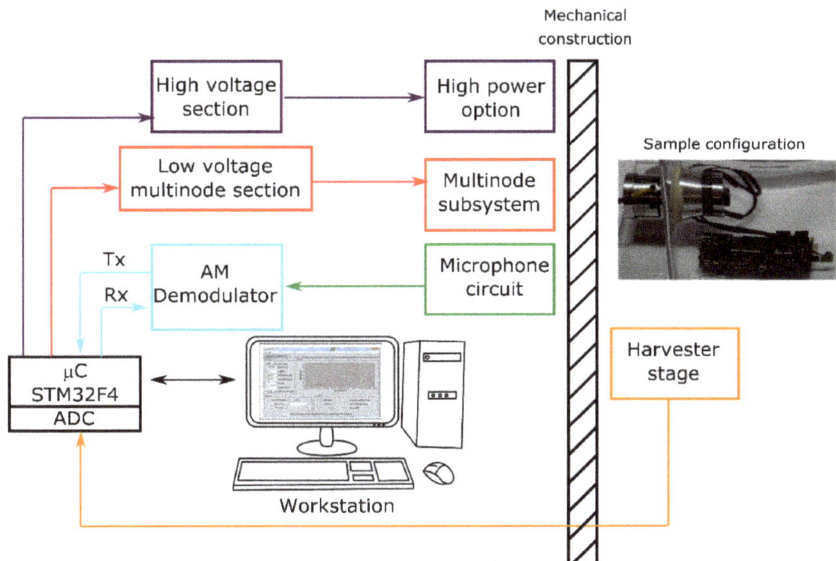

Figure 8. A block diagram of SURPS structure.

The main applications of SURPS encompass:

- service of piezoelectric or magnetic actuators/harvesters;
- scanning of a given frequency range using an actuator–harvester system with a real-time performance readout;
- searching for and generation of the resonance frequency of mechanical constructions;
- data transmission between the actuator and the harvester section in both directions (Tx, Rx), possible vibration generation in the frequency range from 0.1 Hz to 50 kHz with every 0.1 Hz (DDS generator by Analog Devices AD9851 was used);
- possible signal generation for two actuators generating vibration of the same frequency but shifted to each other in phase;
- harvester readout of current RMS voltage.

The results showing the frequency-amplitude characteristics for the circuit shown in Figure 7 are presented in Figure 9. It is worth noting that the highest voltage value (the highest efficiency) was achieved for the frequency located in the above acoustic band (above 20 kHz). The zone defined as SW in the figure defines the acceptable range of resonance frequencies and is approximately 20 kHz. The dashed line shows the voltage value at the level of 2.5 V, above which loss of the microprocessor system occurred. In addition, point A mark the most advantageous frequency ranges in the case of broadcasting information. As can be seen, several frequency ranges could be distinguished that allow the system to be powered and, depending on the conditions and needs, choices could be made between the required ranges of carrier signals. It should also be mentioned that it was possible to connect more microprocessors to the harvester network, but, in that case, it was necessary to follow a strictly defined sequence of their activation. This solution allows the system to be used in SHM (Structural Health Monitoring) applications with numerous sensors.

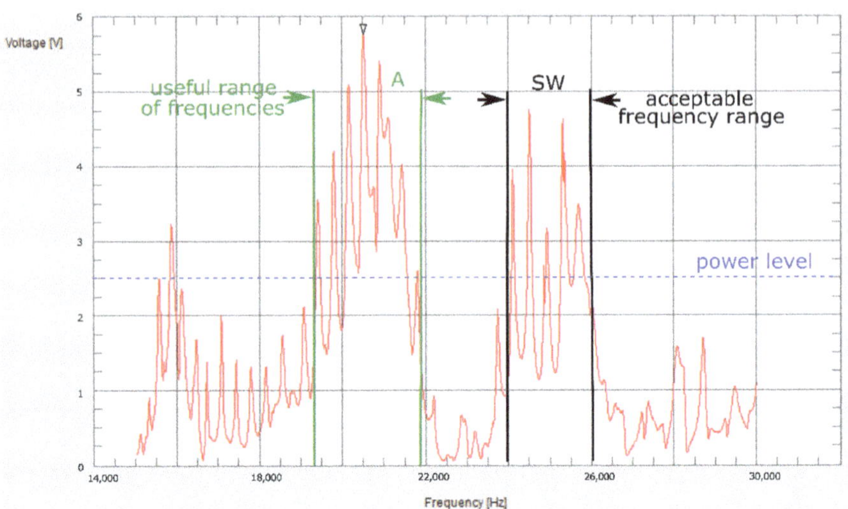

Figure 9. Frequency response of the dual-beam system.

The system was tested using a Gecko microprocessor chip with a 32-bit Cortex-M3 EFM32TG840 processor from Energy Micro. The solution for energy and information transmission proposed in the paper had to allow for powering the Gecko system, as it is a typical system used in industry. Additionally, the proposed system had to ensure a simultaneous half-duplex transmission. This meant that data were sent in one direction and at a certain time but in a two-way channel. The processor software was developed by the authors of the paper and supplemented with numerous useful functions.

The F2F-AM algorithm was used for data transmission. This algorithm guaranteed the stability of the information flow and allowed the system to achieve the transmission value at the level of 1 kbps, which was a sufficient speed in the analyzed case. To obtain higher bit rates, other frequency modulations had to be used. Figure 10 shows the results achieved for the test stand shown in Figure 7. The signal was transmitted by a magnetostrictive actuator, and the nature of the signal was sinusoidal with slight harmonic distortions. The frequency transmitted by this actuator was modulated in the on–off mode, which means that the actuator was turned on and off, which caused temporary shortages in the power supply of the energy harvester. To ensure continuous operation of the microprocessor on the harvester's side, the system was equipped with a set of capacitors, the capacity of which was sufficient for 0.5 s of microprocessor operation. Thanks to this solution, it was possible to continuously power the microprocessor despite the temporary shortages of power supplied by the harvester and to transmit many bits of data encoded in ASCII.

The results obtained during the research showed that the developed proprietary SURPS enables the transmission of energy over distances up to 6 m without the need for wires and using only various types of mechanical structure. This solution allows the use of various types of harvester in many configurations, while the selection of the appropriate harvester system is influenced by the material and form of the medium through which energy and data are transmitted, as well as the ultrasonic wavelength.

To obtain the highest possible efficiency of both information and energy transmission while maintaining a low level of generated noise, the original software was developed. This software allows for the selection of an appropriate actuator type for a given design, as well as modulation and the recommended frequency band, thanks to which it is possible to precisely determine the values of resonance frequencies. Finally, the software selects the resonance range in which the transmission is most effective.

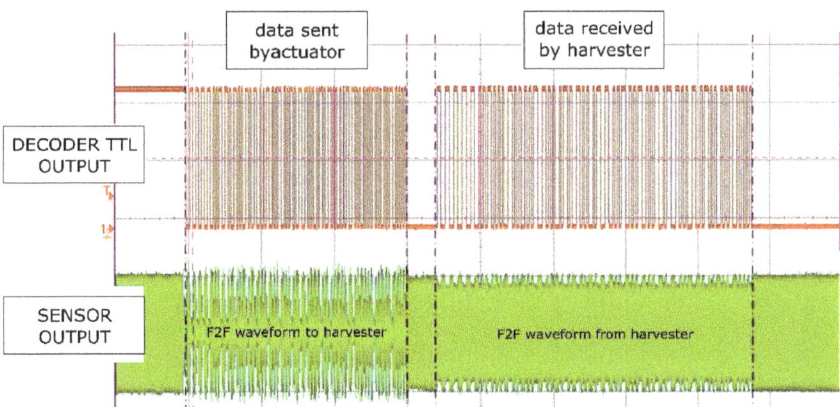

Figure 10. Graphic presentation of the results of power and data transfer.

4. Conclusions

The paper presents the results of works devoted to the transmission of energy and information through various centers. The results achieved include the following:

- A prototype of a multiphase head developed based on four magnetostrictive actuators in a symmetrical system integrated with the accelerometer;
- The possibility of setting objects on the surface of the structure in motion by shaping the appropriate form of vibrations with the aid of the developed head;
- A SURPS allowing the transmission of power and information in long rods using ultrasonic vibration;
- The use of harvesters/actuators based on both magnetostrictive and piezoelectric material and the use of F2F (frequency/double-frequency) procedures, which are a type of frequency modulation, for the transmission of information;
- The development of original software to select the appropriate actuator and modulation type and recommended frequency band for energy and data transfer.

The results presented in this paper are current and constitute the basis for further work in the field of energy and data transmission. One of the main scopes for future research is the usage of controlled vibrations made by the designed actuator in NDE diagnostics. The presented method can be considered as an expansion of the methods known as energy harvesting due to the possibility for it to transform various forms of energy supplied to/collected from the system to perform mechanical work, e.g., object positioning.

Author Contributions: Conceptualization, R.M. and P.W.; methodology, P.W.; software, P.W. and K.W.; validation, R.M. and P.W.; formal analysis, P.W.; investigation, R.M. and P.W.; resources, P.W.; data curation, K.W.; writing—original draft preparation, R.M. and P.W.; writing—review and editing, R.M. and P.W.; visualization, P.W.; supervision, R.M.; project administration, R.M.; funding acquisition, R.M. All authors have read and agreed to the published version of the manuscript.

Funding: This research was funded by The National Centre for Research and Development within the project "Composite magnetostrictive-nanocrystalline materials for use in the field of energy harvesting and transformation", grant number LIDER/21/0082/L-9/17/NCBR/2018.

Institutional Review Board Statement: Not applicable.

Informed Consent Statement: Not applicable.

Data Availability Statement: Data supporting reported results can be provided upon request. Currently, these data are collected as part of the ongoing project and only after its completion will the data be made available to the public.

Conflicts of Interest: The authors declare no conflict of interest.

References

1. Chandrakasan, A.; Amirtharajah, R.; Goodman, J.; Rabiner, W. Trends in low power digital signal processing Circuits and Systems. In Proceedings of the 1998 International Symposium on Circuits and Systems (ISCAS), Monterey, CA, USA, 31 May–3 June 1998; Volume 4, pp. 604–607.
2. Mohri, K. Review on recent advances in the field of amorphous-metal sensors and transducers. *IEEE Trans. Magn.* **1984**, *20*, 942–947. [CrossRef]
3. Mori, K.; Horibe, T.; Ishikawa, S.; Shindo, Y.; Narita, F. Characteristics of vibration energy harvesting using giant magnetostrictive cantilevers with resonant tuning. *Smart Mater. Struct.* **2015**, *24*, 125032. [CrossRef]
4. Loreti, P.; Catini, A.; De Luca, M.; Bracciale, L.; Gentile, G.; Di Natale, C. The design of an energy harvesting wireless sensor node for tracking pink iguanas. *Sensors* **2019**, *19*, 985. [CrossRef] [PubMed]
5. Ait Aoudia, F.; Gautier, M.; Magno, M.; Berder, O.; Benini, L. Leveraging energy harvesting and wake-up receivers for long-term wireless sensor networks. *Sensors* **2018**, *18*, 1578. [CrossRef] [PubMed]
6. Lai, Y.C.; Hsiao, Y.C.; Wu, H.M.; Wang, Z.L. Waterproof fabric-based multifunctional triboelectric nanogenerator for universally harvesting energy from raindrops, wind, and human motions and a self-powered sensors. *Adv. Sci.* **2019**, *6*, 1801883. [CrossRef]
7. Shi, S.X.; Yue, Q.Q.; Zhang, Z.W. A self-powered engine health monitoring system based on L-shaped wideband piezoelectric energy harvester. *Micromachines* **2018**, *9*, 629. [CrossRef]
8. Wang, L.; Yuan, F.G. Vibration energy harvesting by magnetostrictive material. *Smart Mater. Struct.* **2008**, *17*, 045009. [CrossRef]
9. Roundy, S.; Wright, P.K.; Rabaey, J. A study of low level vibrations as a power source for wireless sensor nodes. *Comput. Commun.* **2003**, *26*, 1131–1144. [CrossRef]
10. Scheidler, J.J.; Asnani, V.M. A review of noise and vibration control technologies for rotorcraft transmissions. In Proceedings of the INTER-NOISE and NOISE-CON Congress and Conference Proceedings, InterNoise16, Hamburg, Germany, 21–24 August 2016.
11. Exner, W. *NVH Phenomena in Light Truck Drivelines*; SAE Technical Paper 952641; SAE International: Warrendale, PA, USA, 1995.
12. El-Hami, M.; Glynne-Jones, P.; White, N.W.; Hill, M.; Beeby, S.; James, E.; Brown, A.D.; Ross, J.N. Design and fabrication of a new vibration-based electromechanical power generator. *Sens. Actuators A Phys.* **2001**, *92*, 335–342. [CrossRef]
13. Glynne-Jones, P.; Tudor, M.J.; Beeby, S.P.; White, N.W. An electromagnetic, vibration-powered generator for intelligent sensor systems. *Sens. Actuators A Phys.* **2004**, *110*, 344–349. [CrossRef]
14. Despesse, G.; Jager, T.; Jean-Jacques, C.; Léger, J.; Vassilev, A.; Basrour, S.; Charlot, B. Fabrication and characterization of high damping electrostatic micro devices for vibration energy scavenging. In Proceedings of the Design, Test, Integration and Packaging of MEMS and MOEMS, Montreux, Switzerland, 1–3 June 2005; pp. 386–390.
15. Meninger, S.; Mur-Miranda, J.O.; Amirtharajah, R.; Chandrakasan, A.P.; Lang, J.H. Vibration-to-electric energy conversion. *IEEE Trans. Very Large Scale Integr. (VLSI) Syst.* **2001**, *9*, 64–76. [CrossRef]
16. Mitcheson, P.D.; Green, T.C.; Yeatman, E.M.; Holmes, A.S. Architectures for vibration-driven micropower generators. *J. Microelectromech. Syst.* **2004**, *13*, 429–440. [CrossRef]
17. Moss, S.; Barry, A.; Powlesland, I.; Galea, S.; Carman, G.P. A broadband vibro-impacting power harvester with symmetrical piezoelectric bimorph-stops. *Smart Mater. Struct.* **2011**, *20*, 045013. [CrossRef]
18. Anton, S.R.; Sodano, H.A. A review of power harvesting using piezoelectric materials (2003–2006). *Smart Mater. Struct.* **2007**, *16*, R1. [CrossRef]
19. Mitcheson, P.D.; Yeatman, E.M.; Rao, G.K.; Holmes, A.S.; Green, T.C. Energy harvesting from human and machine motion for wireless electronic devices. *Proc. IEEE* **2008**, *96*, 1457–1486. [CrossRef]
20. Viktor, B. Vibration energy harvesting using Galfenolbased transducer. In Proceedings of the SPIE Smart Structures and Materials + Nondestructive Evaluation and Health Monitoring, San Diego, CA, USA, 10–14 March 2013.
21. Deng, Z. Nonlinear Modeling and Characterization of the Villari Effect and Model-Guided Development of Magnetostrictive Energy Harvesters and Dampers. Ph.D. Thesis, The Ohio State University, Columbus, OH, USA, 2015.
22. Deng, Z.; Dapino, M. Magnetic flux biasing of magnetostrictive sensors. *Smart Mater. Struct.* **2017**, *26*, 055027. [CrossRef]
23. Zhang, H. Power generation transducer from magnetostrictive materials. *Appl. Phys. Lett.* **2011**, *98*, 232505. [CrossRef]
24. Viola, A.; Franzitta, V.; Cipriani, G.; Dio, V.; Raimondi, F.M.; Trapanese, M. A magnetostrictive electric power generator for energy harvesting from traffic: Design and experimental verification. *IEEE Trans. Magn.* **2015**, *51*, 8208404. [CrossRef]
25. Liu, H.; Wang, S.; Zhang, Y.; Wang, W. Study on the giant magnetostrictive vibration-power generation method for battery-less tire pressure monitoring system. *Proc. Inst. Mech. Eng. Part C J. Mech. Eng. Sci.* **2014**, *229*, 1639–1651. [CrossRef]
26. Yan, B.; Zhang, C.; Li, L.; Zhang, H.; Deng, S. Design and construction of magnetostrictive energy harvester for power generating floor systems. In Proceedings of the 2015 18th International Conference on Electrical Machines and Systems (ICEMS), Pattaya, Thailand, 25–28 October 2015; pp. 409–412.
27. Yan, B.; Zhang, C.; Li, L. Design and fabrication of a high-efficiency magnetostrictive energy harvester for highimpact vibration systems. *IEEE Trans. Magn.* **2015**, *51*, 8205404. [CrossRef]
28. Nair, B.; Nachlas, J.A.; Murphree, Z. Magnetostrictive Devices and Systems. U.S. Patent US9438138B2, 14 August 2014.
29. Park, Y.; Kang, H.; Wereley, N.M. Conceptual design of rotary magnetostrictive energy harvester. *J. Appl. Phys.* **2014**, *115*, 17E713. [CrossRef]
30. Zucca, M.; Bottauscio, O.; Beatrice, C.; Hadadian, A.; Fiorillo, F.; Martino, L. A study on energy harvesting by amorphous strips. *IEEE Trans. Magn.* **2014**, *50*, 8002104. [CrossRef]

31. Kita, S.; Ueno, T.; Yamada, S. Improvement of force factor of magnetostrictive vibration power generator for high efficiency. *J. Appl. Phys.* **2015**, *117*, 17B508. [CrossRef]
32. Ueno, T. Performance of improved magnetostrictive vibrational power generator, simple and high power output for practical applications. *J. Appl. Phys.* **2015**, *117*, 17A740. [CrossRef]
33. Deng, Z.; Dapino, M.J. Multiphysics modeling and design of Galfenol-based unimorph harvesters. In Proceedings of the SPIE Smart Structures and Materials + Nondestructive Evaluation and Health Monitoring, San Diego, CA, USA, 8–12 March 2015.
34. Ueno, T.; Yamada, S. Performance of energy harvester using iron-gallium alloy in free vibration. *IEEE Trans. Magn.* **2011**, *47*, 2407–2409. [CrossRef]
35. Lawry, T.J.; Wilt, K.R.; Ashdown, J.D.; Scarton, H.A.; Saulnier, G.J. A high-performance ultrasonic system for the simultaneous transmission of data and power through solid metal barriers. *IEEE Trans. Ultrason. Ferroelectr. Freq. Control* **2012**, *60*, 194–203. [CrossRef]
36. Risquez, S.; Woytasik, M.; Coste, P.; Isac, N.; Lefeuvre, E. Additive fabrication of a 3D electrostatic energy harvesting microdevice designed to power a leadless pacemaker. *Microsyst. Technol.* **2018**, *24*, 5017–5026. [CrossRef]
37. Liang, Y.; Zheng, X. Experimental researches on magneto-thermo-mechanical characterization of Terfenol-D. *Acta Mech. Solida Sin.* **2007**, *20*, 283–288. [CrossRef]
38. Kaleta, J.; Mech, R.; Wiewiórski, P. *Development of Resonators with Reversible Magnetostrictive Effect for Applications as Actuators and Energy Harvesters*; IntechOpen: London, UK, 2018.
39. Claeyssen, F.; Lhermet, N.; Maillard, T. Magnetostrictive Actuators Compared to Piezoeletric Acutators. In Proceedings of the European Workshop on Smart Structures in Engineering and Technology, Giens, France, 21–23 May 2002; p. 4763. [CrossRef]

Article

Rapid Demagnetization of New Hybrid Core for Energy Harvesting

Rafał Mech *, Przemysław Wiewiórski and Karol Wachtarczyk

Faculty of Mechanical Engineering, Wroclaw University of Science and Technology, 50-370 Wroclaw, Poland; przemyslaw.wiewiorski@pwr.edu.pl (P.W.); karol.wachtarczyk@pwr.edu.pl (K.W.)
* Correspondence: rafal.mech@pwr.edu.pl; Tel.: +48-71-320-2899

Abstract: This paper presents the results obtained using the rapid demagnetization method in the case of an NdFeB magnet and a new hybrid core. The developed core consists of three basic elements: an NdFeB magnet, Terfenol-D, and a specifically developed metallic alloy prepared by means of a suction casting method. The main goal of proposing a new type of core in the event of rapid demagnetization is to partially replace the permanent magnet with another material to reduce the rare-earth material while keeping the amount of generated electricity at a level that makes it possible to power low-power electrical devices. To "capture" the rapid change of magnetic flux, a small number of coils were made around the core. However, the very low voltage level at very high current required the use of specialized electronic transducers capable of delivering a voltage level appropriate for powering a microprocessor system. To overcome this problem, a circuit designed by the authors that enabled voltage processing from low impedance magnetic circuits was used. The obtained results demonstrated the usefulness of the system at resonant frequencies of up to 1 MHz.

Keywords: SMART materials; magnetostriction; Terfenol-D; energy harvesting

Citation: Mech, R.; Wiewiórski, P.; Wachtarczyk, K. Rapid Demagnetization of New Hybrid Core for Energy Harvesting. *Sensors* **2022**, *22*, 2102. https://doi.org/10.3390/s22062102

Academic Editors: Zdenek Hadas, Saša Zelenika and Vikram Pakrashi

Received: 10 January 2022
Accepted: 4 March 2022
Published: 9 March 2022

Publisher's Note: MDPI stays neutral with regard to jurisdictional claims in published maps and institutional affiliations.

1. Introduction

Interest in energy harvesting (EH) is continuously increasing from year to year. The primary impulse for development in the discussed area is related to the shortage of electricity and the search for new possible sources of energy. In this context, the idea of EH would also include the generation of energy from wind, solar plants, and other natural power sources on a large scale. In the literature, the discussion of energy harvesting is limited to small electrical and electronic devices that can operate in a self-sufficient manner [1–3]. On the basis of this limitation, it has become clear that EH cannot be considered as a feasible power source in high-power applications. However, as could be predicted, increasing the number of devices that can be powered from harvested energy would make it possible to relieve other energy sources [4–8]. Obtaining energy for devices from the environment has found possible applications in many different markets, and the number of possible applications will grow continuously [9–13]. Currently, the consumer device market is not large, but it is predicted that it will increase continuously, and will have an influence on the direction of future research. Important features and attributes have had, and still have, an influence on the directions of EH development and potential target markets, including:

- Small size, portability;
- Capability of performing in difficult-to-access environments;
- Fewer cables and reduced usage of other materials, devices, and systems;
- Provision of real-time information;
- Decreased cost;
- Low maintenance needs;
- Reduction of batteries—green energy;
- Great market potential, and the absorptiveness of the market.

Due to their small size and lack of need for complicated powering assemblies, these devices can be used in difficult-to-access environments [14–16], such as in complicated machines, structures, etc. Research on the use of EH devices in very fragile environments such as the human body was performed at Southampton University Hospital in the UK. A prototype of an electrodynamics harvester was created that was able to generate enough energy from heartbeats to supply power to a pacemaker for the entire life of a patient. The output was 4.3 micro joules per heartbeat; with the use of new and better polymer materials, that value is expected to double [1].

Additionally, self-sufficient devices mean less cable. This applies mainly to sensors, and is a very important advantage, making monitoring constructions, machines, and different types of elements much easier and safer to use, while providing better information for decision-making processes like with respect to the maintenance needs of an object [17]. These advantages were noticed by the American Society of Civil Engineers, which together with the University of Texas (UT), began testing a new wireless bridge monitoring system [18]. Another system for monitoring the condition of railway bridge structures was presented in Ref. [19]. Sensors that do not need a separate powering system make it possible to test bridges more quickly and receive more detailed information in real time. The importance of such kinds of systems is evidenced by the fact that in America in 2008, over 17,000 bridges did not meet the requirements necessary for operation.

The research and development on generating energy is progressing very quickly, opening new markets for a growing range of different devices. Consumers and the military, along with third-world markets, are the main groups, and will create the biggest market for devices by value. Their share accounted for around 88% of all value predicted for such kinds of device in 2014. As projects are run mainly for profit (including both present and future profits, both financial and strategic), it seems to be significant to realize the importance and the full potential of the EH concept and technology. According to the report presented by IDTechEx Ltd. (Cambridge, UK) [20], these potential markets could represent an income of about USD 1.156 trillion per year.

The main idea of the presented research was to determine whether the chosen types of harvesting device and the proposed type of excitation could be useable for mechanical purposes. The main goal of this investigation was to determine whether the new proposed hybrid core structure, when applied in dedicated harvesters, is able to provide sufficient energy to supply the chosen ATMEL microcontroller. The tests were carried out on a dedicated stand, allowing the impulse load to be applied with a fixed value. The results of the impulse investigations are discussed in detail to provide information about the usefulness of the proposed hybrid core in the area of energy harvesting.

2. Materials and Methods

The study presented in this paper was based on a well-known giant magnetostriction material, Terfenol-D, and on a material prepared by the authors, produced by the suction casting method. In addition, commercially available NdFeB neodymium magnets were also used in the investigation.

2.1. Suction Casting Alloy

The newly created material in the form of a metallic alloy described in this paper was developed based on literature data describing a patented soft, magnetic, amorphous material known as FINEMET [21,22]. It should be noted that the material with the composition $Fe_{57}Co_{10}B_{20}Si_5Nb_4V_4$ described in this paper was developed in the opposite direction to what has been the case in many studies showing materials based on FINEMET. The material presented in this work has a lower content of the basic elements, such as Fe and Co. It should also be emphasized that when describing the material, its composition is given in atomic notation, not mass notation. An arc furnace (arc-melter) was used to produce the material. To ensure the highest possible representation of the developed chemical composition of the alloy, very high-purity elements—up to 99.999%—were used for its production. Each of the alloying elements was weighed out to 4 decimal places, and then these components were placed in the arc furnace chamber. Next, the furnace chamber was sealed, and then the air was removed from it by pumping to a vacuum of 5×10^{-5}, after which the entire chamber was flushed with argon and pumped down again. After several chamber flushing processes, the alloy was prepared by melting all the elements several times to homogenize them in the entire volume. Finally, the liquid alloy was sucked into a specially prepared mold, which made it possible to obtain rods with a diameter of 3 mm and a length of 150 mm. The material prepared in this way was then subjected to mechanical treatment in order to be able to place it in devices for energy harvesting.

2.2. Ferro Magnetic Generator (FMG)

The tests performed in this work are mainly based on the principles of Ferro Magnetic Generators (FMG). As is generally known, the working principle is based on ferromagnets that are demagnetized either by shock loading or from the impact of high-speed flyers. The same effect might also occur as a result of the detonation of high explosives. The team at the Agency for Defense Development in Korea presented the output characteristics of annulus- and cylinder-type explosion-driven devices based on NdFeB magnets [23]. In this paper, the authors decided to use similar principles to FMG to obtain energy that could possibly be used to supply microcontrollers, but without the use of an explosion-driven device, and only with a force impulse. The force impulse causes a change in the magnetization around the material, then this change in the magnetic field is picked up by the coil and converted into electric energy.

In the case of a magnet, during an impact, it will be demagnetized (partially), and with a sufficiently large number of impacts, it could be fully demagnetized. In addition, in the event of an impact with sufficiently high energy, it could even be destroyed, generating a high amount of energy. On the other hand, the results showed that the use of magnets alone as a source of energy may be insufficient, as can be seen from the results shown in this paper.

3. Results

This section is divided into appropriate subsections, allowing the presentation of the course of the processes and the evolution of subsequent solutions, leading to the achievement of the aims of the study. The first part of this section describes the devices for energy harvesting that were then used in the experimental activities described in the second part.

3.1. Harvesters

The research object is a unique basic Energy Harvester Device (EHD) model, shown in Figure 1. The model, called TCCM (Top Core Coil Magnet), is a construction consisting of four major elements: the Top, the role of which is to transfer the shock to the core, and the Coil, Magnet, and Core, which process the impact energy passed from the top, transforming it into electricity. For the aforementioned shock, a reference shock with specific repeatable parameters was used, thanks to which it was possible to test the harvester cores under the same working conditions. However, it should be noted that these devices can also operate in the case of various oscillating systems (with vibrations) of appropriate amplitude.

- A permanent NdFeB was used as a magnet.
- The material used for the core was important, because this is the part of the device responsible for the generation of current of sufficient value to operate as a pulse power supply. A few different core arrangements are proposed. The main material used was Terfenol-D and its powder. Additionally, pieces of the newly produced alloy were introduced to the core arrangement.
- The coil with a plastic body (the number of turns was 980, the resistance was 14Ohm, and the inductance was 1.1 mH) was fixed to the neodymium magnet.

As can be seen from the cross-section (Figure 1), TCCM harvesters can be considered to be one of the simplest models, especially since they do not have a system allowing the introduction of prestress, which makes it a structure with relatively low stiffness, especially when the force impulse is given. It should be noted that due to the simplicity of this structure, the device allows a large number of different materials to be tested relatively quickly for suitability in the field of energy recovery.

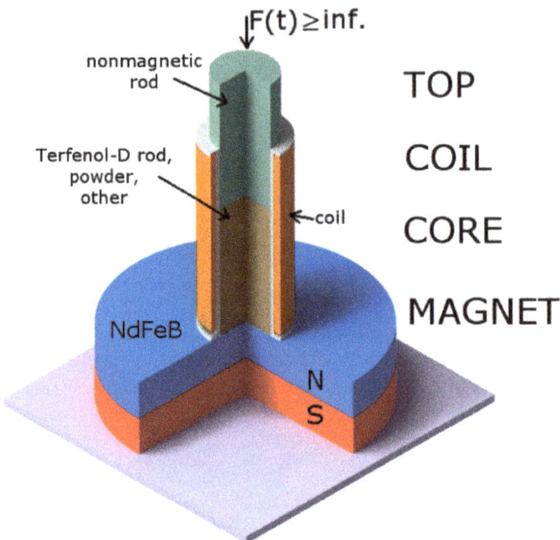

Figure 1. TCCM harvester scheme.

The construction of a more advanced version of TCCM called Double Top Core Coil Magnet (DTCCM) is shown in Figure 2. This is a modified version of the TCCM harvester in which two magnetoelectric circuits are applied. The magnetic circuit of this device is also based on neodymium magnets, and it has low dispersion of magnetic field on the outside due to the permendur plates.

- Two coils used in the shown construction are the same, and have the same parameters, with static resistance $R_c = 11$ kΩ.
- In this type of harvester, it is possible to apply and control prestress. This is due to the thin-walled steel tubes, in which screw-bolts are placed, but the strength of both the screws and the tubes acts as a limitation.

Additionally, different arrangements of cores were also used.

3.2. Test Stand and Measurements

Figure 3 shows the scheme of the stand developed to perform impact tests for energy harvesting devices, in particular, TCCM devices. A horizontally mounted PZT sensor was used to measure the impact force. The sensor was placed on a rigid, non-deformable surface of non-magnetic plate. To generate the impulse force in the device, an aluminum rod was used, which was mounted so that it always hit the center of the upper part of the device. The velocity of the rod movement, and thus the value of the impact force, was regulated. A fast MOSFET transistor interacting with a linear motor was used for regulation, which made it possible to set a precisely defined velocity. Additionally, the impact energy can be changed by changing the load applied to the moving element using weights of a known mass. The mass of each weight was determined with an accuracy of four decimal places. A maximum of 2 kg can be placed on the movable element, but it was decided to apply 0.5 kg due to the small size of the part with the aluminum rod. Thanks to the applied solutions, it was possible to obtain repeatable values of impact energy E_k.

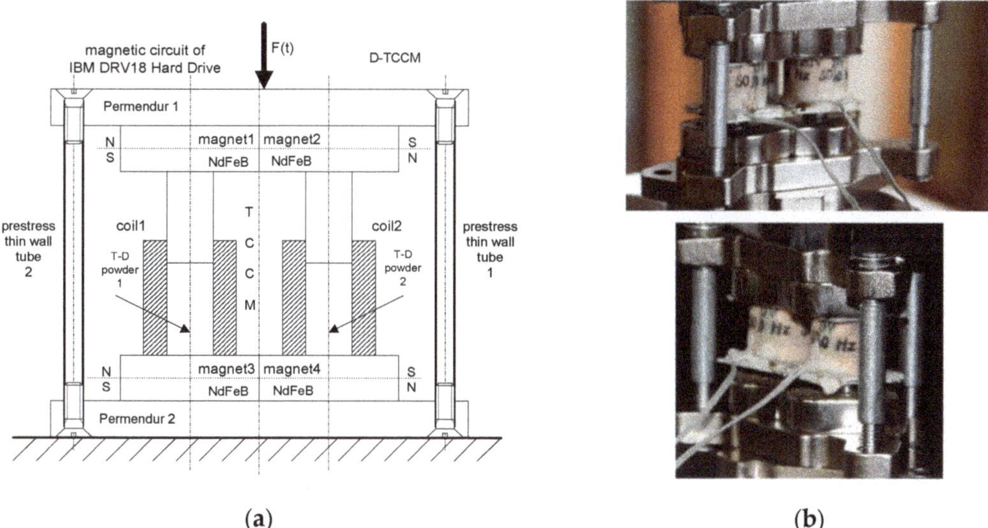

(a)

(b)

Figure 2. DTCCM harvester: (**a**) scheme, (**b**) real device.

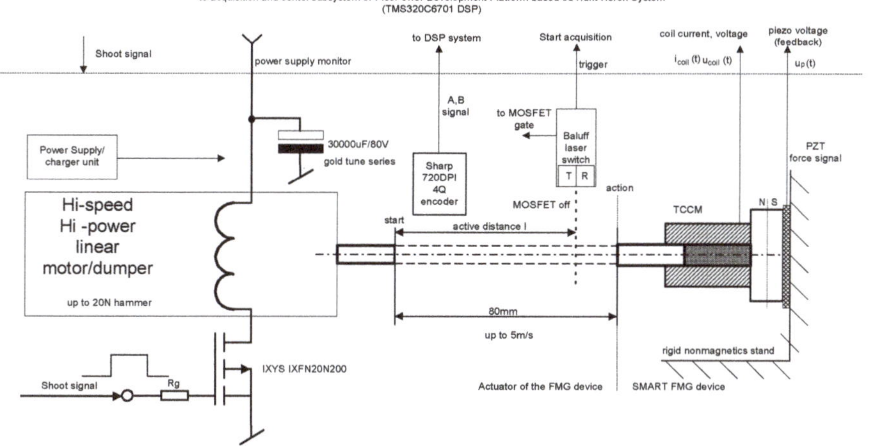

Figure 3. Test stand for TCCM harvester investigations.

The prepared system made it possible to test the cores placed inside the coil of the TCCM device. The impulse force resulting from the movement of the hammer in the form of the aluminum rod is transmitted to the core of the device through its upper part (Top). Figure 4 shows the response of the system during the experiment. Performing the analysis with respect to the experimental process, it is possible to specify individual phases within the experiment. The first phase is the phase just before it hits the device, where the voltage can be observed to rise. This change in voltage is probably caused by the movement of the ram in the magnetic field of the applied neodymium magnet. The next phase is the impact, which causes a shock wave to go through the consecutive elements of the harvester, to the magnet located at the bottom. This changes the magnetic flux, thus inducing a voltage on the used coil.

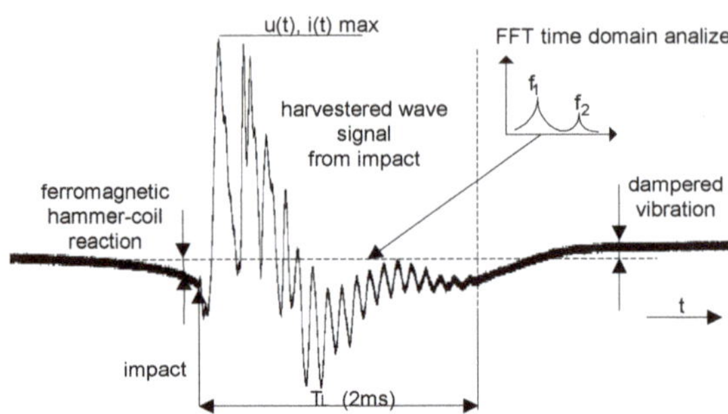

Figure 4. Example of a waveform obtained from TCCM.

In the case of the TCCM harvester, only the top and the core of the system affect the resonance frequency of the system. The shock wave generated by the impact can circulate in the core–magnet system until it is completely suppressed, or exit the system through a magnet that is in contact with another rigid surface. However, it should be noted that this wave is not transferred to the coil itself. It can be observed that the mechanical resonance of the device resulting from the impact is a decisive factor for the signal obtained from the device.

In the case of the energy harvesting device used in the experiment, it was found that the prestress system did not need to be used. It was observed that the prestress effect was obtained during the first moments of impact. This was also the moment at which the maximum response values of the tested system were obtained. It should be noted that the values of the generated voltage were dependent on the applied electromagnetic system, which was in the form of a coil. When applying the same impact energy, the obtained voltage values, apart from the materials used for the core of the device, were also dependent on the number of turns in the coil.

The main goal of this investigation was to show the differences between the materials used to produce the core of the harvester in terms of the current values obtained for each of them, as well as with respect to the reaction of the PZT sensor and TCCM for impact. The final research step was to perform the core selection, which is a significant element of the harvesting device.

The original current signals from the coil were windowed using the Hamming windowing function. All current measurements were performed using a sampling rate of 1 MHz. In line with this, FFT analysis was performed for a spectrum of 500 kHz. Due to the applied windowing, the analysis of the spectrum was narrowed to 5 kHz and 10 kHz when using Terfenol-D as the core material.

As an output of the impact (the applied shock force), the velocity and voltage were determined, and their values are presented in Figure 5, where a correlation between the output signals and the actual test phases is presented.

The first phase of the test encompassed the acceleration of the aluminum ram up to a velocity of 1.1 m/s, E_k = 1.21 J, as shown in Figure 5. The second phase was the moment of impact, upon which the ram rapidly slowed down, and its reflection moment was controlled and damped. In the worst scenario, when the ram was not aligned with the harvester axis, vibration occurred throughout the whole device. On the graphs in Figure 5, the moment of impact is visible. A rapid increase in the current value appears at the moment of impact.

In Figure 6, the results of an experiment with a series of four impacts with the same parameters and a TCCM harvester are presented. Analysis of the signal from the PZT piezo patch revealed the nature of the impact, as well as the reaction of the surface to which the harvester was fixed. However, according to the fact that loading the PZT with an impedance value of 1 MΩ decreases the surface signal one hundred times to the order of a few µV, the practical influence of that signal is negligible. By registering the signal from the PZT, it was possible to control the repeatability of the shock tests, except that, in the future, it will be possible to use that signal as a coupling to change the stiffness of the next generation of harvesters, with the intention of absorbing the maximum amount of energy from the impact with minimal transmission of that energy to the surface. However, the repeatability of the signal in the PZT did not have a place in the current signal of the coil. The reverse magnetostriction effect (Villari's effect), and ensuring that the system has the appropriate operating parameters are very important. Even small changes in the impact conditions, mainly differences in the ram or the core, resulted in significant changes in the current values obtained. The operating parameters of the system are also greatly influenced by the external bias field, the pre-stress, and the impact energy.

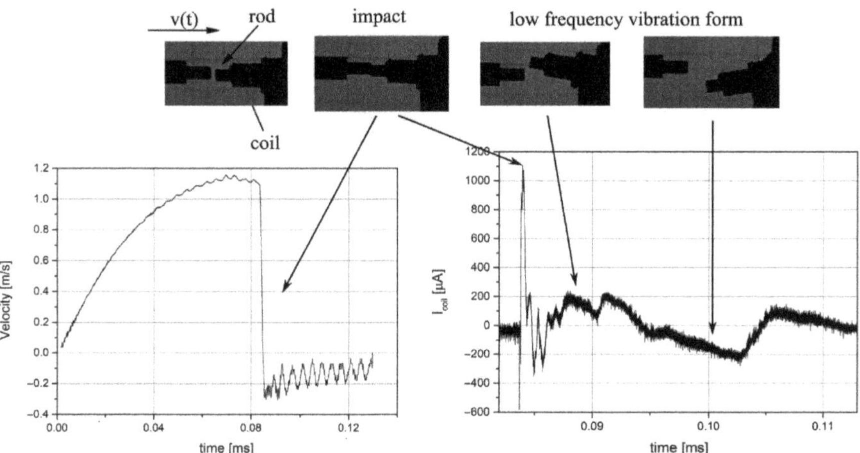

Figure 5. Impact correlated with graphs of velocity and output voltage.

It was important to assess the influence of the state of aggregation (solid, powder) of the core material on the output values. The main criterion for analyzing this influence was the absolute value (the ideal looseness straightens the AC signal) of the current during the 1 ms following the impact moment.

The fundamental frequencies of mechanical resonance are the second parameter describing the core composition. The coil resistive load value $R_c = 125\ \Omega$ was determined using the math library package Numerix SIGLIB v6.0, implemented in the Agilent VEE Pro software.

During the impact tests, the couple–core–magnet system reacts rapidly, while the coil's body is not expanded by the core material. When the impact energy becomes critical, E_k max, an irreversible change in system parameters occurs, resulting in the deformation of the coil body. This becomes an important issue when the core is in a powder state, because the powder adapts itself to the shape of the coil.

During the investigation, it could be seen that the core material had a crucial influence on the results of the tests. These results are shown in Table 1, where the differences between four measured parameters can be seen. It can be observed that the highest value of I_{max} was obtained for the Terefenol-D rod with small pieces of the prepared $Fe_{57}Co_{10}B_{20}Si_5Nb_4V_4$ alloy used as the core. However, due to the high cost of Terfenol-D and its brittleness, the number of tests was limited. As an alternative for a solid rod made using that material, a powder form that was capable of withstanding impact energies greater than those achievable for steel and NdFeB was used. Additionally, the Terenol-D powder was partially mixed with powdered $Fe_{57}Co_{10}B_{20}Si_5Nb_4V_4$ alloy. The results of the impact tests for the powder core are shown in Figure 7. The only limitations of the experiment were the performance of the body and the durability of the coil windings. However, these two parameters were taken under consideration.

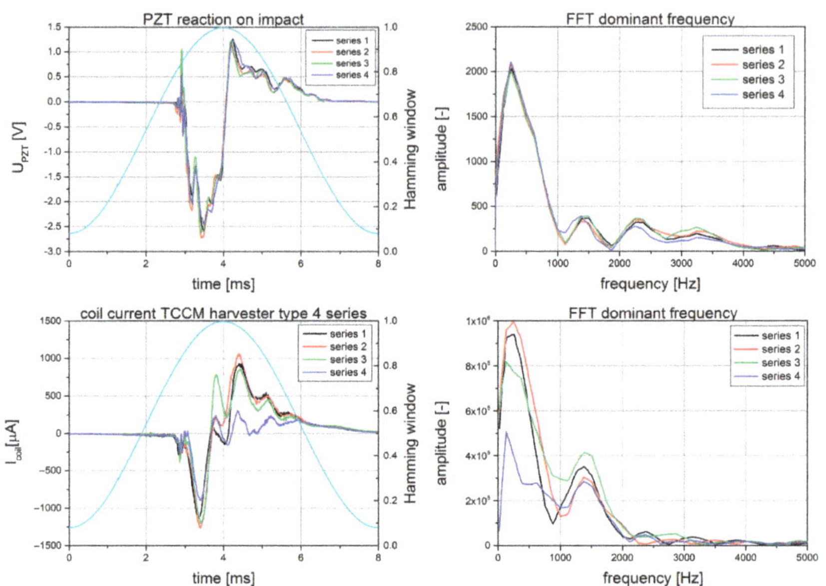

Figure 6. Reaction on impact: the PZT sensor signal, current of coil for TCCM, and their FFT spectrums.

Table 1. Comparison of different harvesting cores with parameters as follows: $Total_{Load}$ = 125 Ω, E_k = 5.6 J, R_c = 180 Ω for V = 6.7 m/s, and m = 0.5 kg.

Core Type	I_{max} [μA]	ABS I_{max} (1 ms) [μA]	FFT Dominate Frequency	
			F1 [Hz]	F2 [Hz]
Cu	1200	600	375	1500
Fe	1600	500	500	1750
Ferrite	500	200	375	1500
NdFeB	1200	400	875	1500
Terfenol-D powder mixed with $Fe_{57}Co_{10}B_{20}Si_5Nb_4V_4$	2500	800	250	1625
Terfenol-D rod with $Fe_{57}Co_{10}B_{20}Si_5Nb_4V_4$ pieces	35,000	8300	500	1875

Where:

- $Total_{Load}$ = 125 Ω is the resistive load on the coil, which is a source of generated energy due to an impact.
- R_c = 180 Ω is the static coil resistance (in this case quite a high value). In the case of harvesting energy from the reverse magnetostrictive effect, the coil parameters, i.e., how many Ohms and how many turns the coil has, are an important factor. In this paper, the aim was to generate the highest possible voltage increase so that the microprocessor power rectifier system was able to work under optimal conditions.
- I_{max} is the parameter that indicates the ability to power the microprocessor system from a single hit. This is the maximum current obtained during the impact phenomenon and the frequency response of the natural vibrations of the harvester's core. In the case of using magnetostrictive cores, the I_{max} parameter is two orders higher than in the case of using other materials.

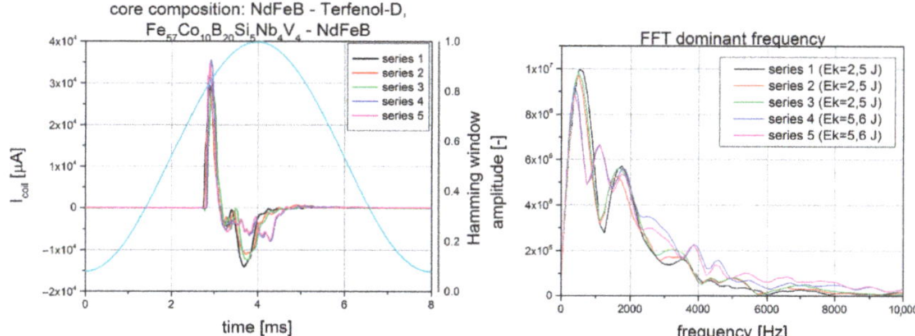

Figure 7. Example of the coil current measured for solid Terfenol-D with $Fe_{57}Co_{10}B_{20}Si_5Nb_4V_4$ alloy pieces, as well as its FFT.

The test performed using powdered Terfenol-D mixed with $Fe_{57}Co_{10}B_{20}Si_5Nb_4V_4$ alloy as the harvester core, even without to the application of prestress, showed that with this type of core, it was possible to achieve higher results for the investigated parameters than when using the other listed materials. The lowest results were obtained for the sintered ferrite-type core.

The tests carried out using the TCCM devices made it possible to perform a comparison of the different types of materials used for the harvester core. However, it should be noted that regardless of the material used, the amount of energy obtained from the device was insufficient to power the microcontroller. Therefore, in the next step, it was decided to modify the test stand to carry out tests with the DTCCM harvester type. For this purpose, the PicoPower platform was specifically developed by the authors, based on the Hereon Hunt system. The block diagram for the research stand is shown in Figure 8. The data acquisition system that was used played a significant role in the stand. This system made it possible to obtain data with a very low level of noise in real time from all sources plugged into it.

The main component of data transfer between the Heron DSP chip and the host was one FIFO buffer that was connected to the PCI interface on the motherboard. The HEDG12 module was used to acquire the coil parameters in the form of current and voltage. This module is a 16-bit ADC converter with eight channels. Thanks to the use of this module, it was possible to acquire data on all eight channels using sigma-delta conversion.

The sigma-delta conversion technique used made it possible to eliminate analog anti-aliasing filters from the measurement path. Simply put, this technique oversamples the waveform that appears at the input eight times. The next step is to digitally filter the over-sampled data. Finally, the resulting samples are reduced eightfold.

The devices presented in this paper are not intended to serve as continuous energy sources for microcontrollers. The main purpose of the developed harvesting devices is to provide a sufficiently large energy pulse to be able to charge a large-capacity capacitor. However, this does not mean that they cannot act as such a source of energy. For the devices shown in the paper, the manner in which they work will depend on the system specification. If the system is intended to send signals every few milliseconds, then the amount of energy supplied may not be large enough to accomplish such tasks. However, if the signal is to be sent every few hours, days, etc., then such a system could be regarded as a continuous source of energy. Figure 9 shows the principle of the system's operation. Additionally, the capacitors would have to act fast enough to capture the current impulse in tens of µm.

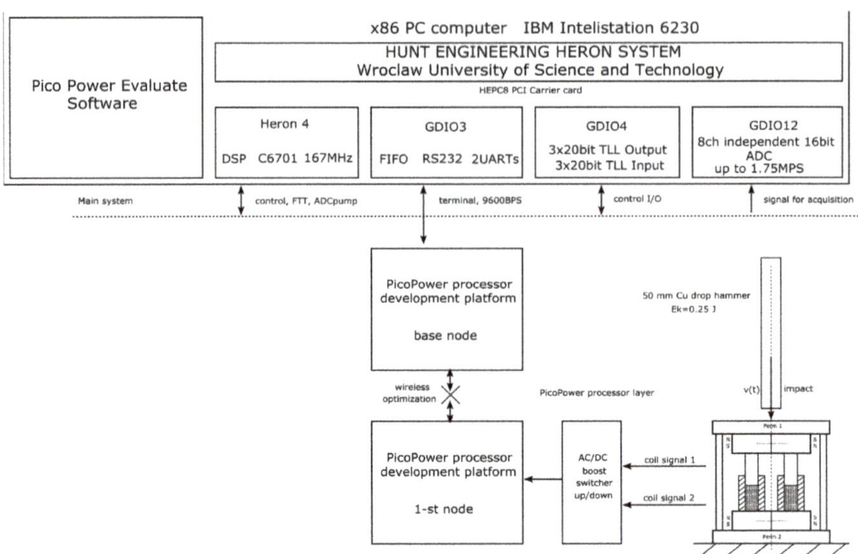

Figure 8. The PicoPower Development Platform was applied as a system to construct a new type of harvesting power supply.

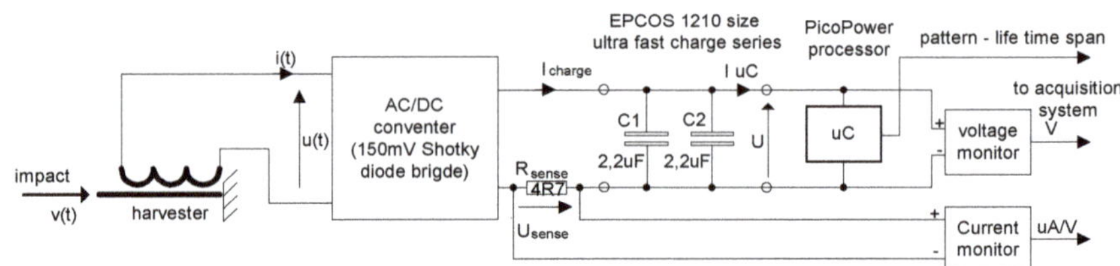

Figure 9. Supply current and voltage measurement scheme for detecting the lifetime of a powered microcontroller.

Figure 10 shows the implementation of the developed energy harvesting device in practice, i.e., with the use of a microcontroller. The voltage drop that can be noticed on the measuring resistor, $R_{sense} = 4.7$ (Figure 9), can be read as the current consumed by the capacitor–microcontroller system. Because the capacitors C1 and C2 (Figure 9) are charged in the first phase, a significant increase in current can be noticed. These capacitors represent a large load for the developed harvester. In the moment at which current consumption decreased, the voltage reached a value of $U_c = 3$ V (Figure 9), making it possible to initiate microcontroller operation. In the first step, the microcontroller performed a reset, and then started to implement the saved algorithm. The microcontroller was supposed to wait until the capacitors were charged to about 5 V before performing the next actions. In the moment at which this value was reached, the microcontroller was to start generating signal sequences, which were then counted and forwarded by the RS232 HEGD3 module.

The lifetime algorithm of the program worked from $U_{max} = 5$ V to $U_{min} = 1.8$ V. During this voltage drop, the microprocessor was able to send about 50 pulses, giving a microcontroller life of about 3 ms. By selecting various mechanical excitation sources, values of up to 200 pulses were obtained, which extended the lifetime of the microcontroller to 8 ms. Finally, an Energy Harvesting Device platform was developed that was able to supply a popular microcontroller, realizing its code for a lifetime of 3 ms with a low impact energy of $E_k = 0.25$ J. The presented device was based on a core made of Terfenol-D powder mixed with $Fe_{57}Co_{10}B_{20}Si_5Nb_4V_4$ alloy powder.

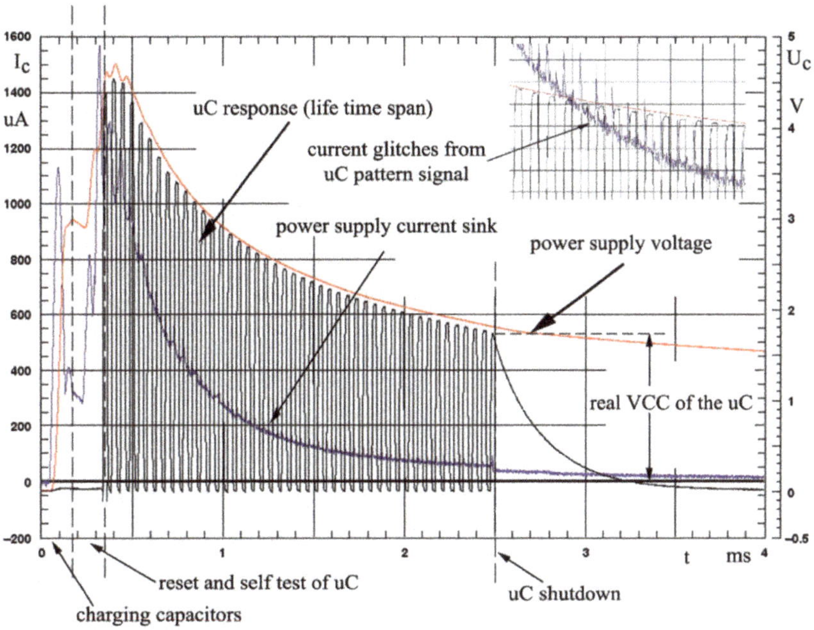

Figure 10. The lifetime of the ATMEGA48V microcontroller circuit powered by the DTCCM harvesting device at a low impact energy of $E_k = 0.25$ J.

In the case of the proposed system, it seems difficult to estimate its efficiency. It cannot be easily calculated, because the electrical efficiency of the system is dependent on the efficiency of each of its components, and the measurement path is complicated. It is obvious that the higher the efficiency of converting energy from one form to another, the better it will be. However, attention should be paid to the fact that the proposed solution needs to work in places where energy is not processed in any way, but only lost in the form of mechanical vibrations. In such cases, recovering or changing even a small part of the energy into another form of energy (in this case electric) seems to make the most sense, even though the efficiency of this transformation is not at a high level. The device shown in this work is based on the reverse magnetostriction effect, but of course, work is also underway using inverse pizoelectric effects, the Farraday effect, and electrostatic charges. In the case of the presented solution, the goal was to develop a device that would not have any moving parts, thus limiting the possibility of defects to some extent. In further research work, it is planned to modify both the measurement system and the harvester in order to increase the amount of electricity generated.

4. Conclusions

Two types of simple energy harvesting devices were developed and investigated using dedicated test stands. The first type of harvester device was used to perform basic research on the influence of the core type on the resulting values of electric energy. On the basis of this series of tests, it is possible to conclude that:

- The proper choice of the coil (the number of coils and its impedance) and the core material for the harvesting device is very important, and has a significant influence on the results.

- Providing low-frequency vibration might have significant implications for the practical realization of energy harvesting, and application in industrial objects, as it could support the optimal operation (generation of current) of the harvester for a long time. However, it should be noted that the value of current generated as a result of impacts will be much higher than that generated from vibrations.
- In the coil and PZT sensor, the same frequencies were dominant, although the PZT spectrum had further harmonics that did not occur in the current signal of the coil. It was found that the two frequencies characterizing the material of the harvester core are sufficient.
- In the low-energy range of impulses, Ek $\leq$ 5.6 J, the application of Terfenol-D rod as the core results in the device to being able to supply about ten 8-bit RISC microelements in a time of about 1 ms.
- To ensure optimal prestress, while also increasing the resonance frequency and the lifetime of waves passing through the harvester device, the TCCM structure was modified. The developed DTCCM harvester structure made it possible to apply and tune the prestress value to the core of the device.
- Finally, an Energy Harvesting Device platform was developed that was able to supply a popular microcontroller that realizes its code within a timeframe of 3 ms with a low impact energy of Ek = 0.25 J.

Moreover, the results obtained in the presented work may constitute a base for further application work with the developed materials, particularly in the field of energy harvesting.

The solution proposed in this paper may find application in various areas, including the transport and manufacturing industries. These devices could be placed under or in the vicinity of railway tracks or on railway carriages. Another example would be trucks or loading bases, where devices could be placed in the ground (e.g., around speed bumps). In addition, devices of this type could find application in the manufacturing industry, especially in steel mills and pressing plants, near presses, or hammers. Another interesting application is the use of the proposed device in ballistics equipment. As additional work, cooperation has been initiated with a group of scientists from the Department of Mechanics, Mechanical and Biomedical Engineering, researching ballistic shields. When testing such shields, large amounts of energy are generated and lost at the same time. Work is currently underway on the effective use of the proposed devices to recover at least a small part of the energy supplied to ballistic shields.

Author Contributions: Conceptualization, R.M. and P.W.; methodology, P.W.; software, K.W.; validation, R.M. and P.W.; formal analysis, P.W.; investigation, R.M., P.W. and K.W.; resources, P.W.; data curation, K.W.; writing—original draft preparation, R.M. and P.W.; writing—review and editing, R.M.; visualization, P.W.; supervision, R.M.; project administration, R.M.; funding acquisition, R.M. All authors have read and agreed to the published version of the manuscript.

Funding: This research was funded by The National Centre for Research and Development within the project: "Composite magnetostrictive-nanocrystalline materials for use in the field of energy harvesting and transformation", grant number LIDER/21/0082/L-9/17/NCBR/2018.

Institutional Review Board Statement: Not applicable.

Informed Consent Statement: Not applicable.

Data Availability Statement: Data supporting reported results can be provided upon request. Currently, these data are collected as part of the ongoing project, and only after its completion will the data be made available to the public.

Conflicts of Interest: The authors declare no conflict of interest.

References

1. Harrop, P. *An Introduction to Energy Harvesting*; IDTechEx Ltd.: Cambridge, UK, 2009.
2. Narita, F.; Fox, M. A Review on Piezoelectric, Magnetostrictive, and Magnetoelectric Materials and Device Technologies for Energy Harvesting Applications. *Adv. Eng. Mater.* **2017**, *20*, 1700743. [CrossRef]

3. Leung, C.M.; Li, J.-F.; Viehland, D.; Zhuang, X. A review on applications of magnetoelectric composites: From heterostructural uncooled magnetic sensors, energy harvesters to highly efficient power converters. *J. Phys. D Appl. Phys.* **2018**, *51*, 263002. [CrossRef]
4. Zucca, M.; Mei, P.; Ferrara, E.; Fiorillo, F. Sensing Dynamic Forces by Fe–Ga in Compression. *IEEE Trans. Magn.* **2017**, *53*, 1–4. [CrossRef]
5. Palumbo, S.; Rasilo, P.; Zucca, M. Experimental investi gation on a Fe-Ga close yoke vibrational harvester by matching magnetic and mechanical biases. *J. Magn. Magn. Mater.* **2019**, *469*, 354–363. [CrossRef]
6. Narita, F.; Katabira, K. Stress-rate dependent output voltage for Fe29Co71 magnetostrictive fiber/polymer composites: Fabrication, experimental observation and theoretical prediction. *Mater. Trans.* **2017**, *58*, 302–304. [CrossRef]
7. Yoffe, A.; Shilo, D. Themagneto-mechanical response of magnetostrictive composites for stress sensing applications. *Smart Mater. Struct.* **2017**, *26*, 065007. [CrossRef]
8. Deng, Z. Explicit and efficient discrete energy-averaged model for Terfenol-D. *J. Appl. Phys.* **2017**, *122*, 043901. [CrossRef]
9. Leon-Gil, J.A.; Cortes-Loredo, A.; Fabian-Mijangos, A.; Martinez-Flores, J.J.; Tovar-Padilla, M.; Cardona-Castro, M.A.; Morales-Sánchez, A.; Alvarez-Quintana, J. Medium and Short Wave RF Energy Harvester for Powering Wireless Sensor Networks. *Sensors* **2018**, *18*, 768. [CrossRef]
10. Loreti, P.; Catini, A.; De Luca, M.; Bracciale, L.; Gentile, G.; Di Natale, C. The Design of an Energy Harvesting Wireless Sensor Node for Tracking Pink Iguanas. *Sensors* **2019**, *19*, 985. [CrossRef]
11. Aoudia, F.A.; Gautier, M.; Magno, M.; Berder, O.; Benini, L. Leveraging Energy Harvesting and Wake-Up Receivers for Long-Term Wireless Sensor Networks. *Sensors* **2018**, *18*, 1578. [CrossRef]
12. Lai, Y.C.; Hsiao, Y.C.; Wu, H.M.; Wang, Z.L. Waterproof fabric-based multifunctional triboelectric nanogenerator for uni-versally harvesting energy from raindrops, wind, and human motions and a self-powered sensors. *Adv. Sci.* **2019**, *6*, 1801883. [CrossRef] [PubMed]
13. Shi, S.; Yue, Q.; Zhang, Z.; Yuan, J.; Zhou, J.; Zhang, X.; Lu, S.; Luo, X.; Shi, C.; Yu, H. A Self-Powered Engine Health Monitoring System Based on L-Shaped Wideband Piezoelectric Energy Harvester. *Micromachines* **2018**, *9*, 629. [CrossRef] [PubMed]
14. Alrashdan, M.H.S.; Hamzah, A.A.; Majlis, B. Design and optimization of cantilever based piezoelectric micro power generator for cardiac pacemaker. *Microsyst. Technol. Micro Nanosyst. Inf. Storage Process. Syst.* **2014**, *21*, 1607–1617. [CrossRef]
15. Deng, Z.; Dapino, M.J. Review of magnetostrictive vibration energy harvesters. *Smart Mater. Struct.* **2017**, *26*, 103001. [CrossRef]
16. Backman, G.; Lawton, B.; Morley, N.A. Magnetostrictive Energy Harvesting: Materials and Design Study. *IEEE Trans. Magn.* **2019**, *55*, 1–6. [CrossRef]
17. Deterre, M.; Lefeuvre, E.; Zhu, Y.; Woytasik, M.; Boutaud, B.; Molin, R.D. Microblood pressure energy harvester for intra cardi-acpacemaker. *J. Microelectromechanical Syst.* **2014**, *23*, 651–660. [CrossRef]
18. Whelan, M.; Fuchs, M.P.; Gangone, M.V.; Janoyan, K.D. Development of a wireless bridge monitoring system for condition assessment using hybrid techniques. *Proc. SPIE* **2007**, *6530*, 65300H. [CrossRef]
19. Song, Y. Finite-Element Implementation of Piezoelectric Energy Harvesting System from Vibrations of Railway Bridge. *J. Energy Eng.* **2019**, *145*, 04018076. [CrossRef]
20. Harrop, P.R. *Energy Harvesting and Storage for Electronic Devices 2009–2019*; IDTechEx Report; IDTechEx Ltd.: Cambridge, UK, 2009.
21. Herzer, G. Grain size dependence of coercivity and permeability in nanocrystalline ferromagnets. *IEEE Trans. Magn.* **1990**, *26*, 1397–1402. [CrossRef]
22. Suzuki, K.; Cadogan, J.M. Random magnetocrystalline anisotropy in two-phase nanocrystalline systems. *Phys. Rev. B* **1998**, *58*, 2730–2739. [CrossRef]
23. Shkuratov, S.I.; Talantsev, E.F.; Dickens, J.C.; Transverse, K.M. Shock wave demagnetization of NdFeB high-energy hard ferromagnetic. *J. Appl. Phys.* **2002**, *92*, 3007–3009. [CrossRef]

Article

Kinetic Electromagnetic Energy Harvester for Railway Applications—Development and Test with Wireless Sensor

Zdenek Hadas *, Ondrej Rubes, Filip Ksica and Jan Chalupa

Faculty of Mechanical Engineering, Brno University of Technology, 616 69 Brno, Czech Republic; Ondrej.Rubes@vut.cz (O.R.); Filip.Ksica@vutbr.cz (F.K.); chalupa@fme.vutbr.cz (J.C.)
* Correspondence: hadas@fme.vutbr.cz

Abstract: This paper deals with a development and lab testing of energy harvesting technology for autonomous sensing in railway applications. Moving trains are subjected to high levels of vibrations and rail deformations that could be converted via energy harvesting into useful electricity. Modern maintenance solutions of a rail trackside typically consist of a large number of integrated sensing systems, which greatly benefit from autonomous source of energy. Although the amount of energy provided by conventional energy harvesting devices is usually only around several milliwatts, it is sufficient as a source of electrical power for low power sensing devices. The main aim of this paper is to design and test a kinetic electromagnetic energy harvesting system that could use energy from a passing train to deliver sufficient electrical power for sensing nodes. Measured mechanical vibrations of regional and express trains were used in laboratory testing of the developed energy harvesting device with an integrated resistive load and wireless transmission system, and based on these tests the proposed technology shows a high potential for railway applications.

Keywords: energy harvesting; train; electromagnetic transducer; model; vibration; test; wireless sensor

Citation: Hadas, Z.; Rubes, O.; Ksica, F.; Chalupa, J. Kinetic Electromagnetic Energy Harvester for Railway Applications—Development and Test with Wireless Sensor. *Sensors* **2022**, *22*, 905. https://doi.org/10.3390/s22030905

Academic Editor: Amir H. Alavi

Received: 12 December 2021
Accepted: 23 January 2022
Published: 25 January 2022

Publisher's Note: MDPI stays neutral with regard to jurisdictional claims in published maps and institutional affiliations.

1. Introduction

Modern railways are required to provide an improved quality of service and high levels of safety. Reliable trackside infrastructure maintained in good condition is important for smooth transportation of goods and passengers. To accomplish that, preventive maintenance and scheduled maintenance techniques are currently being used for trackside infrastructure, which can reveal critical wears, defects or failures. However, continuous condition monitoring and long-time sensing using modern electronics could detect incipient wears, failures and degradation that could affect safe railway operation.

Monitoring of trackside systems is important in order to reveal significant changes in functional parameters (e.g., deformation, vibration and temperature). This type of monitoring and diagnostics is widely known as condition-based maintenance, and its main goal is to provide significant savings in infrastructure operational costs. Predictive maintenance techniques require detailed trackside monitoring and the employment of many sensing systems. Reliable and low-maintenance power supplies are essential prerequisites to reliable predictive maintenance results.

Electrical power for these monitoring systems could be delivered from a catenary that is a part of the trackside infrastructure electrical grid. The catenary, however, could be difficult to access due to tight restrictions set up by the infrastructure management and operation in order to maintain the reliability of the track systems. Furthermore, even in the case of a modern railway network, many electrical systems access points are still remote or quite difficult to access due to poor infrastructure and a lack of foresight in regard to modern wireless sensing system power management. Cables and wires are an expensive part of the infrastructure, are often subject to theft, and are difficult to maintain, especially when the layout of railway tracks is changed. Auxiliary railway systems are ready to accept alternative power sources and achieve economically efficient operation by using alternative and energy harvesting sources to power them. Renewable energy sources, such as solar panels or wind turbines, could be used for remote applications that are quite demanding in terms of their power consumption (e.g., warning and signal lights, track switches, grade crossing signals, point machines, positive train control systems and train positions, communication access points etc.). Other energy harvesting sources are widely discussed for embedded monitoring systems in railways. Energy harvesting has been used for wireless sensor nodes and low-power autonomous systems for over 20 years [1]. In general, energy harvesting is based on the conversion of ambient energy into useful electricity. In trackside environment, the passing train by itself could deliver a wide variety of ambient mechanical energy (e.g., mechanical vibration, rail deformation, the sag of sleepers or rails etc.) that could be utilized for such systems.

Individual trackside energy harvesting technologies are summarized in this paper, and the physical principle of a kinetic energy harvesting solution based on converting track vibration into electricity is proposed. On the basis of a mathematical model, a design for a maintenance-free kinetic energy harvester is developed and described, including experimental results and the testing of a complete system with a sensor node.

2. Energy Harvesting Technologies for Trackside Applications

Recent developments in wireless technologies have resulted in a significantly smaller size, lower price and decreased energy consumption of these systems. For this reason, in railways applications wired sensors are often being abandoned and replaced by wireless alternatives. Their main advantages, on top of the abovementioned ones, are their easy installation and simplified maintenance. The primary battery source and operation in low-power mode could assure the reliable operation of these sensors for more than a year [2]. However, such a period is still close to the required maintenance period of the sensor itself. For this reason, energy harvesting technologies are investigated in order to achieve several years of maintenance-free operation of these sensor nodes.

Wireless sensor nodes with autonomous energy harvesters could find their way into various engineering applications, as they could operate autonomously in maintenance-free mode for long periods of time. As an example, these solutions are currently used in heavy industry applications, structural health monitoring systems [3], aerospace [4] and transportation [5]. Current energy harvesting technologies have been investigated as a possible source of power for wireless applications, which would otherwise be difficult to connect to the existing power grid.

Track condition monitoring applications are developed on the basis of acceleration sensors [6] or strain gauges [7], mainly for the condition monitoring of a crossing [8–10]. Bridge monitoring solutions have also been widely discussed in recent publications. Paper [11] discussed the feasibility of a bridge monitoring system in terms of its operational life. It illustrated how the traffic on a bridge over time could accentuate the identification of damage, which was necessary to know the state and health of the structure. A segmental prefabrication and assembly of the bridge on the Guangzhou Metro was presented in [12]. Passing vehicles induced vibrations used for energy harvesting, and using appropriate modelling and dynamic analyses of the bridge system a new type of electromagnetic vibration energy harvester was proposed. This device was designed in a way that could power strain-collection units for a bridge health monitoring system. These typical sensing applications, such as crossings and bridges, provide a measurable dynamic response of the track infrastructure to the passing train. In this case, the measured response from the passing train could serve as a suitable source of energy for monitoring applications.

Many published papers dealing with trackside energy harvesting solutions showed that harvesting energy from passing trains has a great potential for wireless sensing applications in railways. Authors of paper [13] investigated the possibility of establishing a self-powered wireless sensor network by integrating the ZigBee stack protocol together with an energy harvesting power source. This system is used for the condition monitoring of urban rail transit utilizing localized energy harvesting. Authors of the previous article also present another complex system for the smart monitoring of an underground railway by local energy generation in paper [14].

A study and the results of a portable electromagnetic energy harvesting system are presented in papers [15,16]. Their proposed solution consists of a mechanical transmission and an electrical regulator that converts sags in the rail into electricity, providing a peak voltage of 58 V at 1 Hz with a displacement of 2.5 mm. Authors from Stony Brook presented a preliminary prototype of a mechanical-rectifier-based harvester [17]. This study illustrated that sufficient power can be harvested by the device, which is based on a motion rectifier design. A novel direct-motion-driven harvester was described in publication [18], where the authors describe how an anchorless mounting results in a higher power capacity without the requirement of any special preparation during its installation. An installation and test under a fully loaded freight train running at 64 km/h was also presented in this paper. Paper [19] presented a design, modelling, in-lab experiment and field-test results of a mechanical motion rectifier mechanism which is based on a compact ball-screw-based electromagnetic energy harvester. A theoretical study of a cam mechanism was presented by the University of Nebraska in publication [20], where it was used to exploit the contact between a train wheel and a harvester mechanism to drive an electromagnetic generator. A solution based on a direct load piezoelectric harvesting device was proposed in paper [21], offering a structurally simple solution in the form of a piezoelectric drum device placed under sleepers. Piezoelectric solutions for strain-based energy harvesting have been widely discussed, where piezoceramic patches or piezo stacks transduce deformation into electricity [22].

The abovementioned technology mainly converts a direct train load in form of direct contact, deformation or element strain. These devices have the potential to provide peak output power of several watts; however, they are not suitable for high-speed rail applications due to their necessity for a mechanical contact. In contrast, the subsequently presented drum and patch type piezoelectric element solutions are suitable for high-speed operation at the cost of a lower power output in the range of several microwatts. These piezoelectric elements provide a very high voltage but a low current. This disadvantage could be eliminated by multilayer piezoelectric composites; however, the manufacturing of such materials is a very expensive process and for this reason it is not suitable for cost-effective wireless sensor nodes.

Kinetic energy harvesting solutions capable of transducing kinetic energy from vibrations under the passing train into electricity serve as maintenance-free sources of energy. Authors of publication [23] investigated the possibility of harvesting energy from the vertical vibrations of sleepers generated by passing trains at various speeds. A model combining the track structure and the energy harvesting system was used. Results indicated the generated power was around 100 mW, assuming a 2 mm rail displacement amplitude at a frequency of 6 Hz. The presented track model was validated with UK network experimental data. Testing of a piezoelectric vibration cantilever harvester in publication [24] was focused on energy harvesting at a frequency of 5 to 7 Hz. An output power of 4.9 mW and a peak-to-peak voltage of 22.1 V were achieved on a rail vibrating with amplitudes of 0.2 to 0.4 mm at a frequency of 7 Hz. A design of a resonant electromagnetic harvester was published in papers [13,25]. An approach based on magnetic levitation was capable of energy harvesting at a broadband low-frequency vibration in range of 3 to 7 Hz. This device induced a peak-to-peak voltage 2.32 V and an output power of 119 mW when subjected to vibrations with 1.2 mm amplitudes and with an optimal resistive load of 44.6 Ohm.

An innovative approach was presented in paper [26], where authors deployed piezoelectric energy harvesting devices for monitoring a full-scale bridge structure undergoing forced dynamic testing by passing trains. A similar approach was used in paper [27], where the damage detection and structural health monitoring of a laboratory-scaled bridge was observed using a vibration energy harvesting device, in particular a cantilever-based piezoelectric energy harvesting device. The published approach had an advantage over the conventional accelerometer-based method in terms of power requirements, because energy storage and data transmission units were the only power-consuming parts of the system.

3. Model Based Design of Electromagnetic Trackside Energy Harvester

A passing train provides mechanical vibrations in rails and sleepers. The vertical deflection of a sleeper is depicted in Figure 1 using the variable z. This sag in a sleeper depends on the passing train's mass, velocity and the quality of the rail subgrade. The proposed energy harvesting system is based on a principle of kinetic energy harvesting which can convert the kinetic energy from sleeper oscillations into useful electricity. A mechanical resonator is used for the transfer of input kinetic energy into the free oscillation of a seismic mass. A design of this kinetic energy harvester is based on a mass m which is suspended on two steel cantilevers with a known stiffness k_1 and known mechanical damping d_m. This longitudinal design could be placed on the top of a sleeper, or it could be embedded inside a new generation of innovative sleepers.

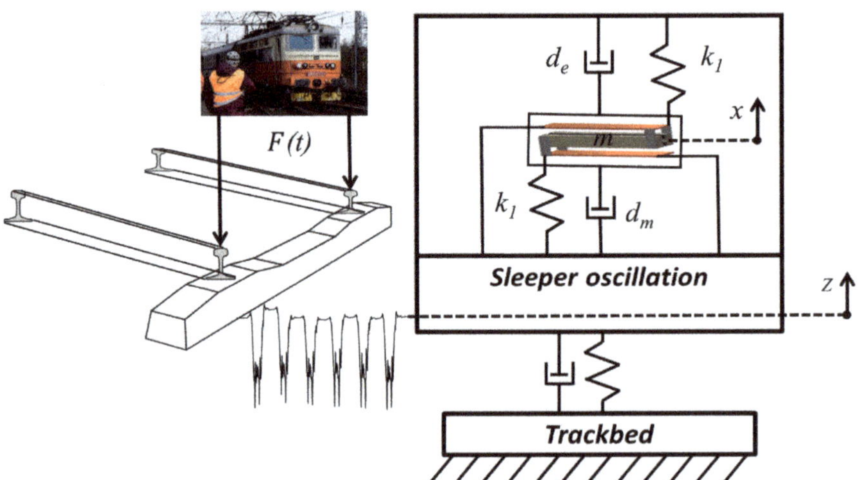

Figure 1. Physical principle of kinetic energy harvester under passing train vibrations.

On the basis of the previously published analysis, the electromagnetic energy transducer provides an effective harvesting power for this application. The oscillating mass is a part of a magnetic circuit, and its free oscillation x against a fixed coil generates useful electricity, which provides electromagnetic damping forces in the form of electrical damping d_e.

3.1. Mathematical Model of One Degree of Freedom System

The kinetic electromagnetic energy harvesting system could be described by a multidomain model in the form of coupled mechanical and electrical systems, as depicted in Figure 2. The mechanical resonator is excited by ambient mechanical shocks z to the oscillating sleeper and this results in the relative movement x. The relative movement x of the mass m in the magnetic circuit consisting of the frame and a fixed coil is inversely proportional to the mechanical damping d_m. Due to Faraday's law, the relative movement of the magnetic circuit results in a change in the magnetic field of the coil, inducing an electromotive voltage u_i. A model of an electromagnetic coupling coefficient was used for the description of the interaction between both the mechanical and electrical domains. The induced voltage depends on the design of the electromagnetic transducer (the electromagnetic coupling coefficient c_{EH}) and its relative velocity. When a resistive electrical load R_L is connected to a coil, then a current flows through the coil and electrical power is extracted from the system. The electrical power extracted from this system provides electromechanical feedback in a form of an electrical damping, which is depicted as a damper d_e. This electrical damping feedback is proportional to the electromagnetic coupling coefficient c_{EH}. A derived mathematical model with one degree of freedom was used for predicting the harvested power in a resonance operation.

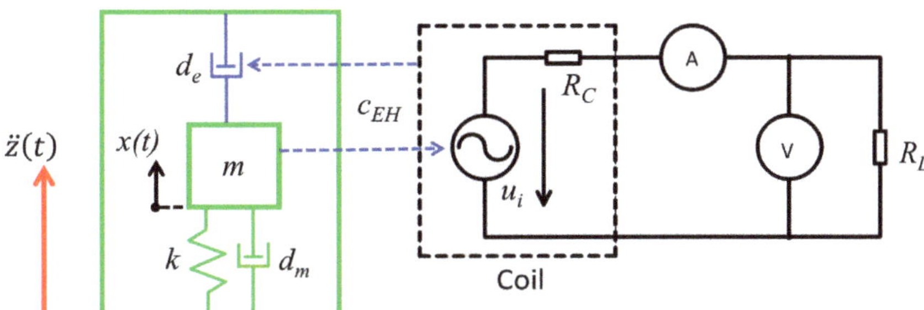

Figure 2. Coupled mechanical and electromagnetic models of the proposed kinetic energy harvester.

A second-order equation according to the mechanical model in Figure 2 describes mechanical oscillations of a seismic mass as a response to the kinetic excitation of the sleeper:

$$m\ddot{x} + d_m\dot{x} + d_e\dot{x} + kx = -m\ddot{z} \tag{1}$$

where x is the relative displacement of the oscillating mass, z is the absolute displacement of the vibrating sleeper, m is the moving mass, d_m is the mechanical damping, d_e is the electrical damping, and k is the mechanical stiffness.

The mechanical stiffness combines stiffness of both cantilevers. The stiffness of a single beam k_1 can be calculated using this equation:

$$k_1 = \frac{3 \cdot E \cdot J}{l^3} \tag{2}$$

where E is Young's modulus of the used material (steel), J is a second moment of the area, and l is the length of the cantilever. The mechanical stiffness k is then simply $2 \cdot k_1$ for this double suspended system.

The mechanical damping d_m can be calculated using this relation:

$$d_m = \frac{1}{2Q_m} 2m\Omega \tag{3}$$

where Q_m is the mechanical quality factor, either estimated or calculated from an experiment. The natural frequency Ω can be calculated using a commonly known formula for a single degree of freedom system:

$$\Omega = \sqrt{\frac{k}{m}} \tag{4}$$

The electrical damping d_m of the electromechanical system can be calculated using this equation:

$$d_e = \frac{(BNl)^2}{R_C + R_L} = \frac{(c_{EH})^2}{R_C + R_L} \tag{5}$$

where B is the magnetic flux density in the coil, N is the number of turns, l is the active length of one turn, R_C is the coil resistance, R_L is the load resistance, and c_{EH} is the electromagnetic coupling coefficient of the energy harvester, where $c_{EH} = BNl$.

The induced voltage on the coil u_i, can be calculated using equation:

$$u_i = BNl \cdot \dot{x} = c_{EH} \cdot \dot{x} \tag{6}$$

The first-order electric differential equation of the electrical circuit in Figure 2 is then:

$$L \cdot \frac{di}{dt} + i \cdot (R_C + R_L) = u_i \tag{7}$$

where L is the inductance of the coil, and i is the electric current. By design, the inductance of the coil in our harvester is very small (for a coil with an air core) and the current change is very slow, therefore the first term is irrelevant and the equation can simplified:

$$i = \frac{c_{EH} \cdot \dot{x}}{R_C + R_L} \tag{8}$$

The coupled mechanical equation can modified, where Equations (5) and (8) provide the electrical damping as a function of the electric current:

$$m\ddot{x} + d_m\dot{x} + c_{EH}i + kx = -m\ddot{z} \tag{9}$$

The fundamental performance of the energy harvester is the equation for the output power:

$$p_{out} = i^2 \cdot R_L \tag{10}$$

The displacement amplitude (peak values) could be simply calculated from these equations assuming a resonance operation. The mechanical amplitudes of both the displacement and velocity follow these relations:

$$x_A = z_A Q_T = \frac{\ddot{z}_A}{\Omega^2} Q_T \rightarrow \dot{x}_A = \frac{\ddot{z}_A}{\Omega} Q_T \tag{11}$$

where the term Q_T is the total quality factor of both the mechanical and electrical damping. This quality factor is a compound on the basis of the following relation:

$$Q_T = \frac{1}{2\frac{d_m + d_e}{2m\Omega}} = \frac{m\Omega}{d_m + d_e} \tag{12}$$

The calculation of the velocity amplitude can be used for the calculation of the amplitude of the induced voltage:

$$u_{i_A} = c_{EH} \cdot \dot{x}_A \tag{13}$$

and the amplitude of the output voltage on the resistive load is:

$$u_{L_A} = u_{i_A} \frac{R_L}{R_C + R_L} \tag{14}$$

The output power amplitude can then be expressed using either voltage or current:

$$p_{out_A} = i_A^2 \cdot R_L = \frac{u_{L_A}^2}{R} \tag{15}$$

3.2. Design of Energy Harvesting Device

The proposed design of the electromagnetic kinetic energy harvester could be capable of converting sleeper vibrations into useful electricity. Resonance operation is not possible due to the pulse excitation characteristics produced by the passing train. However, the free oscillation response to the passing train provides a relative oscillation of the suspended seismic mass against the fixed base with a coil, which has the potential to generate satisfactory levels of useful electrical power. In the case of the longitudinal design of the device mounted on top of the sleeper, the suspension system consists of a pair of steel cantilevers with dimensions of $400 \times 30 \times 3$ mm^3. The mechanical resonator is by design tuned up to have a natural frequency of 12 Hz, which provides a sufficient relative movement. The long steel cantilever design results in a mechanical resonator with one degree of freedom in the vertical direction, which makes it sensitive to the train induced vibrations. The relative movement amplitude is important for a correct design of the magnetic circuit, which is fixed inside the seismic mass. A concept of a sleeper kinetic energy harvester design for trackside application is shown in Figure 3.

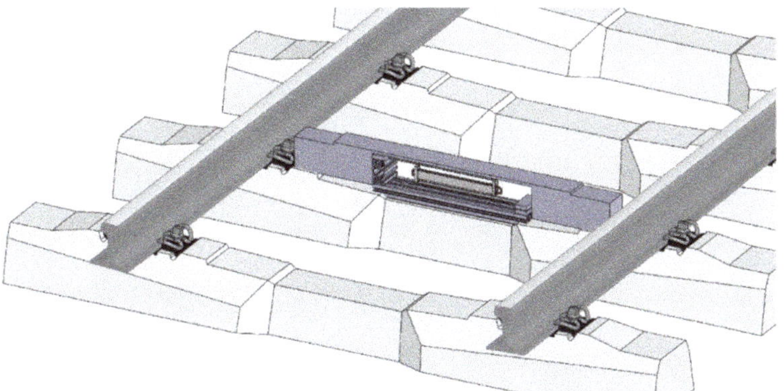

Figure 3. Proposed integration of KEH design for railway applications.

The fundamental part of the seismic mass is a magnetic circuit with 16 rare earth FeNdB magnets and ferromagnetic holders. A pair of ferromagnetic holders with permanent magnets moves freely in the air gap of a fixed coil. A planar finite element analysis of this magnetic circuit was conducted in order to calculate the average magnetic flux density in the area of the coil for a given relative movement. The coil was designed to have an air core and wound up around a plastic frame fixed to the base. All active turns of the coil were placed in the air gap of the magnetic circuit. A relative position of the magnetic circuit and coil was set with a minimal air gap to achieve efficient electro-mechanical energy conversion. This analysis was done in an FEMM environment and the calculated magnetic field is shown in Figure 4.

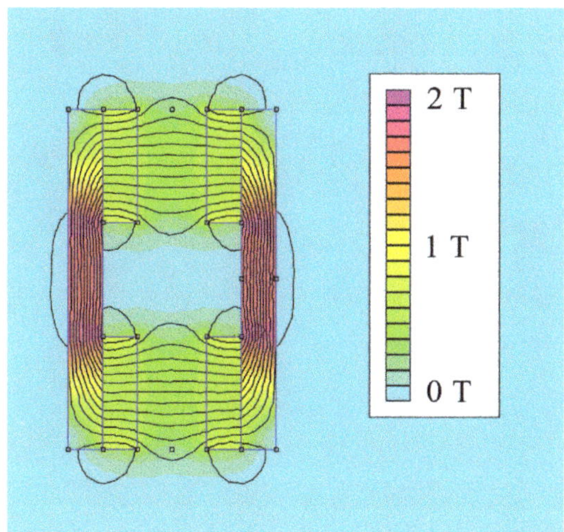

Figure 4. Planar FEMM model of magnetic circuit; analysis of magnetic flux density B.

The developed and assembled kinetic electromagnetic device for trackside application is shown in Figure 5 and it consists of:

1. A base (1) fixed on a vibrating structure,
2. The flexible suspension of a resonator (2)—its stiffness is provided by a pair of steel cantilevers,
3. A resonator mass (3) with a magnetic circuit inside,
4. A self-bonded air coil with a plastic coil holder (4).

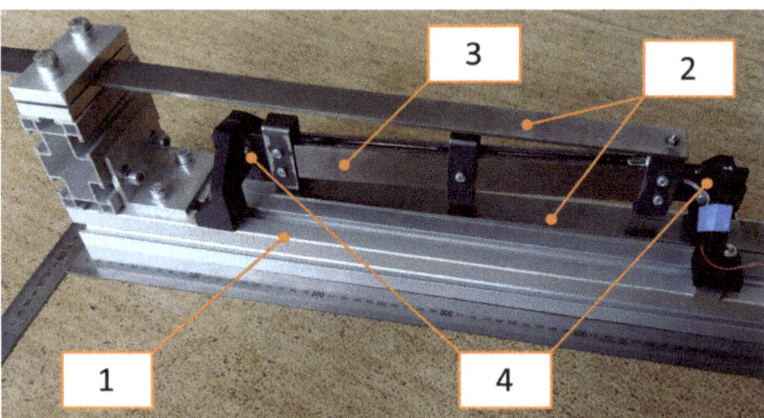

Figure 5. Design of proposed kinetic energy harvester.

The model from the previous chapter was used for the design of the individual parameters with respect to the required harvested power. The parameters of the final model and the assembled device (see Figure 5) are summarized in Table 1.

Table 1. Parameters of individual harvesters used in experiments.

Parameter	Symbol	Value
Total weight	-	3.9 kg
Total dimensions	-	$600 \times 160 \times 90$ mm^3
Moving mass	m	0.8 kg
Resonance frequency	Ω	12 Hz
Mechanical quality factor	Q_M	150
Coil dimensions	-	$210 \times 25 \times 3$ mm^3
Coil wire diameter	-	0.15 mm
Coil turns	N	300
Coil resistance	R_C	150 Ω
FeNdB magnetic circuit dimensions	-	Two pairs, $3 \times 10 \times 180$ mm^3
Air gap	-	6 mm
Average magnetic flux density	B	0.3 T

4. Electromagnetic Kinetic Energy Harvester Testing with Resistive Load

4.1. Resonance Operation: Model Results and Experiment

The designed parameters of the model are used to predict the output voltage and power in a resonance operation. The presented electro-mechanical equations in combination with the model of peak voltage and peak power described in Section 3.1 were used for output calculations for a variable resistive load. The experiment was conducted on a laboratory shaker, an RMS SW8142–SWH600APP, connected to its auxiliary measurement instruments and depicted in Figure 6. The harvested voltage was measured on an oscilloscope, a Rigol MSO 5204. Both the model and experiment were excited in a resonance frequency with an acceleration amplitude of 1 ms^{-2}.

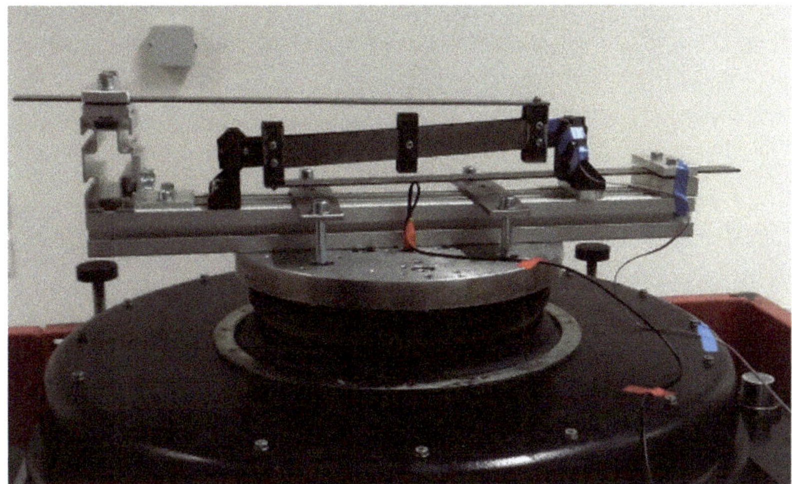

Figure 6. Shaker lab test of kinetic energy harvester.

The calculated output voltage and power are shown in Figure 7, and both outputs are compared with values obtained from the experiment. The correlation between the model and experiment is very good for low values of the resistive load. Based on the harvester model, the maximal power was expected with a resistive load of 3 kΩ. However the experimental results showed that the maximal power was harvested for a resistive load of 2 kΩ, but the harvested power was very similar across a wide range of resistive loads, 2–3 kΩ. The experimentally measured voltage and power for a higher resistance were lower than the theoretical values and it seemed that the real damping was higher compared to the damping model at higher speeds, causing a less pronounced increase in the output voltage due to the voltage being proportional to the speed.

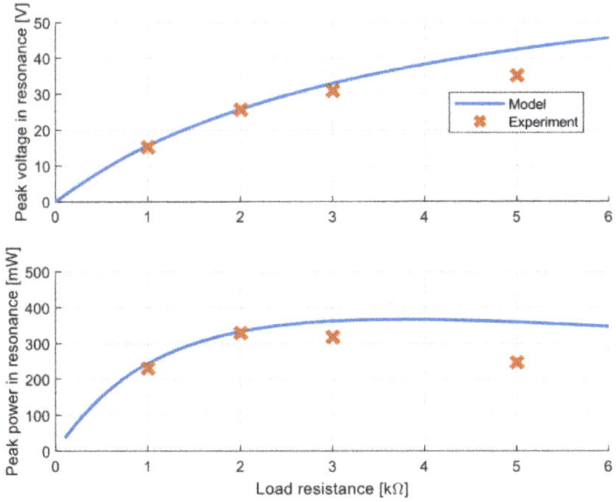

Figure 7. Voltage and power responses in resonance operation vs. load resistance—simulation results and measurements with excitation acceleration amplitude of 1 ms^{-2}.

4.2. Test of Kinetic Energy Harvester—Harmonic Vibration

The frequency response is very important for characterizing an energy harvester's performance. A shaker test with sinusoidal vibrations was used to measure the frequency response around resonance. Both sweep up and sweep down tests with a slowly changing input frequency of vibration were used for voltage and output power response measurements in the frequency domain. The input acceleration with an amplitude of 1 ms^{-2} and frequency rate of change of 0.1 Hz/s were used in the tests. These sweep tests were realized around the resonance frequency with different load resistance.

The measured peak voltage and peak power are depicted in Figure 8. The lab experiment was made with four different load resistances: 1, 2, 3 and 5 kΩ. The voltage response was proportional to the load resistance. It was caused by an increase in the velocity due to decreasing electrical damping. However, the output power peaks at around 2 kΩ, and below and above that value the power amplitudes decreased. This power maximum was observed in a resonance operation, but outside of the resonance operation the device harvested higher power with the lowest load resistance of 1 kΩ. The kinetic energy harvester with a lower resistance provided higher power outside of a resonance operation. Based on this fact, it is suitable to use a resistive load of 2 kΩ in a resonance operation but use a lower value of 1 kΩ for a non-resonance operation. This fact is discussed further in the next section.

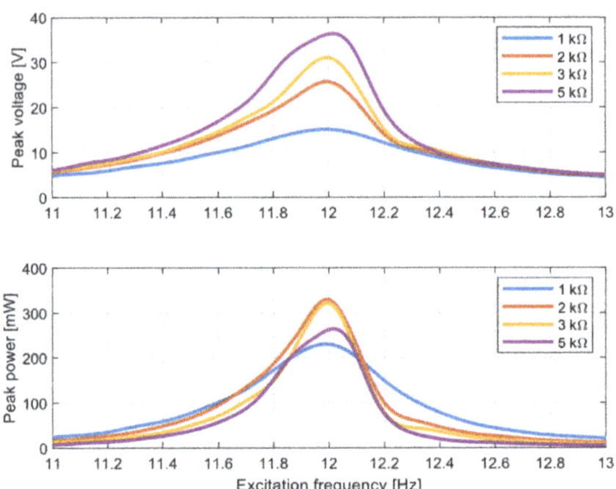

Figure 8. Voltage and output power depending on excitation frequency with different load resistance. Video S1.

4.3. Test of Kinetic Energy Harvester—Train-Induced Vibrations

The typical transient dynamic response of a kinetic energy harvester was provided with real train-induced vibrations as inputs. Real vibrations in the trackside sleeper, where the vibrations were not pure sinusoidal, but rather had the characteristic of a series of mechanical pulses, are shown in Figure 9. The acceleration and displacement of the sleeper was measured on a trackside in the Czech Republic using an inertial accelerometer on the sleeper and a capacitive displacement sensor mounted between the sleeper and a fixed point. This detail is for a single bogie consisting of two train wheels. The movement of the sleeper had general characteristics based on many parameters of the track subgrade. For this particular sleeper movement, the shown input vibrations for whole trains are used for the kinetic energy harvester test. Experiments with different electric loads were made with the aim of finding an optimal resistive load for a maximal energy harvesting potential.

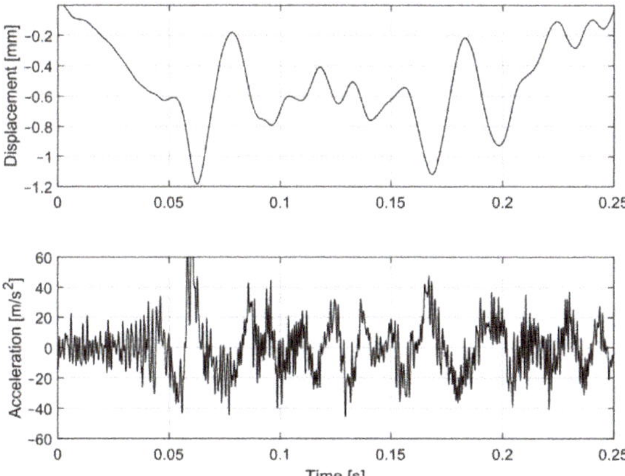

Figure 9. Typical vibrations of sleeper under passing train—detail of bogie/two wheels.

Real acceleration and displacement measurements of sleeper movement for two typical trains were used for lab tests of the developed kinetic energy harvesters:

- Train 1—Regional train travelling at 80 km/h,
- Train 2—Express train travelling at 130 km/h.

As Figure 10 shows, the maximal average power was measured with a resistive load of 150 Ω, which was the same as the coil resistance. The output power was lower at higher resistive loads due to lower damping and lower energy harvesting from trackside vibrations. The power was lower at lower resistive loads due to higher energy dissipation on the coil resistance rather than the load resistance.

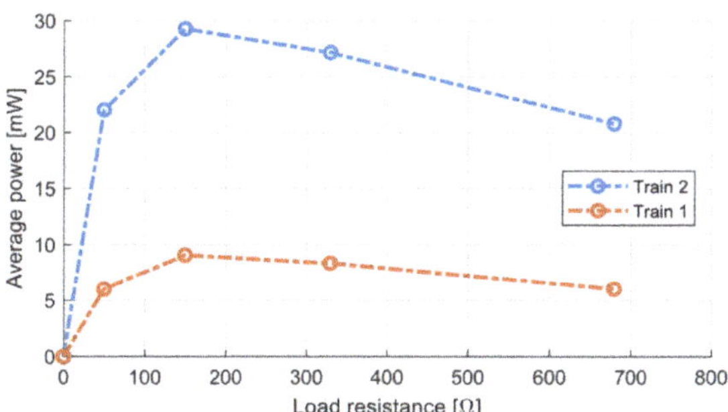

Figure 10. Average power depending on load resistance for Train 1 and Train 2.

Different weights of the seismic mass of this device were tested in order to achieve the optimal design of this energy harvesting device. The default weight of the moving mass, 800 g, was increased using an additional external weight. Unfortunately, decreasing the weight below the default value without significant structural modifications was not possible. A relation of this mass value to the generated power was experimentally verified, and the results of his test are shown in Figure 11. It was possible to add an extra mass to the initially calculated seismic mass of 800 g; however, it is evident that the output power would decrease with higher values of the seismic mass, and it was verified that the modelled mass of 800 g provided the most effective energy harvesting operation with the given train vibration data. On the other hand, it was not possible to decrease the mass in the current design due to the nature of the magnetic circuit, which needs a minimal cross-section area of the core to function properly.

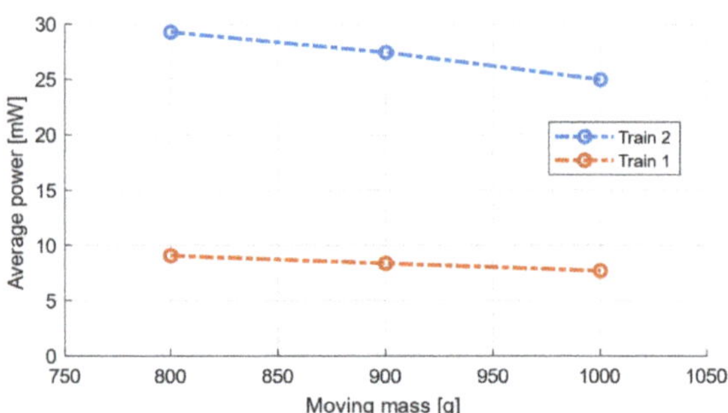

Figure 11. Average power depending on moving mass for Train 1 and Train 2.

Time-domain measurements of the voltage and power with the excitation acceleration and displacement for both testing trains are shown in Figure 12. These measurements were done with the optimal resistive load of 150 Ω. The average harvested power and total harvested energy for a passing train are presented in Table 2.

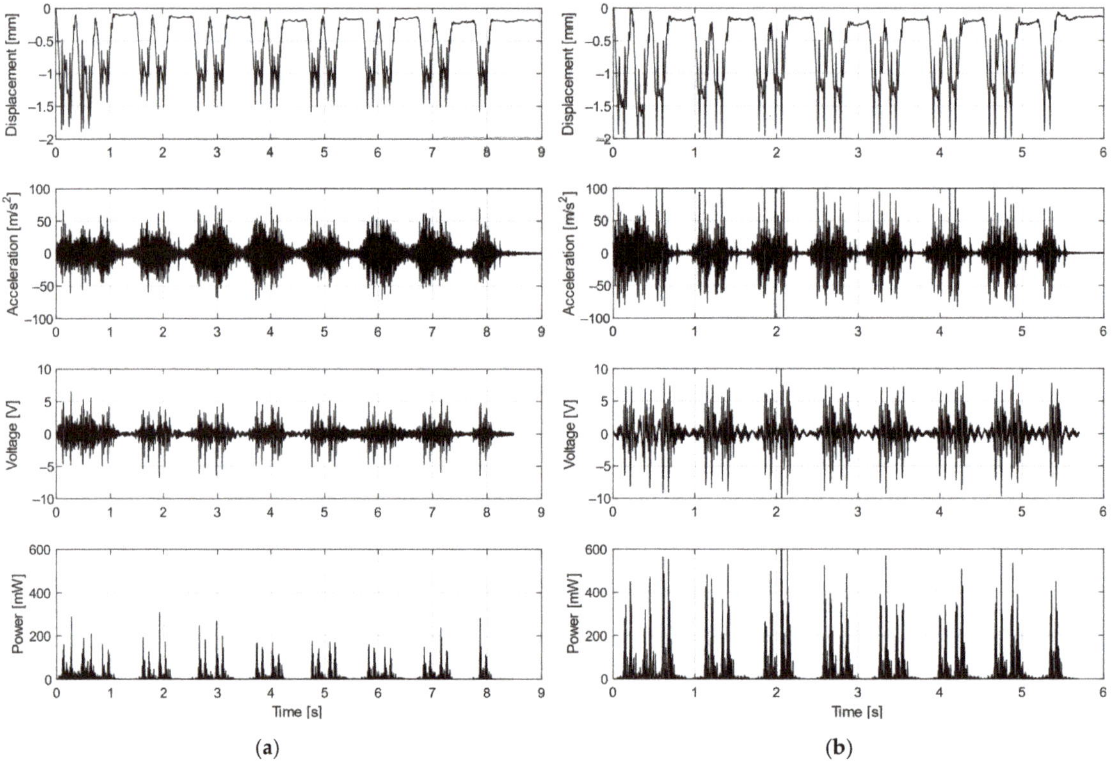

(a)   **(b)**

Figure 12. Lab shaker test with real acceleration data from regional Train 1 (**a**) and from express Train 2 (**b**); measured voltage and power with load resistance of 150 Ω for input mechanical vibrations of the shaker.

Table 2. Harvested power and energy with optimal load resistance.

Acceleration Source	Average Power	Harvested Energy
Regional Train 1	9.1 mW	77 mJ
Express Train 2	29.3 mW	167 mJ

Both the voltage and power strongly depended on the input acceleration during the train's passage. While the average power was 9.1 mW for the regional Train 1, the maximum power during much more pronounced acceleration peaks was up to 300 mW. In the case of the express Train 2, this vibration could generate average power of 29.3 mW and several peaks above 600 mW could be observed. Given all the information it is necessary to mention that the variable nature of the output power must be considered during the design of power management electronics for railway applications.

5. Kinetic Energy Harvester as Source of Energy for Wireless Sensing

The previous chapter shows that the kinetic energy harvester could generate useful electric power from passing trains. However, a test with power management electronics, a sensing node, and a communication module is required to fully assess the feasibility of this energy harvesting technology. For this reason, operational tests of the kinetic energy harvester utilized as a source of energy for a wireless sensor node were conducted for both types of trains. Our lab successfully tested autonomous wireless sensor nodes with the vibration energy harvester in lab conditions, and these results, including the design of a power management, sensing unit and communication module, were successfully published in our research paper [28].

The electric diagram of an energy harvesting system with a sensor node and communication module is shown in Figure 13. The developed kinetic energy harvester is connected to previously published electronic systems. It consists of a rectifier, a storage capacitor with a capacity of 493 µF, an LTC 3588 power management circuit, an analogue front end for the sensing signal, and a communication module based on the NRF24L01 chip from Nordic Semiconductor.

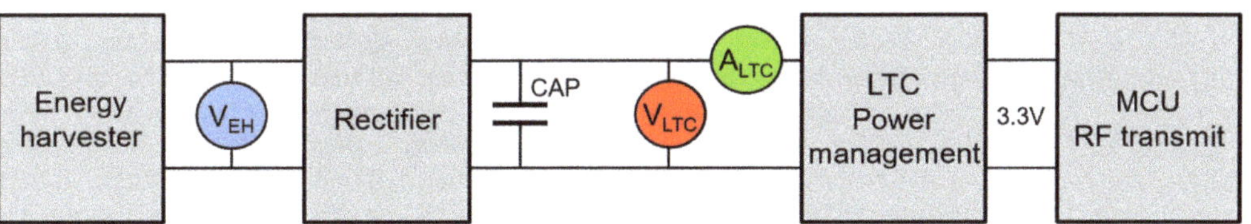

Figure 13. Schematic diagram of energy harvesting system with highlighted electrical parameters measurement.

Measured vibration signals for both types of passing trains (the same as those in the previous section) were used for the test of this autonomous source of energy for the sensing node and communication module. Measurements of electrical signals highlighted in Figure 13 (the voltage induced by the energy harvester, the input voltage and the current into the LTC power management circuit of the wireless sensor node) were used for the harvester performance assessment.

Experimental results of the autonomous wireless sensing node operation are depicted in Figure 14 for both train types. The voltage induced on the developed kinetic energy harvester was very similar to the response with a resistive load. The input voltage into the LTC circuit rose quickly during the first second of the train's passage and then began to fluctuate around a constant value for the regional Train 1, or slowly increase for the express Train 2, which provided more power in general. After a train passed, the voltage decreased slowly and transmission continued. The current was discontinuous due to characteristics of the LTC power electronics, and this pattern reflected the transmission of the measured data.

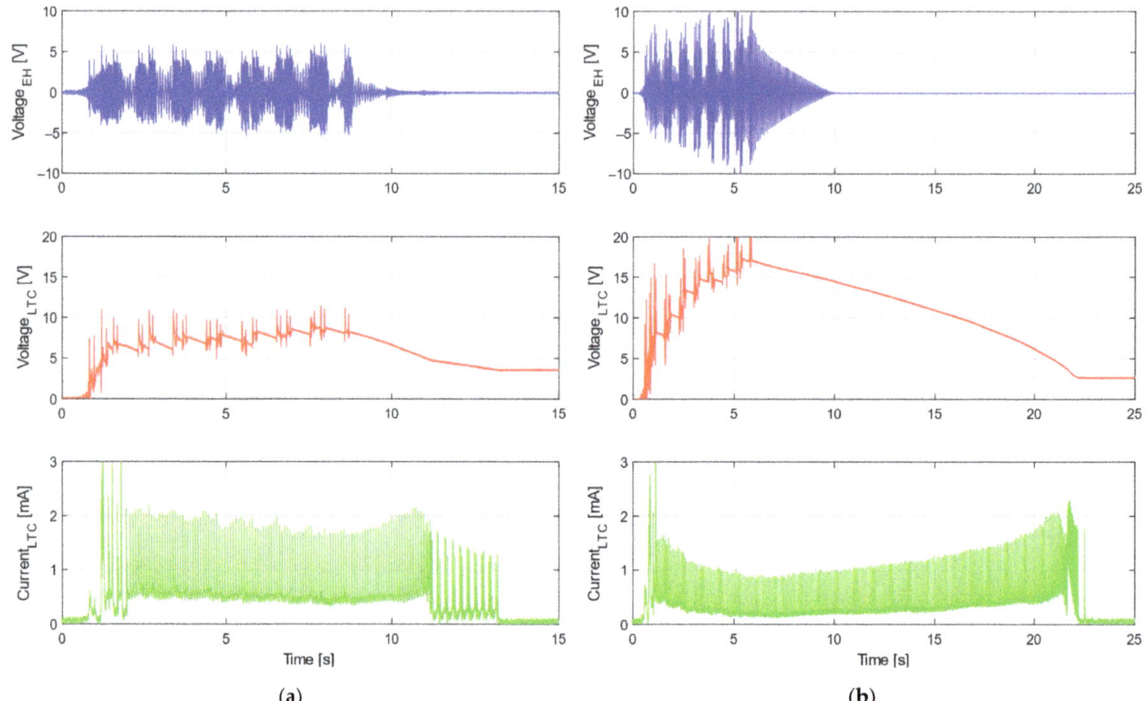

Figure 14. Shaker test with real acceleration data from Train 1 (**a**) and Train 2 (**b**) with connected power electronics; measured voltage and current according to electric schema. Video S2.

During this experiment, the average power consumption of the radio transmission unit was around 4.5 mW, which was less than the average power acquired from each train. The power harvested during a train's passage was stored in the capacitor, and after the passage it was used to continue the data transmission. This energy could also be stored for later if the transmission was no longer active after the respective train passed. However, it would strongly depend on the sensor setup and monitoring requirements for any given application. Nevertheless, experiments with both types of trains confirmed that the proposed system was able to provide enough power for continuous transmission, as is evident in Figure 14.

Measurement with different capacitor values are depicted in Figure 15. The excitation acceleration data were acquired from Train 1, similarly to Figure 14a. For a better understanding of the energy storage capabilities, the graph of the voltage is accompanied by a graph depicting the calculated energy stored in the capacitor. Initially, the lower capacity caused the voltage to increase rapidly; however, after a few seconds of operation the stored energy was lower compared to the case with higher capacity. Furthermore, with lower capacity values the voltage tended to fluctuate significantly.

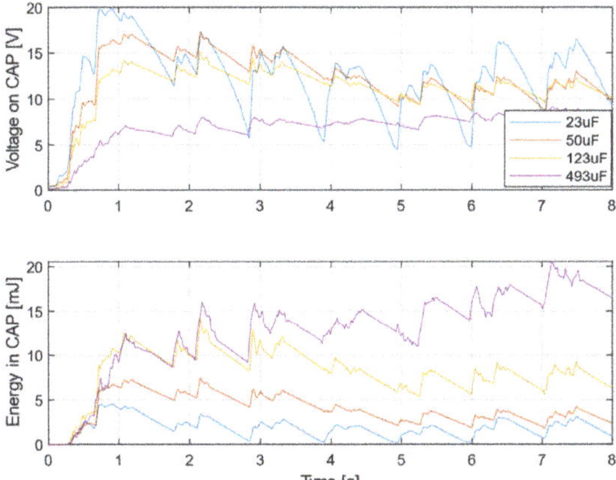

Figure 15. Charging process of capacitor of different value with continuous power consumption of wireless sensor.

6. Potential Applications

Several manufacturers of railway infrastructure systems (e.g., sleepers/bearers, railway switches, point machines, axle counters etc.) have begun to seriously consider the development of new equipment with energy harvesters for predictive maintenance applications of their products. There are two potential workstreams: retrofitting existing products, and novel design of products with embedded energy harvesting. Both emphasize the main advantages of energy harvesting devices, which would result in a reduction in wiring and cables (in both communication and power), reduction of losses due to cable theft, decreased costs of the energy power supply and battery replacement etc.—and all of these aspects contribute to a significant reduction in maintenance costs. This fact is amplified by the subsequent application of predictive maintenance methods, which replace periodic inspections and, compared to the conventional methods, can reveal potential defects and abnormal degradation before a fatal system failure occurs.

6.1. Smart Railway Monitoring Application

A new generation of smart sleepers could detect overloaded trains, trains with abnormal wheel wear or problems with suspension systems, all of which contribute to a significantly higher load on the track and a rapidly increasing wear of railways. The railway wear or significant changes in subgrade and ballast properties—mainly the gap under a sleeper—could also be detected by smart monitoring systems embedded in sleepers. The proposed concept of autonomous application is shown in Figure 16. The kinetic energy harvester could be integrated or embedded inside innovative sleepers and provide electricity for autonomous monitoring. The sensing node could be integrated into rails in the form of a piezoelectric layer that generates an active voltage signal and does not consume power. This system could be even more affordable if PVDF piezopolymers for structural monitoring were used.

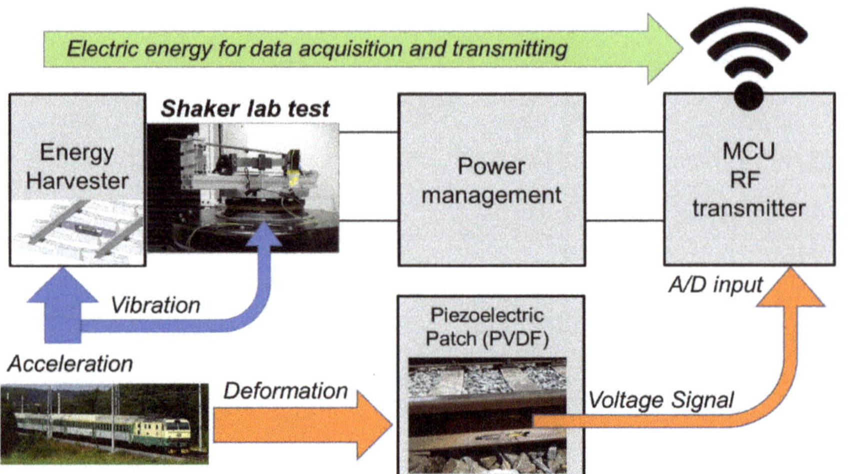

Figure 16. Proposed autonomous railway application of autonomous piezoelectric sensing. Video S3.

This maintenance system could be interesting for infrastructure management and freight train providers interested in detecting critical wear or damage of both the railway and trains. The comprehensive monitoring system could solve the sensitive question of whether freight cars are subjected to excessive wear due to poor track quality, or conversely whether damaged freight cars in operation are causing excessive wear of railways.

6.2. Concept of Smart Turnouts

Switches and crossings are the parts of a railway track most impacted by the dynamic forces applied by trains. From a maintenance point of view, it is important to recognize faults or degradation processes at an early stage, before any significant limitation in operability occurs. The best way to monitor the conditions of crossings is with condition monitoring systems capable of measuring and evaluating the dynamic impacts on crossings over longer periods of time. The concept of an autonomous sensing node with a vibration energy harvester could represent a suitable solution for this system, and kinetic energy harvesters could provide sufficient energy, especially if piezopolymer materials were utilized as the active piezoelectric sensors (e.g., PVDF), mainly because they do not require an external power source to operate. The concept of an autonomous wireless node depicted in Figure 17 can transmit signals from the turnout structure to a close trackside IoT point. There is also the option to process signals on-site and transmit only the results of embedded data analyses. Energy harvesting could mostly be useful in the case of short-distance wireless communication between the track structure and the IoT point, which would use electricity from the power grid.

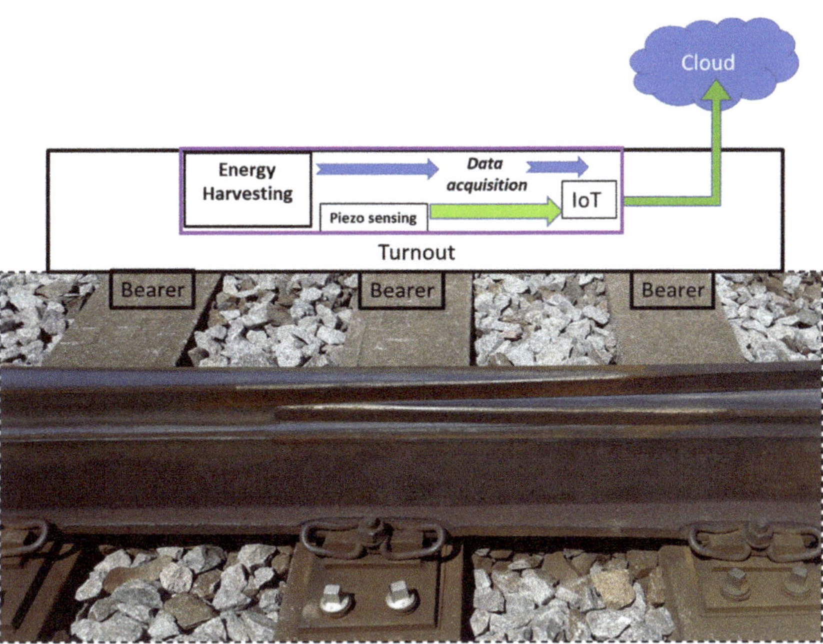

Figure 17. Proposed concept of smart turnout based on energy harvesting.

7. Conclusions

The main design goal of all energy harvesting devices should be to deliver sufficient power for the operation of a given system. In the case of railway applications, replacing cables, cutting off railway systems from the power grid, and using energy harvesting sources for every system is not always an optimal solution, and in many cases is nearly impossible to implement. It is important to keep in mind that energy harvesting devices are utilized as power sources for specific tasks and must be developed and designed with respect to the system they are used to power in order to achieve reliable, low-power and maintenance-free operation over long periods of time. Only such an approach would inevitably lead to a long-term deployment of energy harvesting devices with a new generation of smart railway systems and parts for sustainable rail transportation.

The maintenance of cable systems could result in damage to the wires used in wired sensing systems. For this reason, using embedded kinetic energy harvesters as a source of energy for autonomous wireless sensing nodes is an advantageous approach for long-term railway sensing and monitoring. Two potential applications are presented in this paper for smart rail monitoring and a turnout predictive maintenance system.

The main aim of this paper was to present the development of a kinetic energy harvesting device for rail track applications, a device that is able to provide sufficient power for short-distance communication. The developed device was tested under lab conditions with the input vibration signals of two different trains, regional and express. In lab tests, vibrations obtained from a real rail track used as an input provided enough energy for the communication module and transmission of the sensing signal, with the results presented in this paper. The concept it uses, of placing a kinetic energy harvester on the top of sleepers, could generate an average output power in range of 5–35 mW, depending on the train speed. In this case, this technology could be attractive for the retrofitting of the existing railway infrastructure and for innovative products.

Supplementary Materials: The following are available online at https://www.mdpi.com/article/10.3390/s22030905/s1, Video S1: Resonance, Video S2: Train, Video S3: Communication.

Author Contributions: Conceptualization, Z.H.; data curation, O.R., F.K. and J.C.; formal analysis, O.R. and F.K.; funding acquisition, Z.H.; investigation, Z.H., O.R., F.K. and J.C.; methodology, Z.H., O.R., F.K. and J.C.; project administration, Z.H.; resources, Z.H.; software, J.C.; supervision, Z.H.; validation, O.R. and J.C.; visualization, O.R. and F.K.; writing—original draft, Z.H., O.R. and F.K. All authors have read and agreed to the published version of the manuscript.

Funding: The presented energy harvesting research and development were supported by H2020 projects ETALON S2R-OC-IP2-02-2017 and I2T2 S2R-OC-IP2-02-2020. The sensing part was supported by the Czech Science Foundation project GA19-17457S 'Manufacturing and analysis of flexible piezoelectric layers for smart engineering', Czech Republic.

Institutional Review Board Statement: Not applicable.

Informed Consent Statement: Not applicable.

Data Availability Statement: Not applicable.

Conflicts of Interest: The authors declare no conflict of interest. The funders had no role in the design of the study; in the collection, analyses or interpretation of data; in the writing of the manuscript; or in the decision to publish the results.

References

1. Roundy, S.; Wright, P.K.; Rabaey, J. A study of low level vibrations as a power source for wireless sensor nodes. *Comput. Commun.* **2003**, *26*, 1131–1144. [CrossRef]
2. Hodge, V.J.; O'Keefe, S.; Weeks, M.; Moulds, A. Wireless Sensor Networks for Condition Monitoring in the Railway Industry: A Survey. *IEEE Trans. Intell. Transp. Syst.* **2015**, *16*, 1088–1106. [CrossRef]
3. Aktakka, E.E.; Najafi, K. A Micro Inertial Energy Harvesting Platform With Self-Supplied Power Management Circuit for Autonomous Wireless Sensor Nodes. *IEEE J. Solid-State Circuits* **2014**, *49*, 2017–2029. [CrossRef]
4. Zelenika, S.; Hadas, Z.; Bader, S.; Becker, T.; Gljušćić, P.; Hlinka, J.; Janak, L.; Kamenar, E.; Ksica, F.; Kyratsi, T.; et al. Energy Harvesting Technologies for Structural Health Monitoring of Airplane Components—A Review. *Sensors* **2020**, *20*, 6685. [CrossRef]
5. Yoon, Y.-J.; Park, W.-T.; Li, K.H.H.; Ng, Y.Q.; Song, Y. A study of piezoelectric harvesters for low-level vibrations in wireless sensor networks. *Int. J. Precis. Eng. Manuf.* **2013**, *14*, 1257–1262. [CrossRef]
6. Shen, C.; Li, Z.; Dollovoet, R. A Novel Method for Railway Crossing Monitoring Based on Ambient Vibration Caused by Train-Track Interaction. In *Advances in Dynamics of Vehicles on Roads and Tracks. IAVSD 2019. Lecture Notes in Mechanical Engineering*; Springer: Cham, Switzerland, 2020; pp. 133–141. [CrossRef]
7. Oßberger, U.; Kollment, W.; Eck, S. Insights towards Condition Monitoring of Fixed Railway Crossings. *Procedia Struct. Integr.* **2017**, *4*, 106–114. [CrossRef]
8. Liu, X.; Markine, V.L. Correlation analysis and verification of railway crossing condition monitoring. *Sensors* **2019**, *19*, 4175. [CrossRef]
9. MBS Vehicle–Crossing Model for Crossing Structural Health Monitoring. *Sensors* **2020**, *20*, 2880. [CrossRef]
10. Liu, X.; Markine, V.; Wang, H.; Shevtsov, I. Experimental tools for railway crossing condition monitoring (crossing condition monitoring tools). *Measurement* **2018**, *129*, 424–435. [CrossRef]
11. Cahill, P.; Hanley, C.; Jaksic, V.; Mathewson, A.; Pakrashi, V. Energy Harvesting for Monitoring Bridges over their Operational Life. In Proceedings of the 8th European Workshop on Structural Health Monitoring, EWSHM 2016, Bilbao, Spain, 5–8 July 2016; Volume 4, pp. 1–11.
12. Hou, W.; Li, Y.; Guo, W.; Li, J.; Chen, Y.; Duan, X. Railway vehicle induced vibration energy harvesting and saving of rail transit segmental prefabricated and assembling bridges. *J. Clean. Prod.* **2018**, *182*, 946–959. [CrossRef]
13. Gao, M.; Wang, P.; Wang, Y.; Yao, L. Self-Powered ZigBee Wireless Sensor Nodes for Railway Condition Monitoring. *IEEE Trans. Intell. Transp. Syst.* **2017**, *19*, 900–909. [CrossRef]
14. Gao, M.; Lu, J.; Wang, Y.; Wang, P.; Wang, L. Smart monitoring of underground railway by local energy generation. *Undergr. Space* **2017**, *2*, 210–219. [CrossRef]
15. Zhang, X.; Zhang, Z.; Pan, H.; Salman, W.; Yuan, Y.; Liu, Y. A portable high-efficiency electromagnetic energy harvesting system using supercapacitors for renewable energy applications in railroads. *Energy Convers. Manag.* **2016**, *118*, 287–294. [CrossRef]
16. Zhang, X.; Pan, H.; Qi, L.; Zhang, Z.; Yuan, Y.; Liu, Y. A renewable energy harvesting system using a mechanical vibration rectifier (MVR) for railroads. *Appl. Energy* **2017**, *204*, 1535–1543. [CrossRef]
17. Wang, J.J.; Penamalli, G.P.; Zuo, L. Electromagnetic energy harvesting from train induced railway track vibrations. In Proceedings of the 2012 8th IEEE/ASME International Conference on Mechatronic and Embedded Systems and Applications, MESA, Suzhou, China, 8–10 July 2012; Volume 11787, pp. 29–34.
18. Lin, T.; Pan, Y.; Chen, S.; Zuo, L. Modeling and field testing of an electromagnetic energy harvester for rail tracks with anchorless mounting. *Appl. Energy* **2018**, *213*, 219–226. [CrossRef]

19. Pan, Y.; Lin, T.; Qian, F.; Liu, M.; Yu, J.; Zuo, J.; Zuo, L. Modeling and field-test of a compact electromagnetic energy harvester for railroad transportation. *Appl. Energy* **2019**, *247*, 309–321. [CrossRef]
20. Pourghodrat, A.; Nelson, C.A.; Hansen, S.E.; Kamarajugadda, V.; Platt, S.R. Power harvesting systems design for railroad safety. *Proc. Inst. Mech. Eng. Part F J. Rail Rapid Transit* **2014**, *228*, 504–521. [CrossRef]
21. Tianchen, Y.; Jian, Y.; RuiGang, S.; Xiaowei, L. Vibration energy harvesting system for railroad safety based on running vehicles. *Smart Mater. Struct.* **2014**, *23*, 125046. [CrossRef]
22. Wang, J.; Shi, Z.; Xiang, H.; Song, G. Modeling on energy harvesting from a railway system using piezoelectric transducers. *Smart Mater. Struct.* **2015**, *24*, 105017. [CrossRef]
23. Cleante, V.G.; Brennan, M.J.; Gatti, G.; Thompson, D.J. Energy harvesting from the vibrations of a passing train: Effect of speed variability. *J. Phys. Conf. Ser.* **2016**, *744*, 12080. [CrossRef]
24. Gao, M.Y.; Wang, P.; Cao, Y.; Chen, R.; Liu, C. A rail-borne piezoelectric transducer for energy harvesting of railway vibration. *J. Vibroeng.* **2016**, *18*, 4647–4663. [CrossRef]
25. Gao, M.; Wang, P.; Cao, Y.; Chen, R.; Cai, D. Design and Verification of a Rail-Borne Energy Harvester for Powering Wireless Sensor Networks in the Railway Industry. *IEEE Trans. Intell. Transp. Syst.* **2016**, *18*, 1–14. [CrossRef]
26. Cahill, P.; Hazra, B.; Karoumi, R.; Mathewson, A.; Pakrashi, V. Vibration energy harvesting based monitoring of an operational bridge undergoing forced vibration and train passage. *Mech. Syst. Signal Process.* **2018**, *106*, 228–265. [CrossRef]
27. Fitzgerald, P.C.; Malekjafarian, A.; Bhowmik, B.; Prendergast, L.J.; Cahill, P.; Kim, C.-W.; Hazra, B.; Pakrashi, V.; Obrien, E.J. Scour Damage Detection and Structural Health Monitoring of a Laboratory-Scaled Bridge Using a Vibration Energy Harvesting Device. *Sensors* **2019**, *19*, 2572. [CrossRef]
28. Rubes, O.; Chalupa, J.; Ksica, F.; Hadas, Z. Development and experimental validation of self-powered wireless vibration sensor node using vibration energy harvester. *Mech. Syst. Signal Process.* **2021**, *160*, 107890. [CrossRef]

Article

On Theoretical and Numerical Aspects of Bifurcations and Hysteresis Effects in Kinetic Energy Harvesters

Grzegorz Litak [1,*], Jerzy Margielewicz [2], Damian Gąska [2], Andrzej Rysak [1] and Carlo Trigona [3]

[1] Faculty of Mechanical Engineering, Lublin University of Technology, 20-618 Lublin, Poland; a.rysak@pollub.pl
[2] Faculty of Transport and Aviation Engineering, Silesian University of Technology, 40-019 Katowice, Poland; jerzy.margielewicz@polsl.pl (J.M.); damian.gaska@polsl.pl (D.G.)
[3] Dipartimento di Ingegneria Elettrica Elettronica e Informatica, University of Catania, 95125 Catania, Italy; carlo.trigona@dieei.unict.it
* Correspondence: g.litak@pollub.pl

Abstract: The piezoelectric energy-harvesting system with double-well characteristics and hysteresis in the restoring force is studied. The proposed system consists of a bistable oscillator based on a cantilever beam structure. The elastic force potential is modified by magnets. The hysteresis is an additional effect of the composite beam considered in this system, and it effects the modal solution with specific mass distribution. Consequently, the modal response is a compromise between two overlapping, competing shapes. The simulation results show evolution in the single potential well solution, and bifurcations into double-well solutions with the hysteretic effect. The maximal Lyapunov exponent indicated the appearance of chaotic solutions. Inclusion of the shape branch overlap parameter reduces the distance between the external potential barriers and leads to a large-amplitude solution and simultaneously higher voltage output with smaller excitation force. The overlap parameter works in the other direction: the larger the overlap value, the smaller the voltage output. Presumably, the successful jump though the potential barrier is accompanied by an additional switch between the corresponding shapes.

Keywords: vibration energy-harvesting system; hysteretic effect; bistable oscillator; bifurcation

Citation: Litak, G.; Margielewicz, J.; Gąska, D.; Rysak, A.; Trigona, C. On Theoretical and Numerical Aspects of Bifurcations and Hysteresis Effects in Kinetic Energy Harvesters. *Sensors* **2022**, *22*, 381. https://doi.org/10.3390/s22010381

Academic Editor: Chelakara S. Subramanian

Received: 22 November 2021
Accepted: 29 December 2021
Published: 5 January 2022

Publisher's Note: MDPI stays neutral with regard to jurisdictional claims in published maps and institutional affiliations.

1. Introduction

Mechanical vibrations typically induced during machine operation are a disadvantageous phenomenon. In most practical applications, the influence of mechanical vibrations is limited by means of vibration-reduction systems, while the complete elimination of vibration is practically impossible. On the other hand, there are devices in technology in which vibrations are intentionally induced: vibrating conveyors, compactors, pneumatic hammers, etc. Currently, there is growing interest from both scientists and industry in the harvesting of this irretrievably lost energy resulting from vibrations [1–3]. This is possible because of energy harvesters that use, among others, the piezoelectric effect.

Piezoelectric energy harvesting from ambient vibration sources has been widely studied in recent years [1–11]. The applied devices are made up of a vibration resonator formed as a beam-mass resonator structure and a piezoelectric transducer [1,2]. To improve efficiency, a nonlinear oscillator was proposed [3–12], which is characterized by inclinations of resonance curves [13] and additional resonances defined by the multiplied rational and fractions of the main resonance frequencies [11,12,14]. Simultaneously, multiple coexisting solutions are present in such a system and depend on initial conditions [13–15]. Among the main disadvantages of such systems, one can distinguish the difficulty of controlling particular solutions [16]. In recent years, several studies have been conducted to investigate, among others, the hysteretic effects of elastic beam materials and related structures, as well as piezo elements [17–19]. In this study, we continue the investigation of the dynamics of a nonlinear bistable energy harvester with hysteresis induced by the snap-through phenomenon of a bistable beam [17,19]. Our study was inspired by a recent work [17], wherein composite bistable beams were used for energy harvesting, and by [20], where bistability was used in combination with an additional magnetic field to model the potential shape. In [17], the authors proposed a piezoelectric energy harvester that stores the potential energy induced by the mutual self-constraint of the subbeams and harvests the large energy released during the rapid shape transition.

Regarding nonlinear systems, the assessment of the impact of individual parameters is difficult to implement without detailed numerical experiments [8,12]. This claim is justified because even a small change in the value of any parameter in a nonlinear system can lead to drastic changes in the dynamics of the system [21]. In this publication, we are considering cutting the potential barrier and shifting the cut parts so that they overlap. This can be modelled by an elastic cantilever beam that is subjected to initial elastic deformation. An example of such an arrangement would be a typical hairpin. Another possibility is to use a shape-memory material.

2. Mathematical Model

The starting point for the numerical experiments performed is the piezoelectric beam resonator system (Figure 1a) [22], which is the subject of many publications (see [23]). In general, the analyzed system consists of a flexible beam I with piezoelectric transducers II attached to its flat surfaces. The flexible beam I is rigidly fixed in the frame IV, which is bolted to the vibrating subassembly of the mechanical system, from which energy is recovered by means of screws III. In our research, the impact of the hysteresis loop (presumably from the composite beam) on the efficiency of the energy harvested was assessed. For simplicity, we used the initial configuration of the system without hysteresis (Figure 1a) accompanied by the interacting permanent magnets to support the double-well solution (Figure 1b). The hysteresis loop caused by the beam complexity is introduced to the above system potential as an effect of cut and sift the left- and right-hand sides of potential characteristic (Figure 1c) with respect to the central point at the top of the potential barrier by the distance of d (Figure 1c).

(**a**)

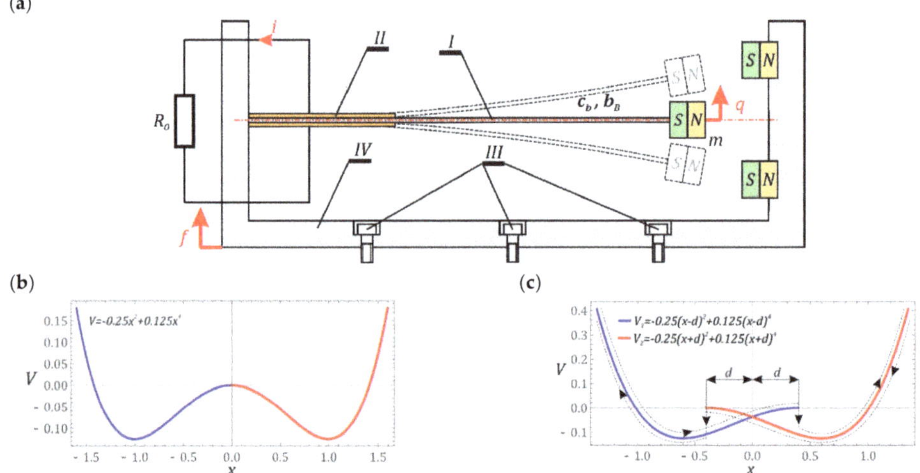

(**b**)   (**c**)

Figure 1. System and its potential characteristics: (**a**) schematic diagram without indicating the origin of hysteresis, modelling of the hysteretic branches: (**b**) potential without hysteresis loop, (**c**) with hysteresis loop, *d* indicates the overlap shift, *q* is the displacement of the tip point of the beam. *i* is the current in the electrical circuit.

In numerical simulations, it was assumed that the system was influenced by mechanical vibrations of the frame mapped by the harmonic function having an amplitude A and a frequency ω_W, $f = A\sin(\omega_W t)$. The differential equations of motion, taking into account the electromechanical coupling, have a general form of:

$$\begin{cases} m_1 \frac{d^2 q}{dt^2} + b_B \left(\frac{dq}{dt} - \frac{df}{dt} \right) + c_B(q - f) - c_1(q - f) + c_2(q - f)^3 + k_P U_P = 0, \\ C_P \frac{dU_P}{dt} + \frac{U_P}{R_O} - k_P \left(\frac{dq}{dt} - \frac{df}{dt} \right) = 0, \end{cases} \quad (1)$$

where m_1, b_B, c_B are effective (modal) beam parameters corresponding to the mass, damping and stiffness, respectively. c_1 and c_2 denote the coefficients of the magnetic force. C_P and R_O are parameters in the electrical circuit indicating capacity and resistance, respectively. Finally, k_p is the coupling parameter and U_p is the voltage output on the resistor [4,12,22].

Considering quantitative and qualitative computer simulations, the system of Equation (1) was transformed into a dimensionless form. A new coordinate y, $y = q - f$, was introduced, defining the difference between the displacement of the free end of the flexible beam q and the point of its attachment to the rigid frame IV f (see Figure 1a). The differential equations of motion, considering dimensionless time and dimensionless displacement, finally take the form:

$$\begin{cases} \ddot{x} + 2\delta\dot{x} - \alpha(x + a\cdot d) \cdot \left[1 - (x + a\cdot d)^2 \right] + \theta u = \omega^2 p \sin(\omega\tau), \\ \dot{u} + \sigma u - \vartheta\dot{x} = 0. \end{cases} \quad (2)$$

where:

$$\omega_0^2 = \frac{c_1 - c_B}{m}, \quad \delta = \frac{b_B}{m\omega_0}, \quad \alpha = \frac{c_2 x_0^4}{c_1 - c_B}, \quad \theta = \frac{k_P}{\omega_0^2 x_0 m}, \quad p = \frac{A}{x_0},$$
$$\omega = \frac{\omega_W}{\omega_0}, \quad x = \frac{y}{x_0}, \quad \vartheta = \frac{k_P x_0}{C_P}, \quad \sigma = \frac{1}{\omega_0 C_P R_O}, \quad \tau = \omega_0 t.$$

In the model tests, x_0 represents the scaling parameter equal to the absolute value of the coordinate defining the position of the minimum potential barrier. On the other hand, the variable a occurring in the mathematical model reflects control quantity responsible for the shift of the operating point from one half of the potential barrier to the other. In general terms, the variable a takes the value of 1 or -1 depending on the hysteretic branch. Here, we limit ourselves to listing the numerical values of the mathematical model coefficients, based on which quantitative and qualitative computer simulations were carried out, i.e.,: $\delta = 0.05$, $\vartheta = 0.5$, $\theta = 0.05$, $\sigma = 0.05$, $\alpha = 0.5$.

3. The Results of Model Tests

Based on the assumed numerical data characterizing the mathematical model of the analyzed energy-harvesting system, numerical experiments were performed to investigate the impact of the d shift on the efficiency of energy harvesting. In the first stage, the impact of the shift on the location of the zones of chaotic and periodic solutions was assessed. Among the periodic solutions, small and large orbits were found. Areas of chaotic solutions can be identified via different numerical tools, such as bifurcation diagrams [21], 0–1 test [19,24] or by determining the maximal Lyapunov exponent [25]. In our numerical simulations (performed in Mathematica), the areas of chaotic and periodic solutions were visualized in the form of multicolor maps showing the distribution of the maximal Lyapunov exponent (Figure 2). According to the authors, this approach makes it possible to look at system dynamics with a wide range of variability of the parameters characterizing the source of mechanical vibration affecting the energy-harvesting system. The maximal Lyapunov exponent was estimated in a three-dimensional phase space following the numerical procedure proposed by Wolf et al. [26]. The phase space was spanned by the computed coordinates (x, $\dot{x}$, u). All the included two-dimensional, multicolor distribution maps were plotted for the assumed nodal initial conditions: $x(0) = 0$, $\dot{x}(0) = 0$ and $u(0) = 0$. Furthermore, two control parameters, ω and p, were selected to describe changes in the external excitation source. Technically, the small distance of the initial conditions between the tested trajectory and the reference trajectory was assumed to be $\varepsilon(0) = 10^{-5}$. To obtain a satisfactory resolution, the ranges of the control parameters ω and p were divided into 500 subintervals.

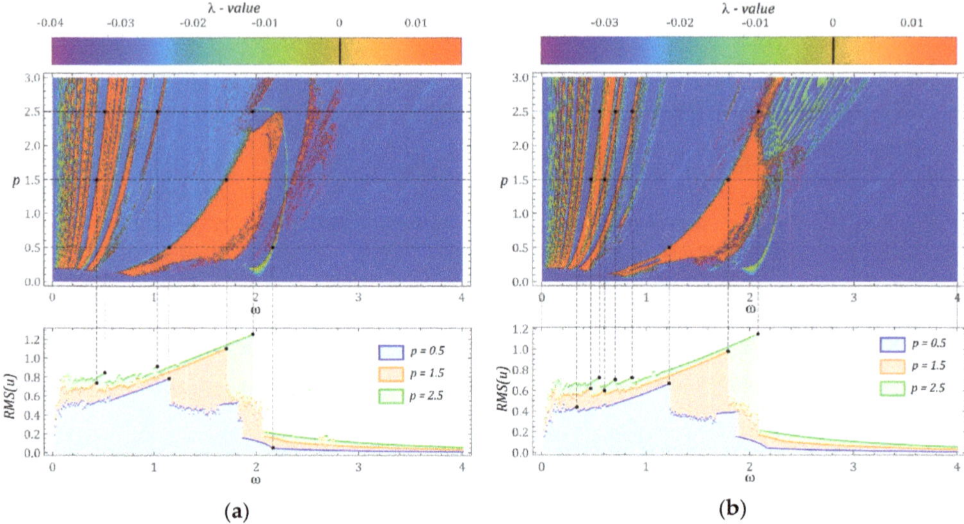

Figure 2. Influence of the parameter d and damping on the distribution of the largest Lyapunov exponent: (**a**) $d = 0$, (**b**) $d = 0.3$ with nodal initial conditions.

It is worth mentioning that the positive values of λ relate to the chaotic dynamic response of the system, otherwise (negative, $\lambda < 0$) the system response is regular with the corresponding phase trajectories tending to stable points or periodic orbits. However, when λ approaches values equal to zero, we are dealing with the so-called bifurcation points (or quasiperiodic solutions). In a low frequency range ω, narrow repetitive zones of chaotic solutions are arranged along the curves that can be approximated by functions with a fairly high exponential growth. Regardless of the parameter d value, the largest area of chaotic solutions is located in the central parts of the multicolor maps showing the distribution of the maximal Lyapunov exponent. For the case of $d = 0$, this region is located in the band ω [1,2] (see Figure 2a). However, for $d = 0.3$, this zone stretches towards higher values of the dimensionless frequency ω [1, 2.4] (see Figure 2b). An indicator expressed as the effective value of the voltage induced on the piezoelectric electrodes was taken as a measure of energy-harvesting efficiency. To determine the effect of the solution on the efficiency of energy harvesting, the multicolor maps of the maximal Lyapunov exponent distribution were compared with the diagrams of effective voltage values. The results clearly indicate that when the solution changes from periodic to chaotic, the voltage induced on the piezoelectric electrodes is limited. Examples of these landmarks are marked in black. The RMS(u) diagrams (RMS—root mean square) were identified in relation to every value of the dimensionless excitation frequency from the range ω [0,4], based on a time sequence of 50 periods of the mechanical vibration signal affecting the energy-harvesting system.

With regard to the zero initial conditions, it can be observed that—irrespective of the displacement value of the "cut" halves of the potential barrier d (Figure 1b,c) and the amplitude of the excitation source p—periodic solutions with a periodicity of 1T (single excitation period T) dominate in the range of low, dimensionless excitation frequencies of $\omega < 0.5$. It should be noted that increasing the width of the hysteresis loop reduces the energy-harvesting capability. Consequently, the reduction in the voltage induced on the piezoelectric electrodes is particularly noticeable in the range of higher parameter p values. In the range of low amplitudes of forced vibration, $p < 0.1$, particularly in the zone of chaotic solutions, no significant decrease in effective voltage value is observed. As predicted by

RMS(u), the voltage value is directly proportional to the amplitude and average kinetic energy of cantilever beam vibration.

3.1. Influence of the Parameter d on the Geometrical Structure of a Chaotic Attractor

The below graphs (Figure 3) show examples of attractors that occur in the regions of chaotic solutions. It is worth mentioning that in the classic visualization of the Poincaré cross-section, numerical results are presented as a set of points located on the phase plane. Much more information about the geometric structure of chaotic attractors can be obtained by analyzing the density of the distribution of the points of intersection of the trajectory and the control plane. In this way, one obtains information from the areas of the phase plane where the trajectory most often intersects with the control plane. The graphical representations of the Poincaré cross-sections are then plotted against bifurcation diagrams. From a theoretical point of view, bifurcation diagrams can be drawn based on the following: Poincaré cross-section, phase trajectory and time sequence. In our case, these diagrams were plotted based on the local minima (points marked in red) and maxima (points marked in blue) of the analyzed time series (Figure 3).

The plotted Poincaré cross-sections show different geometrical structures of attractors depending on the external excitation frequency value affecting the energy-harvesting system. To quantify the geometric structure of the chaotic attractors, their correlation dimension was identified, and the effective voltage values induced on the piezoelectric electrodes were also given. Based on the model tests, it was found that in the case of the "developed" attractors (a developed attractor is understood as a structure consisting of the Poincaré cross-section points forming an even distribution along the geometric structure), the correlation dimension of the attractor of the system with a displaced characteristic ($\omega = 1.7$ Figure 3b) is comparable to the attractor plotted for the classical potential function ($\omega = 1.7$ Figure 3a). In conclusion, the parameter d does not significantly affect the position of the band of chaotic solutions. Both for $d = 0$ and $d = 0.3$, unpredictable solutions are located in comparable ranges of the variability $\omega \in [1.5, 1.9]$. Similar values of the correlation dimension (DC) also result from the similarity of the geometric structure of the chaotic attractors and the density distribution of the points of intersection between the trajectory and the control plane.

The most important element distinguishing the two discussed cases is the parameter representing the effective value of the voltage induced on the piezoelectric electrodes. In the case of the "split" and shifted characteristic ($\omega = 1.7$ in Figure 3b), the value is lower by approximately 0.08. This RMS voltage value is directly related to the lower amplitude of the displacement of the free end of the flexible beam I (Figure 1). The influence of the displacement of the "cut" halves of the potential barrier is best visible in the bifurcation diagrams. In particular, this is very visible in the range of low values of ω, where increasing the parameter d results in narrowing the zones of chaotic solutions. It is also worth noting that in the range of low values of ω, the zones of unpredictable solutions are additionally shifted, as a result of which it is impossible to directly refer to the Poincaré cross-sections that were plotted for the cases $d = 0$ and $d = 0.3$. The numerical modelling results presented in this section were obtained for zero initial conditions. This approach however does not fully reflect the capability of harvesting energy from vibrating mechanical devices. Therefore, the numerical simulations focused on the assessment of initial conditions from the point of view of coexisting solutions.

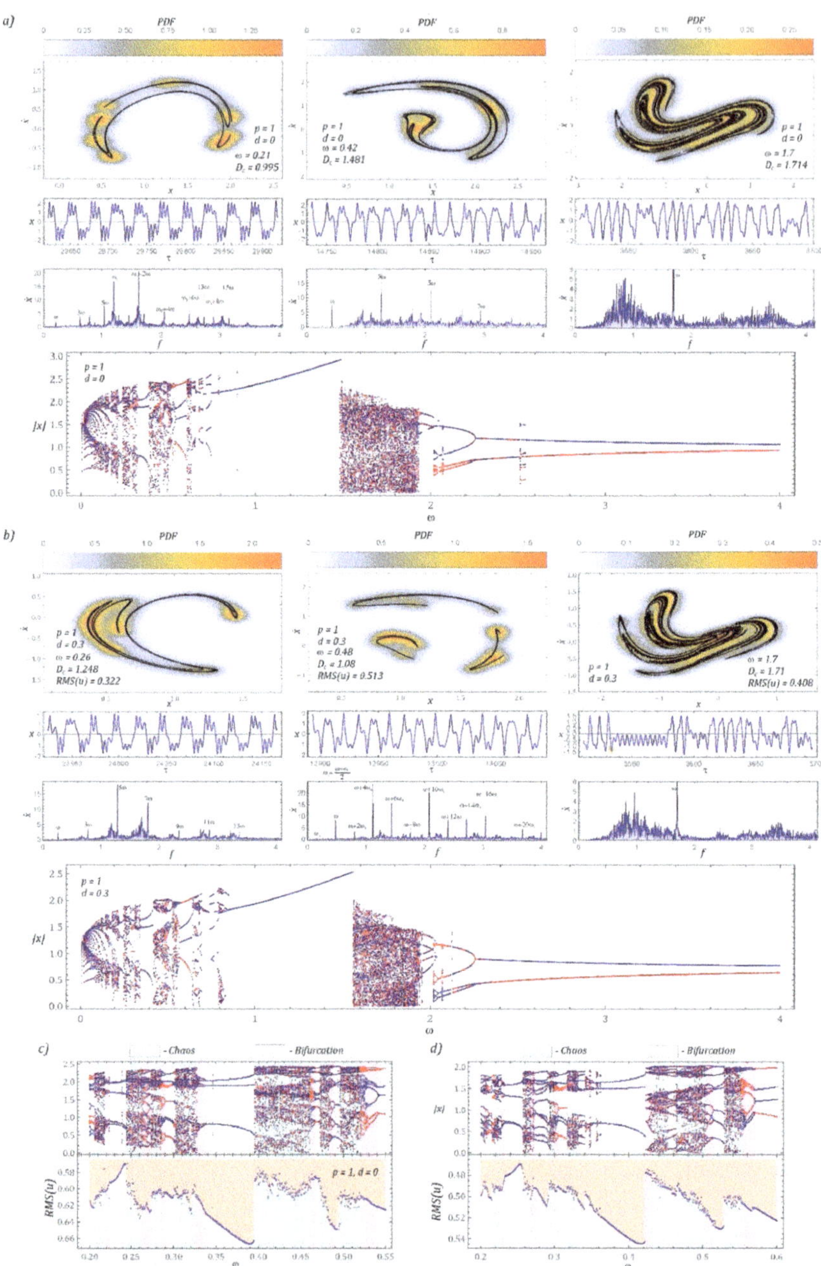

Figure 3. Influence of the parameter d on the geometric structure of the Poincaré cross-section: (**a**) $d = 0$, (**b**) $d = 0.3$. Starting from the top, the horizontal panels correspond to the Poincaré map and the corresponding time series for selected ω and amplitude–frequency spectrum. The bottom panels show the bifurcation diagrams based on the local minima (red) and the local maxima (blue) compared with RMS(u) for $d = 0.3$. To clarify the influence of the parameter d on bifurcations in the selected frequency interval, (**c**,**d**) magnify the difference between the cases $d = 0$ and $d = 0.3$, respectively. The simulation results were obtained assuming zero initial conditions.

With reference to the presented Poincaré sections, Fourier amplitude–frequency spectra were plotted. They provide information on the dominant harmonics in the time series, which constitutes the formal basis for depicting Poincaré maps. Based on the spectra, it is possible to conclude that in the case of developed attractors, for which the correlation dimension $D_C > 1.5$, harmonic components representing the frequency of the source of excitation dominate in the Fourier spectra. We deal with such geometrical structures of Poincaré cross-sections in the widest zone of chaotic solutions, located in the band $\omega \in [1.4, 1.9]$. However, in the range of low ω values, the correlation dimension of the Poincaré cross-sections $D_C < 1.5$. At the same time, in the amplitude–frequency spectra, it is represented by the domination of harmonic components that are a multiple of the frequency of mechanical vibrations affecting the energy-harvesting system. We deal with such a sequence of dominant harmonic components of the Fourier spectrum when the correlation dimension of Poincaré cross-sections is in the D_C range of $[1.1, 1.5]$. At this point, it is worth noting that such spectra occur both in the case of continuous and smooth ($p = 1, d = 0, \omega = 0.42$) and discontinuous ($p = 1, d = 0.3, \omega = 0.26$) characteristics representing the potential barrier. On the basis of the presented results of numerical experiments, it is also possible to state that for correlation dimensions $D_C < 1.1$, in the amplitude–frequency spectra, the dominant harmonic components are the multiples of the combination of the two fundamental frequencies ω and $\omega 1$. As in the previous example, this is the case with both continuous and smooth ($p = 1, d = 0, \omega = 0.21$) and discontinuous ($p = 1, d = 0.3, \omega = 0.48$) characteristics representing the barrier potentials.

The diagrams (the bottom panels Figure 3a,b) show a direct comparison of Feigenbaum's steady state bifurcation diagrams with the diagrams of RMS values of the voltage induced on the piezoelectric electrodes. In the wide range of variability of the dimensionless excitation frequency $\omega \in [0, 4]$, bifurcation diagrams are dominated by one relatively wide band of chaotic solutions, within the ω range of $[1.4, 1.9]$. On the other hand, in the band $\omega < 1$, there are many bands corresponding to unpredictable solutions, separated by pieriodic areas and bifurcation zones. The individual bands characterizing the dynamics of the tested energy-harvesting system were distinguished by colors: the bands of chaotic solutions were depicted with a light cyan color, and the bifurcation zones with a light magenta color. Based on the presented graphs, it is difficult to unambiguously characterize the dynamics of the system, because both in the bands of chaotic solutions and bifurcation zones we deal with an increase and a decrease in the effective voltage induced on piezoelectric electrodes. We are also dealing with both the decrease and increase in the voltage induced on piezoelectric electrodes in the areas of periodic solutions. For example, the voltage drop induced on piezoelectric electrodes in the area of periodic solutions is in the ω band of $[0.225, 0.245]$ (Figure 3c). The same is the case when the halves of the "cut" potential barrier overlap with ω in $[0.23, 0.255]$ (Figure 3d), and the range of variation of the dimensionless excitation frequency is similar. In the remaining bands of periodic solutions, for example $\omega \in [0.337, 0.395]$ (Figure 3c) and $\omega \in [0.355, 0.42]$ (Figure 3d), we deal with the opposite situation, i.e., an increase in the voltage induced on the electrodes is observed. While any detailed inference regarding the correlation of the bifurcation diagram with the diagram of rms voltage values in the bands of chaotic solutions and bifurcation zones is difficult due to the large number of points, in the case of periodic solutions it is possible to formulate a probable diagnosis.

It is possible to formulate the hypothesis that if the mean of the slope coefficients approximating the branches of the bifurcation diagram in the area of periodic solutions is positive, then we are dealing with an increase in the effective value of the voltage induced on the piezoelectric electrodes. On the other hand, when the average value of the slope approximating branches of bifurcation diagrams takes negative values, a decrease in the RMS voltage is observed on the electrodes. If the mean value of the slope of the branch takes values in the vicinity of zero, then the RMS voltage diagram assumes a constant value.

3.2. Identification of Multiple Solutions

It should be emphasized that the results presented in the previous section pertain to selected cases of system dynamics for zero initial conditions. However, one of the most important properties of nonlinear dynamic systems is related to the coexistence of different solutions depending on the initial conditions. For the defined system and excitation parameters, one has to find the system's evolution with various initial conditions. This problem comes down to the study of phase plane orbit topologies [22], the origins of which are located in different places. In the case of systems with a greater number of degrees of freedom, coexisting periodic and chaotic solutions are identified in a multidimensional phase space. The space dimension is a multiple of the number of degrees of freedom. Considering the research problem formulated in this way, it is possible to conduct complex numerical calculations, the results of which can be illustrated in the form of a diagram of solutions (DS) [27,28]. In this approach, the information about the efficiency of harvesting energy from coexisting solutions is neglected. For this reason, this work contributes to the state of the art by proposing a different approach, the essence of which boils down to multiple plotting of voltage effective value diagrams. Every diagram is plotted for randomly chosen initial conditions of the phase space. Such presentation of computer simulation results provides information about the number of coexisting solutions. However, it does not provide information about their periodicity. For this reason, additional detailed computer simulations were performed to identify the periodicity of individual branches appearing in the diagrams of effective voltage induced on the piezoelectric electrodes (Figure 4). The following convention was adopted to define periodicity: the digit before the letter T indicates the periodicity of the solution, while the number of solutions with a given periodicity is denoted by the right subscript.

It should be noted that both the plotted DS diagrams and the diagrams of effective values plotted for randomly selected RMS(u) initial conditions were obtained in a simplified manner. As a result, their accuracy is a compromise between the precision of numerical calculations and the time of computer simulation. Below are some examples of diagrams showing the effective values of the voltage induced on the piezoelectric electrodes. On their basis it was possible to determine the ranges of variability of the dimensionless frequency ω in which energy harvesting is most effective. Examples of numerical results showing the effects of the dimensionless excitation amplitude p and the shift d of the left and right potential halves overlap in the barrier zone are given in the graphs (Figure 4).

Given the identified branches on the plotted diagrams of the RMS values of the voltage induced on the piezoelectric electrodes, additional numerical simulations were performed to identify phase trajectories of the coexisting solutions. The periodicity of individual branches was identified for the frequency $\omega > 0.4$. Irrespective of the d shift value of the "cut" potential barrier, the highest energy-harvesting efficiency was obtained for 1T-periodic solutions in the band $\omega < 2$. A comparison of the numerical results presented in this section with the RMS(u) diagrams (Figure 2) reveals that the highest energy-harvesting efficiency in this band is achieved by assuming zero initial conditions. However, in the zone $\omega > 2$, energy-efficient solutions can be obtained too, for example by initiating appropriate initial conditions. It is also worth noting that the parameter d does not affect the nature (periodicity) of a given solution. Its influence is mainly visible in the shift of individual branches in relation to the frequency axis ω. However, with increasing the parameter d value, this shift is directed towards higher values of the dimensionless excitation frequency ω. Moreover, as the parameter d is increased, the efficiency of energy harvesting is reduced. A detailed examination of the single points and branches in the diagrams (Figure 4) shows that they represent transient solutions which finally become attracted to permanent periodic solutions over long enough time spans. This is the case with, e.g., the $4T_2$ branch that appears in the diagrams identified for $p = 2$.

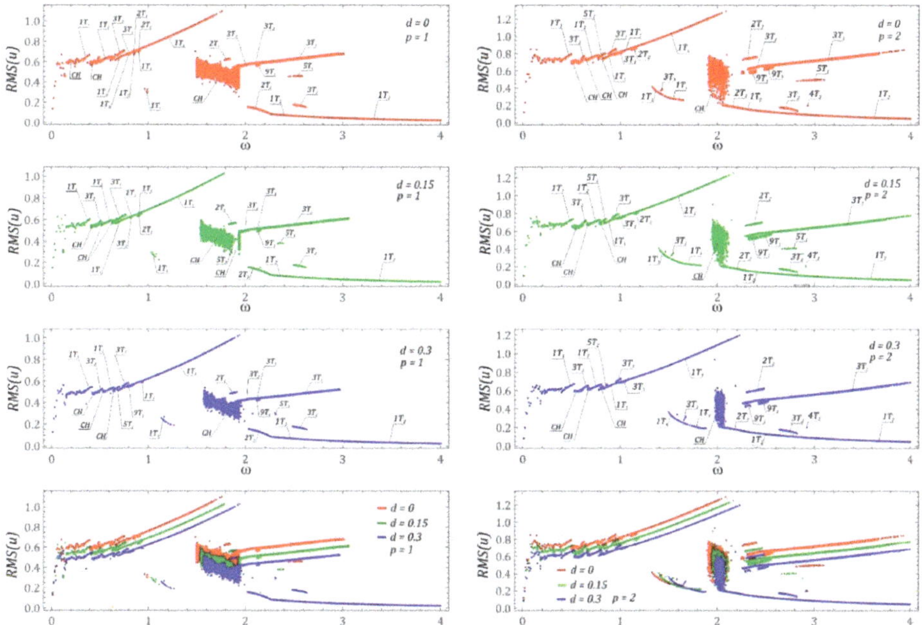

Figure 4. Bifurcation diagram: effect of parameters d and p on energy recovery efficiency with various initial conditions considered simultaneously for the given system parameters. The parameters are indicated in the figures. For better clarity, the solutions are marked by nT_m (n—the response period, m—the number of different solutions with the same response period). The bottom panels are the summery of the upper left and right parametric cases to show the tendencies of the system evolution with d parameter changes.

For clarity, the selected cases were studied with the help of phase plane trajectories (Figure 5). The images are shown against the background of the potential barrier. It should be noted that the multiple solutions, which are marked in different colors depending on the solution periodicity, occur with the evolution of the frequency ω. The Poincaré points are also identified, which confirms the periodicity of the solutions obtained. The multiple solutions of the same period are plotted using the same color. Usually, they are trapped in the left and right potential wells.

The trajectories, which were depicted against the background of the "cut" and shifted halves of the potential barrier, illustrate the possible cases of coexisting solutions. Figure 6 presents selected examples of coexisting solutions. The examples were selected to illustrate the effect of the superimposition of the "cut" halves of the potential barrier, with respect to large (Figure 6a) and small (Figure 6d) orbits of coexisting solutions and unpredictable responses (Figure 6b). The following convention was adopted during the graphical visualization of the results of computer simulations: the colors of the coexisting solutions were assigned to the individual branches of the diagrams of the RMS values of the voltage induced on the piezoelectric electrodes (Figure 4), while the coexisting solutions are plotted with dashed lines.

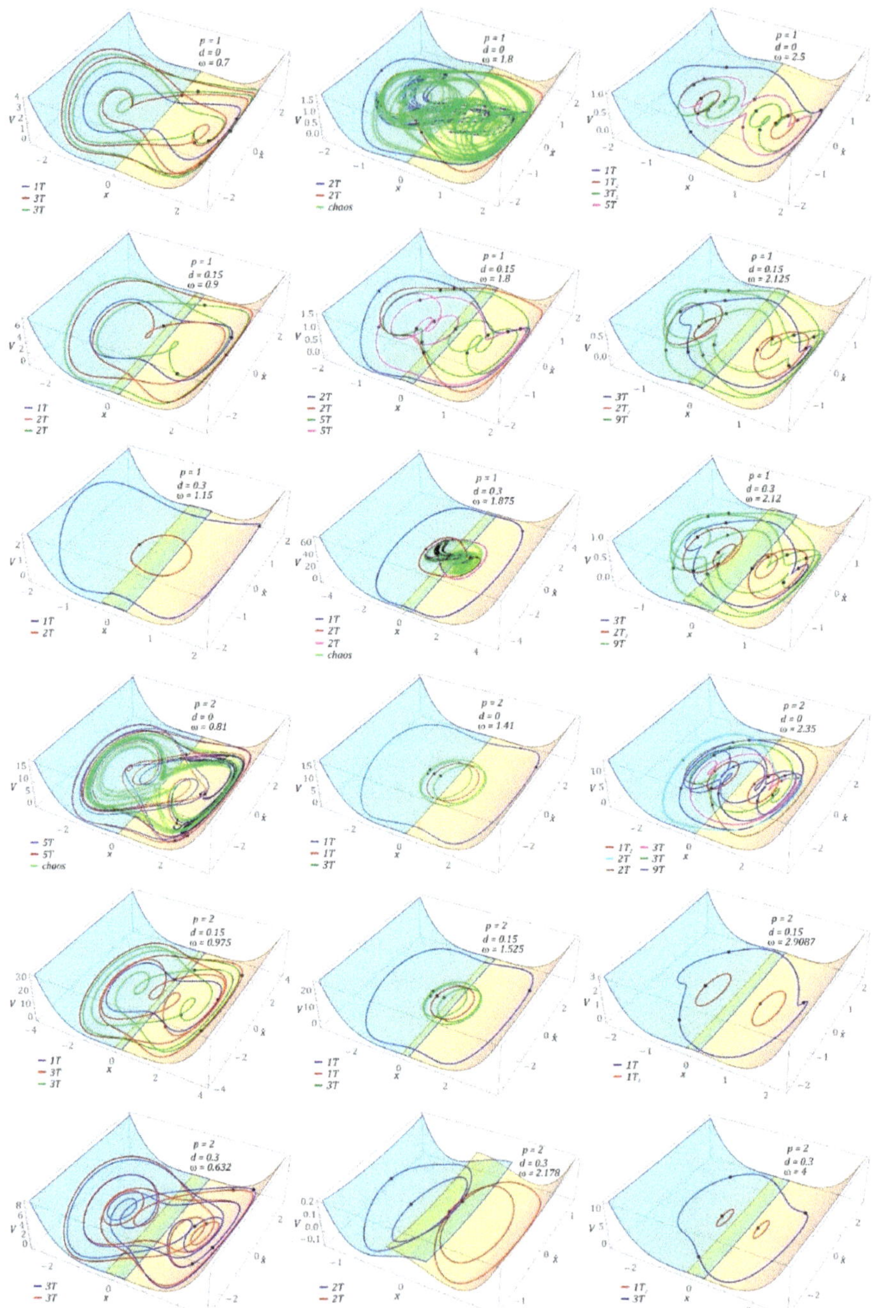

Figure 5. Influence of the parameters d and p on the efficiency of energy harvesting: 3D orbits with the vertical axis indicate the total mechanical energy. For clarity, the corresponding hysteretic potential is plotted with a division into two-color flaps with overlap. The system parameters are indicated in the figures.

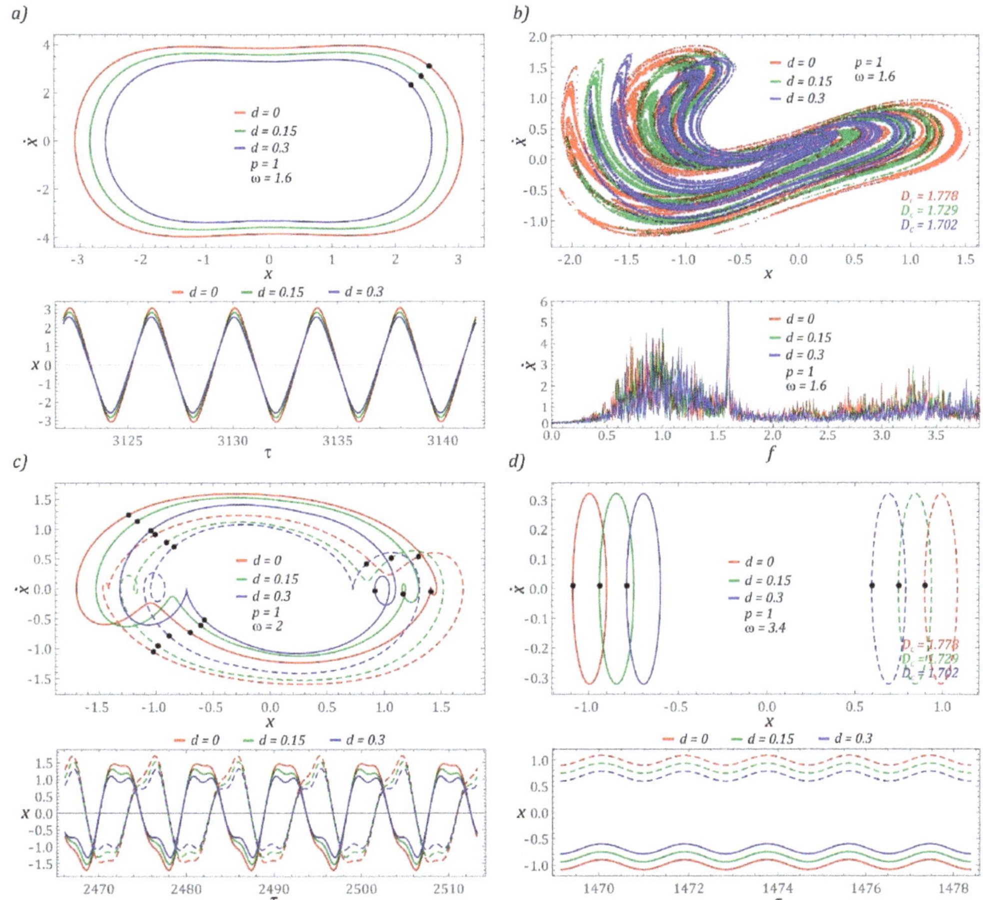

Figure 6. Examples showing the influence of parameter *d* on coexisting solutions, plotted for: (**a**) and (**b**) coexisting solutions for $p = 1$, $\omega = 1.6$, (**c**) $p = 1$, $\omega = 2.0$, (**d**) $p = 1$, $\omega = 3.4$.

Based on the results of computer simulations, it can be concluded that the increase in the value of the parameter *d* does not change the nature of the solution. This behavior of the system is observed both in the case of periodic and chaotic solutions. It is also worth noting that the vibrations of the elastic cantilever beam for periodic solutions do not undergo a phase shift with the increase in the parameter *d* (Figure 6a,c). In the case of large and medium orbits, only a reduction in the amplitude of vibrations is observed. However, amplitude limitations were not observed for solutions whose orbits are located inside the potential well (Figure 6d). If the response of the system is given in the form of a chaotic solution (Figure 6b), the change of parameter *d* does not cause a significant deformation of the geometric structures of the Poincaré sections, and their similarity is preserved. The differences in the plotted Poincaré cross-sections become apparent when the correlation dimensions are identified. With an increase in the value of the parameter *d*, a decrease in the value of the correlation dimension *Dc* is observed. On the other hand, in graphical images of amplitude–frequency spectra, it is manifested by limiting the amplitude of the excited harmonics, with a simultaneous slight shift towards higher values.

4. Conclusions

The influence of parameter d reduces the distance between the external potential barriers, and as a result, the efficiency of energy harvesting changes. On the one hand, the system easier undergoes the potential barrier, on the other, the large orbit size is limited. The small orbit solutions are very similar for any d as they are governed by the linearized equations. In the hysteretic potential case, chaotic solutions may appear easier as a result of the higher competition between unstable small and large orbits. It is worth noting that the hysteretic property of the system can be a side effect of bistable structures such as bistable plates or beams [20,29]. Consequently, the amplitude of resonator oscillations is smaller, and the efficiency of energy harvesting is decreased with respect to the system without hysteresis. In the next step, we would perform experimental verification of the observed tendencies with a suitable metrological characterization. Based on the numerical experiments carried out, it is possible to formulate the following, more detailed conclusions:

- If the mean value of slope coefficients approximating the branches of the bifurcation diagram in the area of periodic solutions assumes positive values, then we deal with an increase in the effective value of the voltage induced on the piezoelectric electrodes. For its negative values, a decrease in the RMS voltage is observed at the piezoelectric electrodes.
- With increasing overlapping of the cut halves of the potential barrier, in a wide range of variability of the dimensionless excitation frequency, a reduction in the efficiency of energy harvesting is observed, which is confirmed by the diagrams of RMS voltage values presented in the graphs (Figure 4). The limitation of energy-harvesting efficiency, caused by the increase in the value of parameter d, is determined by the reduction in the width measured between the potential barriers.

The presented graphs clearly indicate that when designing energy-harvesting systems, the impact of hysteresis caused by overlapping potential barriers should be avoided or minimized. From an engineering point of view, such characteristics of potential barriers can be found in mechanical systems that have been pre-deformed, or in the case of shape-memory materials.

Author Contributions: Conceptualization, G.L., J.M., D.G. and A.R.; methodology, G.L., J.M. and D.G.; calculations, J.M., D.G., A.R.; validation, G.L., C.T.; formal analysis, J.M. and D.G.; original draft preparation, J.M., G.L., review G.L., J.M., D.G., A.R. and C.T.; visualization, J.M.; supervision, G.L. and J.M.; project administration, G.L. and J.M.; funding acquisition, G.L. All authors have read and agreed to the published version of the manuscript.

Funding: This work was supported by the Ministry of Science and Higher Education in Poland under the project DIALOG 0019/DLG/2019/10 during the years 2019–2021.

Institutional Review Board Statement: Not applicable.

Informed Consent Statement: Not applicable.

Data Availability Statement: Data is contained within the article.

Conflicts of Interest: The authors declare no conflict of interest.

References

1. Kaźmierski, T.J.; Beeby, S. *Energy Harvesting Systems, Principles, Modelling and Applications*; Springer: New York, NY, USA, 2011.
2. Priya, S.; Inman, D.J. *Energy Harvesting Technologies*; Springer: Blacksburg, VA, USA, 2009.
3. Erturk, A.; Inman, D.J. *Piezoelectric Energy Harvesting*; John Wiley and Sons: Chichester, UK, 2011. [CrossRef]
4. Erturk, A.; Hoffmann, J.; Inman, D.J. A piezomagnetoelastic structure for broadband vibration energy harvesting. *Appl. Phys. Lett.* **2009**, *94*, 254102. [CrossRef]
5. Cottone, F.; Vocca, H.; Gammaitoni, L. Nonlinear energy harvesting. *Phys. Rev. Lett.* **2009**, *102*, 080601. [CrossRef]
6. Ferrari, M.; Ferrari, V.; Guizzetti, M.; Ando, B.; Baglio, S.; Trigona, C. Improved energy harvesting from wideband vibrations by nonlinear piezoelectric converters. *Sens. Actuators A Phys.* **2010**, *162*, 425–431. [CrossRef]
7. Friswell, M.I.; Ali, S.F.; Bilgen, O.; Adhikari, S.; Lees, A.W.; Litak, G. Non-linear piezoelectric vibration energy harvesting from a vertical cantilever beam with tip mass. *J. Intell. Mater. Syst. Struct.* **2012**, *23*, 1505–1521. [CrossRef]

8.   Andò, B.; Baglio, S.S.; Maiorca, F.; Trigona, C. Analysis of two dimensional, wide-band, bistable vibration energy harvester. *Sens. Actuators A Phys.* **2013**, *202*, 176–182. [CrossRef]

9.   Harne, R.L.; Wang, K.W. A review of the recent research on vibration energy harvesting via bistable systems. *Smart Mater. Struct.* **2013**, *22*, 023001. [CrossRef]

10. Pellegrini, S.P.; Tolou, N.; Schenk, M.; Herder, J.L. Bistable vibration energy harvesters: A review. *J. Intell. Mater. Syst. Struct.* **2013**, *24*, 1303–1312. [CrossRef]

11. Huguet, T.; Lallart, M.; Badel, A. Orbit jump in bistable energy harvesters through buckling level modification. *Mech. Syst. Signal Process.* **2019**, *128*, 202–215. [CrossRef]

12. Litak, G.; Ambrożkiewicz, B.; Wolszczak, P. Dynamics of a nonlinear energy harvester with subharmonic responses. *J. Phys. Conf. Ser.* **2021**, *1736*, 012032. [CrossRef]

13. Daqaq, M.F.; Masana, R.; Erturk, A.; Quinn, D.D. On the role of nonlinearities in vibratory energy harvesting: A critical review and discussion. *Appl. Mech. Rev.* **2014**, *66*, 40801. [CrossRef]

14. Syta, A.; Litak, G.; Friswell, M.I.; Adhikari, S. Multiple solutions and corresponding power output of a nonlinear bistable piezoelectric energy harvester. *Eur. Phys. J. B* **2016**, *89*, 99. [CrossRef]

15. Huang, D.; Zhou, S.; Litak, G. Theoretical analysis of multi-stable energy harvesters with high order stiffness terms. *Commun. Nonlinear Sci. Numer. Simul.* **2019**, *69*, 270–286. [CrossRef]

16. Tan, T.; Wang, Z.; Zhang, L.; Liao, W.-L.; Yan, Z. Piezoelectric autoparametric vibration energy harvesting with chaos control feature. *Mech. Syst. Signal Process.* **2021**, *161*, 107989. [CrossRef]

17. Qian, F.; Hajj, M.R.; Zuo, L. Bio-inspired bi-stable piezoelectric harvester for broadband vibration energy harvesting. *Energy Convers. Manag.* **2020**, *222*, 113174. [CrossRef]

18. Montegiglio, P.; Maruccio, C.; Acciani, G.; Rizzello, G.; Seelecke, S. Nonlinear multi-scale dynamics modeling of piezoceramic energy harvesters with ferroelectric and ferroelastic, hysteresis. *Nonlinear Dyn.* **2020**, *100*, 1985–2003. [CrossRef]

19. Harris, P.; Bowen, C.R.; Kim, H.A.; Litak, G. Dynamics of a vibrational energy harvester with a bistable beam: Voltage response identification by multiscale entropy and "0–1" test. *Eur. Phys. J. Plus* **2016**, *131*, 109. [CrossRef]

20. Harris, P.; Arafa, M.; Litak, G.; Bowen, C.R.; Iwaniec, J. Output response identification in a multistable system for piezoelectric energy harvesting. *Eur. Phys. J. B* **2017**, *90*, 20. [CrossRef]

21. Strogatz, S.H. *Nonlinear Dynamics and Chaos: With Applications to Physics, Biology, Chemistry and Engineering*; Westview Press: Boulder, CO, USA, 2015.

22. Margielewicz, J.; Gąska, D.; Litak, G.; Wolszczak, P.; Yurchenko, D. Nonlinear dynamics of a new energy harvesting system with quasi-zero stiffness. *Appl. Energy* **2021**, 118159. [CrossRef]

23. Zhou, S.; Cao, J.; Inman, D.J.; Lin, J.; Liu, S.; Wang, Z. Broadband tristable energy harvester: Modeling and experiment verification. *Appl. Energy* **2014**, *133*, 33–39. [CrossRef]

24. Bernardini, D.; Litak, G. An overview of 0–1 test for chaos. *J. Braz. Soc. Mech. Sci. Eng.* **2016**, *38*, 1433–1450. [CrossRef]

25. Kantz, H. A robust method to estimate the maximal Lyapunov exponent of a time series. *Phys. Lett. A* **1994**, *185*, 77–87. [CrossRef]

26. Wolf, A.; Swift, J.B.; Swinney, H.L.; Vastano, J.A. Determining Lyapunov Exponens from a time series. *Physica D* **1985**, *16*, 285–317. [CrossRef]

27. Litak, G.; Margielewicz, J.; Gaska, D.; Wolszczak, P.; Zhou, S. Multiple solutions of the tristable energy harvester. *Energies* **2021**, *14*, 1284. [CrossRef]

28. Margielewicz, J.; Gaska, D.; Opasiak, T.; Litak, G. Multiple solutions and transient chaos in a nonlinear flexible coupling model. *J. Braz. Soc. Mech. Sci. Eng.* **2021**, *43*, 471. [CrossRef]

29. Syta, A.; Bowen, C.R.; Kim, H.A.; Rysak, A.; Litak, G. Experimental analysis of the dynamical response of energy harvesting devices based on bistable laminated plates. *Meccanica* **2015**, *50*, 1961–1970. [CrossRef]

sensors

MDPI

Article

Numerical Analysis and Experimental Verification of Damage Identification Metrics for Smart Beam with MFC Elements to Support Structural Health Monitoring

Andrzej Koszewnik [1,*], **Kacper Lesniewski** [1] and **Vikram Pakrashi** [2]

[1] Department of Robotics Control and Mechatronics, Faculty of Mechanical Engineering, Bialystok University of Technology, Wiejska 45C Street, 15-351 Bialystok, Poland; k.lesniewski@doktoranci.pb.edu.pl

[2] UCD Centre for Mechanics, Dynamical Systems and Risk Laboratory, School of Mechanical and Materials Engineering, University College Dublin, D04 V1W8 Dublin, Ireland; vikram.pakrashi@ucd.ie

* Correspondence: a.koszewnik@pb.edu.pl; Tel.: +48-571-443-052

Abstract: This paper investigates damage identification metrics and their performance using a cantilever beam with a piezoelectric harvester for Structural Health Monitoring. In order to do this, the vibrations of three different beam structures are monitored in a controlled manner via two piezoelectric energy harvesters (PEH) located in two different positions. One of the beams is an undamaged structure recognized as reference structure, while the other two are beam structures with simulated damage in form of drilling holes. Subsequently, five different damage identification metrics for detecting damage localization and extent are investigated in this paper. Overall, each computational model has been designed on the basis of the modified First Order Shear Theory (FOST), considering an MFC element consisting homogenized materials in the piezoelectric fiber layer. Frequency response functions are established and five damage metrics are assessed, three of which are relevant for damage localization and the other two for damage extent. Experiments carried out on the lab stand for damage structure with control damage by using a modal hammer allowed to verify numerical results and values of particular damage metrics. In the effect, it is expected that the proposed method will be relevant for a wide range of application sectors, as well as useful for the evolving composite industry.

Keywords: Structural Health Monitoring; piezoelectric; energy harvesting; damage detection; macro fiber composites (MFC); damage sensitive feature; finite element method (FEM)

Citation: Koszewnik, A.; Lesniewski, K.; Pakrashi, V. Numerical Analysis and Experimental Verification of Damage Identification Metrics for Smart Beam with MFC Elements to Support Structural Health Monitoring. *Sensors* **2021**, *21*, 6796. https://doi.org/10.3390/s21206796

Academic Editor: Zdenek Hadas

Received: 19 August 2021
Accepted: 8 October 2021
Published: 13 October 2021

Publisher's Note: MDPI stays neutral with regard to jurisdictional claims in published maps and institutional affiliations.

1. Introduction

Modern mechanical and civil structures are becoming increasingly flexible and complex with time. Beams, pipes, and cables in disparate areas of engineering (e.g., aeronautical, bridge, petroleum, renewable energy) remain critical structural elements which continue to degrade and remain susceptible to both external excitation and internal disturbances, leading to increasing risk and maintenance costs [1–5]. Critical components in these complex structures are typically designed around limit state principles for a certain level of damage tolerance and their maintenance and inspection schedules may not necessarily provide an appropriate evolution of damage due to logistic and procedural aspects, epistemic and aleatory uncertainties around such processes, and due to human effects.

In this context, damage detection or assessment of the current condition of structure can be addressed effectively through Structural Health Monitoring (SHM) by reducing the number and frequency of inspections, uncertainties, and also provide some information about the capacity of a system, which is not possible from the more popular visual inspections, thereby also increasing the value of information from such systems and eventually create more resilient structures. A wide range of SHM systems are widely studied by both academia and the industry, including vibration, optical, thermal, or impedance-based methods [6–8]. In particular, the vibration method is intensively developed by many researchers due to the simplicity of implementation of chosen sensors on real structures and fast detection of damage in the structures with the use of real-time monitoring systems. The instance of these considerations are papers [9,10], where the behavior of the intact and damaged mechanical structures has been assessed based upon natural frequencies, mode shapes, frequency response functions, and also power spectral density or based on spatial wavelet analysis [11].

Many investigations in this field have shown that one-layer piezoelectric patches or piezo harvesters with lightweight fiber materials allow identifying changes in different kinds of vibrating structures [12–16]. The relevant properties of these element, especially their fragility or extreme value effects allow to use them in many applications to monitor structure or detect damage [17–21]. A core aspect behind energy harvesting not being a part of SHM is the relative lack of sector specific examples and benchmarks, where the damage detection markers are compared in terms of their performance. Such markers can relate to damage existence, location, extent, and a combination thereof and it is important to discover and distinguish which markers are relevant for which purposes. In recent times, there has been some effort in creating benchmarks yielding sector specific challenges and energy harvesting sensors, along with markers for damage detection. A recent work [22] uses energy harvesting (EH) systems to assess the leak localization in water pipes. Here, the authors investigated several pipes with different widths of leak to propose and calibrate a leak index based on the monitoring voltage from a piezoelectric energy harvester (PEH) and the power spectrum of the output signal generated from particular polyvinylidene fluoride (PVDF) piezoelectric transducers. Subsequently, the use of Pb-free biomolecular piezoelectrics was also used to enhance SHM of water pipes [23]. Similar examples are available for bridge monitoring under operational conditions via Pb-zirconate titanate (PZT) patches [24,25]. Similar investigations have been also performed with piezoelectric PVDF sensors for wireless monitoring of tension conditions in a cable stayed bridge [26]. Similar applications have also been considered for the aerospace sector in terms of component monitoring in airplanes [27], including those harvesting energy from fuselage vibration with one-layer piezo transducers [28].

While the potential of an energy harvesting system for supporting SHM is established in principle, the implementation of it in specific engineering sectors is still fraught with several questions around interpretation and performance of responses, their analyses, and metrics developed for monitoring features of interest. An initial computational model is helpful in this regard to assess sensor location, potential impact of nonlinearities of the energy harvesting element, and the eventual translation of such information into designing appropriate SHM systems. The advantage of an energy harvesting-based monitoring often lies in the low-power aspect of it and the ease of use, which can lead to the possibility of extensive instrumentation. Model updating and digital twinning are also becoming more common in such industries and, consequently, a work like this will also provide a connectivity of harvesters integrated to such updating processes, for operational structures and those which are evolving through varied technological readiness levels.

There exists papers which address elements of the abovementioned challenges. The influence of non-linear geometric responses of a piezoelectric composite plate considering von Karman large strain theories into the classic plate solution has been investigated by using a 3D element model [29], with results indicating that the problem cannot be omitted especially when correct prediction of the stress-strain over the PEH is analyzed. A piezo-electric element modeled as a shell element acting under d31 effect of the crystal [30] noted that the effect of non-linearity is small and can be neglected, especially when commercial piezoelectric patches are used. Other scientific works in this field focused on introducing piezoelectric coupling to the shell element [31,32] and the results indicate the influence of non-linearity for piezoelectric laminated shell is significant and should be further analyzed. As a result, this led to developing investigations and modeling the piezo structure as a higher order layer-wise plate finite element considering piezoelectric coupling [33]. In summary, despite many scientific contributions related to formulations of plates and shells for piezoelectric laminated elements, there is a gap in verification of numerical results considering the shell finite elements of the piezoelectric element. This paper addresses this knowledge gap by carrying out numerical analysis and subsequently validating them against experimental results.

Studies similar to what has been presented in this paper are also of particular relevance for new sensors that are being developed from environmental perspectives to avoid Pb-based systems [23] or multi-functional materials [34]. As their material properties and uncertainties become lower, the possibility of their use in many sectors get higher and the current work can make them better adapted to the composites sector where energy harvesting based SHM still requires significant research. While the presence of damage is often easier to detect with a sensor, the localization of such damages often provides further challenges, especially when the damages are relatively closely spaced. The impact of closely spaced damages have been studied before [11,20], and while there is a natural understanding that indicators of damage at some point would present a coalesced effect of two damages after a certain proximity, there is a paucity of literature in investigating quantitatively what the effects are for patch based systems. Furthermore, the use of composite materials for this topic is important as a benchmark due to their extensive use in new sectors, especially in marine environments [21]. Over time, their degradation, especially in the durability aspects for saline and harsh marine environments [35], will be of particular relevance around this topic. There is thus a need for a detailed numerical and experimental investigation of a relatively generic example which can be used in the future for similar studies, but also as an evidence base for current performance of patch-based energy harvesting and SHM, future adaption to new sensors, and to new structural systems and environments. This will also lend eventually to estimates on their lifetime risk levels [36] and comparative performances [37] around such levels, especially when the exposure is of a stochastic nature.

Therefore, this paper is focused on damage identification in thin structures by using piezo fiber composites. In contrast to [20,21,23], the piezo harvester with three dimensional material in the piezoelectric fiber layer is considered to enhance the harvesting effect, especially in resonance regions with the host structure. To establish this, three different structures have been investigated, where the first one is intact, while the other two are damaged. The damage is introduced by drilling holes in the area of the beam closer to the end of the piezo-composite located closer the fixed end of the beam. The numerical results obtained from the finite element (FE) models of both sensors and also the damage indexes were determined next based on the frequency response functions, which subsequently allows for damage localization assessment. An experimental test on a real structure was finally carried out to verify the numerical results.

The paper is organized as follows: Section 2 describes the methodology of modeling a piezo-harvester based on the First Order Shear Theory (FOST) and also presents the procedure of determining the shell finite element of the laminated structure of the piezo-composite. Section 3 presents the computational models of all structures (intact and damage) with a homogenous model of macro-fiber composite (MFC), which is also a core novelty of this manuscript. In Section 4, experimental investigations were carried out for all structures with the use of an impact modal hammer. The measurement signals from both MFC elements allow calculation of real values of damage matrices and for comparison with numerical results. Additionally, the approach allows assessing the damage location. Section 5 concludes the main findings of this work.

2. Mathematical Model of the Smart Structure with a Laminated Model of the MFC Element

The finite element model designed on the basis of the First Order Shear Theory (FOST) is mainly used in many applications consisting of thin structures like beams or plates to monitor their state via piezoelectric patches like macro-fiber composites [38]. The chosen finite element (FE) model is described by the laminated shell quadratic finite element with eight nodes: five mechanical degrees of freedom (three translations, two rotations) and three electrical DOFs related to the number of the piezoelectric-active layers of MFCs. In addition, this FE model can act as an equivalent single layer model to describe the mechanical behavior and as a layer-wise model to describe the electrical behavior. As a result, all the aforementioned performances of the proposed FE element caused that this element can simulate not only layered materials (e.g., laminated structural composites) with the piezoelectric layers embedded in the structure, but also non-layered materials (e.g., aluminum) with piezoelectric sensors attached to this structure. Another important aspect to note is that only displacements, forces applied to the structure (impulse inputs), and electrical potentials can be enough to obtain the frequency response functions of this structure and calculate the damage metrics properly. This makes the proposed FE model more useful to increase the computation efficiency, especially for real time SHM, without compromising damage identification accuracy.

The shell finite element, capable of simulating a plane, can be implemented in Ansys and solved using the curvature formulation, which must be at least quadratic to describe single or double curved shells with adequate accuracy [39]. Under such circumstances, the problem of interpolation of the curvature field should be considered in the same way as in the case of the issue of variables. In order to do this, it can be assumed that direction 1 is taken to be aligned to the fibers of a given lamina, direction 3 is aligned to the normal laminate direction, while direction 2 is obtained based on the right-hand rule. Indexes of all variables described in this manuscript are then given in the range of $i = 1..3, j = 1..3, k = 1..3, l = 1..3$.

Constitutive Equations and Electrical Assumptions of the Piezo-Electric Lamina and Its Finite Element Model

The piezoelectric MFC, as shown in Figure 1, is a five-layer smart composition with a single active layer of the thickness of t_{MFCa}, two electrode layers of the thickness of t_{MFCe}, and two Kapton layers of the thickness of t_{MFCk}. As a result, this smart MFC element, attached to the host structures, is described in the form of constitutive equations given by Equation (1).

$$\sigma_{ij} = C_{ijkl}^{E} \varepsilon_{kl} - e_{kij} E_k,$$
$$D_i = e_{ijk} \varepsilon_{kl} + d_{ik}^{\varepsilon} E_k \tag{1}$$

where:

σ_{ij}—the laminate stresses;
D_k—the electrical displacement;
E_k—the electrical field;
C_{ijkl}^{E}—the short-circuit elastic properties of the piezo-laminate;
ε_{kl}—the laminate strains;
e_{ijk}—the piezoelectric coupling coefficients;
d_{ik}^{ε}—the dielectric properties of the piezo-laminate.

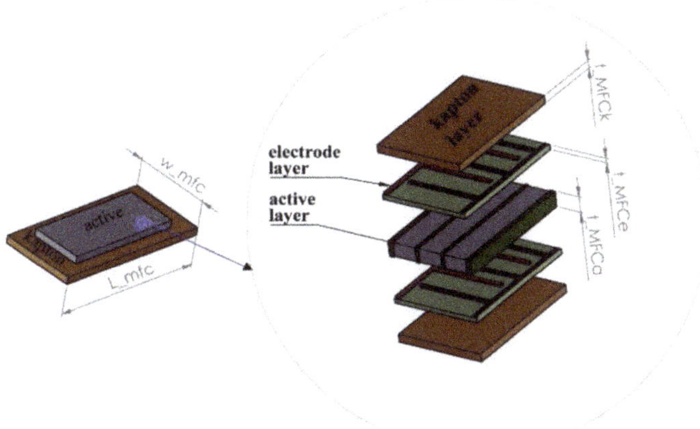

Figure 1. The structure of the macro-fiber composite.

From the measurement point of view, a proper polarization direction of the active element attached to the structure is required. In many applications, especially in those related to SHM, this problem has been solved by choosing the transverse direction of the active element polarization. As a result, Equation (1) can be simplified to the form given by Equation (2) when the Voigt notation is used, and also the normal transversal stress can be considered by omitting ($\sigma_{33} \approx 0$).

$$\begin{Bmatrix} \sigma_{11} \\ \sigma_{22} \\ \sigma_{12} \\ \tau_{23} \\ \tau_{13} \\ D_3 \end{Bmatrix} = \begin{bmatrix} Q_{11} & Q_{12} & 0 & 0 & 0 & -e'_{31} \\ Q_{12} & Q_{22} & 0 & 0 & 0 & -e'_{31} \\ 0 & 0 & Q_{66} & 0 & 0 & 0 \\ 0 & 0 & 0 & Q_{44} & 0 & 0 \\ 0 & 0 & 0 & 0 & Q_{55} & 0 \\ e'_{31} & e'_{31} & 0 & 0 & 0 & d'_{33} \end{bmatrix} \begin{Bmatrix} \varepsilon_{11} \\ \varepsilon_{22} \\ \varepsilon_{12} \\ \gamma_{23} \\ \gamma_{13} \\ E_3 \end{Bmatrix}, \qquad (2)$$

where the material matrix components are

$$Q_{11} = C_{11}^E - \frac{C_{13}^{E2}}{C_{33}^E}, Q_{12} = C_{12}^E - \frac{C_{13}^{E2}}{C_{33}^E}, Q_{22} = C_{22}^E - \frac{C_{13}^{E2}}{C_{33}^E}, Q_{66} = 2\left(C_{11}^E - C_{22}^E\right),$$

$$Q_{44} = C_{44}^E + \frac{e_{15}^2}{d_{11}}, Q_{55} = C_{55}^E + \frac{e_{15}^2}{d_{11}}, e'_{31} = e_{31} - \frac{C_{13}^E e_{33}}{C_{33}^E}, d'_{33} = d_{33}^\varepsilon - \frac{e_{33}^2}{C_{33}^E}.$$

Application of smart elements in the form of macro fiber composites in mechanical structures requires also considering their mechanical and electrical properties. In the case of the mechanical behavior of the laminated structures, the Equivalent Single-Layer (ESL) approach has been used. Then, according to Figure 1, actuating and generalized forces and moments, like: shearing forces (Q), bending moments (M), normal moments (N), and torsion moments (T) in relation to the thickness of the piezo, can be written in the local coordinate system in the following forms:

$$\begin{Bmatrix} N_x \\ N_y \\ N_{xy} \end{Bmatrix} = \int_{-t_{MFCa}/2}^{t_{MFCa}/2} \begin{Bmatrix} \sigma_{xx} \\ \sigma_{yy} \\ \sigma_{xy} \end{Bmatrix} dz + \sum_{op=1}^{nl_p} \int_{-t_{MFCp}/2}^{t_{MFCp}/2} \begin{Bmatrix} \sigma_{xx} \\ \sigma_{yy} \\ \sigma_{xy} \end{Bmatrix} dz + \sum_{ok=1}^{nl_k} \int_{-t_{MFCk}/2}^{t_{MFCk}/2} \begin{Bmatrix} \sigma_{xx} \\ \sigma_{yy} \\ \sigma_{xy} \end{Bmatrix} dz, \qquad (3)$$

$$\begin{Bmatrix} M_x \\ M_y \\ M_{xy} \end{Bmatrix} = \int_{-t_{MFCa}/2}^{t_{MFCa}/2} z \begin{Bmatrix} \sigma_{xx} \\ \sigma_{yy} \\ \sigma_{xy} \end{Bmatrix} dz + \sum_{op=1}^{nl_p} \int_{-t_{MFCp}/2}^{t_{MFCp}/2} z \begin{Bmatrix} \sigma_{xx} \\ \sigma_{yy} \\ \sigma_{xy} \end{Bmatrix} dz + \sum_{ok=1}^{nl_k} \int_{-t_{MFCk}/2}^{t_{MFCk}/2} z \begin{Bmatrix} \sigma_{xx} \\ \sigma_{yy} \\ \sigma_{xy} \end{Bmatrix} dz, \qquad (4)$$

$$\left\{ \begin{array}{c} Q_x \\ Q_y \end{array} \right\} = \int\limits_{-t_{MFCa}/2}^{t_{MFCa}/2} \frac{5}{6}\left(1 - \frac{4s_3^2}{t_{MFCa}^2}\right)\left\{ \begin{array}{c} \tau_{yz} \\ \tau_{xz} \end{array} \right\}dz + \sum_{op=1}^{nl_p} \int\limits_{-t_{MFCp}/2}^{t_{MFCp}/2} \frac{5}{6}\left(1 - \frac{4s_3^2}{t_{total\_MFCp}^2}\right)\left\{ \begin{array}{c} \tau_{yz} \\ \tau_{xz} \end{array} \right\}dz + \sum_{ok=1}^{nl_k} \int\limits_{-t_{MFCk}/2}^{t_{MFCk}/2} \frac{5}{6}\left(1 - \frac{4s_3^2}{t_{total\_MFCk}^2}\right)\left\{ \begin{array}{c} \tau_{yz} \\ \tau_{xz} \end{array} \right\}dz, \tag{5}$$

$$\left\{ \begin{array}{c} T_{rx} \\ T_{ry} \end{array} \right\} = \int\limits_{-t_{MFCa}/2}^{t_{MFCa}/2} \frac{5}{6}z\left(1 - \frac{4s_3^2}{t_{MFCa}^2}\right)\left\{ \begin{array}{c} \tau_{yz} \\ \tau_{xz} \end{array} \right\}dz + \sum_{op=1}^{nl_p} \int\limits_{-t_{MFCp}/2}^{t_{MFCp}/2} \frac{5}{6}z\left(1 - \frac{4s_3^2}{t_{total\_MFCp}^2}\right)\left\{ \begin{array}{c} \tau_{yz} \\ \tau_{xz} \end{array} \right\}dz + \sum_{ok=1}^{nl_k} \int\limits_{-t_{MFCk}/2}^{t_{MFCk}/2} \frac{5}{6}z\left(1 - \frac{4s_3^2}{t_{total\_MFCk}^2}\right)\left\{ \begin{array}{c} \tau_{yz} \\ \tau_{xz} \end{array} \right\}dz, \tag{6}$$

where:

nl_p—the amount of the electrode layers of the MFC element, $op = 1..nl_p$;

nl_k—the amount of the Kapton layers of the MFC element, $ok = 1..nl_k$;

nl—the total amount of layers of the MFC element: $nl = nl_p + nl_k + 1$;

t_{MFCa}—the thickness of the active layer of the MFC element;

t_{MFCp}—the thickness of the passive layer of the MFC element;

t_{MFCk}—the thickness of the Kaption layer of the MFC element;

t_{total_MFCp}—the total thickness of the passive layers of the MFC element;

t_{total_MFCk}—the total thickness of the Kaption layers of the MFC element.

In the case of determining the electrical behavior of this laminate piezo structure, the layer-wise approach has been used. Then, according to this method, electrical displacement of this piezo-composite for the active piezoelectric layer is expressed in the following form:

$$D_3 = \int\limits_{-t_{MFCa}/2}^{-t_{MFCa}/2} \{E_3\}dz, \tag{7}$$

where:

D_3—the electrical displacement of the active layer of the MFC element

E_3—the electrical field of the MFC element with the vertical polarization.

Determining the representative mechanical behavior of the smart structure with the attached laminate to its surface is also required and is relevant for SHM. To do this, the degenerated shell theory with an implicit curvature [38] is used. Displacements, strains, and the electrical field can be written then as a function of the nodal degree of freedom of the finite element in the following form (Equations (8)–(10)), respectively:

$$u_i = u_i^n \phi_n, \quad \theta_k = \theta_k^n \phi_n, \quad \varphi_p = \varphi_p^n \phi_n, \tag{8}$$

$$\left\{ \begin{array}{c} \varepsilon_{11} \\ \varepsilon_{22} \\ \varepsilon_{12} \\ \gamma_{23} \\ \gamma_{13} \end{array} \right\} = \left(\begin{array}{c} B_m + B_{b0} + s_3 B_{b1} \\ B_s + B_{t0} + s_3 B_{t1} \end{array} \right)\widetilde{u}, \tag{9}$$

$$E_3 = B_\phi \widetilde{\varphi}, \tag{10}$$

where: ϕ_n is the shape function for n-th node of the finite element, B_m, B_{b0}, B_{b1}, B_s, B_{t0}, B_{t1} are the curvature-displacement components calculated versus in-plane membrane strains (m), b_0 is the uniform term of in-plane bending strains, b_1 is the linear term of in-plane bending strains, s is the out-of-plane shearing distortions, t_0 is a uniform term of out-of-plane torsions, and t_1 is a linear term of out-of-plane torsions, respectively.

Taking in Equations (8)–(10) to account, the strains, displacements, and electrical potentials of the laminated elements can be expressed in terms of the nodal variables. Subsequently, taking a solution of the elemental equilibrium equation adopted from [38] into account, equations for the piezoelectric problem of laminate structure where w_i and w_j are Gauss' Quadrature weights, can be expressed as

$$\begin{aligned} \mathbf{M}\ddot{u} + \mathbf{C}\dot{u} + \mathbf{K}_{uu}u + \mathbf{K}_{u\varphi}\varphi &= \mathbf{F} \\ \mathbf{K}_{\varphi u}u + \mathbf{K}_{\varphi\varphi}\varphi &= \mathbf{Q} \end{aligned} \tag{11}$$

where:

$$\mathbf{M} = \sum_{i,j=1}^{j=3} \det(J^{-1}) w_i w_j \rho \left[\frac{h}{2} H_0^T H_0 + \frac{h^2}{4} \left(H_1^T H_0 + H_0^T H_1 \right) + \frac{h^3}{12} H_1^T H_1 \right], \tag{12}$$

$$\mathbf{K}_{uu} = \sum_{i,j=1}^{j=3} \det(J^{-1}) w_i w_j \left([B_u^m]^T \begin{bmatrix} A & B \\ B & D \end{bmatrix} [B_u^m] \right) + \sum_{i,j=1}^{j=2} \det(J^{-1}) w_i w_j \left([B_u^t]^T \begin{bmatrix} G & G_h \\ G_h & H \end{bmatrix} [B_u^t] \right), \tag{13}$$

$$\mathbf{K}_{\varphi u} = \sum_{i,j=1}^{j=3} \det(J^{-1}) w_i w_j \left([B_{ut}]^T \begin{bmatrix} \bar{e}_1 \\ \bar{e}_2 \end{bmatrix} [B_{ut}] \right), \tag{14}$$

$$\mathbf{K}_{\varphi\varphi} = \sum_{i,j=1}^{j=3} \det(J^{-1}) w_i w_j B_{u\varphi}^T \bar{d} B_{u\varphi}, \tag{15}$$

$$\mathbf{C} = \alpha\mathbf{M} + \beta\mathbf{K}_{uu}, \tag{16}$$

$$B_u^m = \begin{bmatrix} B_{b0} + B_{b0} \\ B_{b1} \end{bmatrix}, \; B_u^t = \begin{bmatrix} B_{t0} + B_{t0} \\ B_{t1} \end{bmatrix}, \; B_{u\varphi} = \begin{bmatrix} B_{b0} \\ B_{b1} \end{bmatrix}, \; H_{0_{i,5(n+1)+j}} = \delta_{ij}\phi_n, H_{1_{i,5(n+1)+j}} = \delta_{(i,j+3)}\frac{h}{2}\phi_n H_{ij}^n. \tag{17}$$

From a numerical point of view, the electrical and mechanical degree of freedom, as well as the generalized mass, damping and stiffness matrices need to be transformed to modal coordinates in such a way that the nodal variables for a given element can be obtained in a single vector ordered by node numbering. In order to do this, the transforming matrix R should be used. Then, a general mechanical-electrical system in modal coordinates can be express in the following form, respectively:

$$\hat{\mathbf{u}} = \left\{ \begin{array}{cccccccc} u_1^1 & u_2^1 & u_3^1 & \theta_x^1 & \theta_y^1 & \varphi_{p(1)}^1 & \cdots & \varphi_{p(last)}^1 & \cdots \\ u_1^n & u_2^n & u_3^n & \theta_x^n & \theta_y^n & \varphi_{p(1)}^n & \cdots & \cdots & \varphi_{p(last)}^n \end{array} \right\}^T, \tag{18}$$

$$\hat{\mathbf{u}} = \mathbf{R} \left\{ \begin{array}{c} \tilde{u} \\ \tilde{\varphi} \end{array} \right\}, \tag{19}$$

$$\hat{\mathbf{K}} = \mathbf{R} \begin{bmatrix} K_{uu} & K_{u\varphi} \\ K_{\varphi u} & K_{\varphi\varphi} \end{bmatrix} \mathbf{R}^T, \tag{20}$$

$$\hat{\mathbf{M}} = \mathbf{R} \begin{bmatrix} M & 0 \\ 0 & 0 \end{bmatrix} \mathbf{R}^T, \tag{21}$$

$$\hat{\mathbf{C}} = \mathbf{R} \begin{bmatrix} C & 0 \\ 0 & 0 \end{bmatrix} \mathbf{R}^T, \tag{22}$$

$$\hat{\mathbf{F}} = \mathbf{R} \left\{ \begin{array}{c} F \\ Q \end{array} \right\}, \tag{23}$$

$$\hat{\mathbf{M}}\ddot{\hat{u}} + \hat{\mathbf{C}}\dot{\hat{u}} + \hat{\mathbf{K}}\hat{u} = \hat{\mathbf{F}} \tag{24}$$

3. Numerical Analysis of a Smart Beam as a Laminated Structure

The numerical calculations of the cantilever beam of a length of 380 mm, width of 31 mm, and a thickness of 1.8 mm, with two piezo-stripe elements (MFC 8528 P2), consisting of a three-dimensional homogenized material in the active layer, is described in this section. The parameters of a homogenized material for the MFC taken from [40] are collected in Table 1. To assess the values of damage identification metrics, the considered structures (intact and damage structures with first one, two and three drilled holes) were modelled in the Ansys software with the assumption that the first MFC element is located in the distance of 40 mm from the fixed end of the beam, while the second one is at a distance of 15 mm from the free end of the beam. Taking into account the structure of the cantilever beam with MFC attached to its top surface, the host beam structure is modelled by using an 8-node coupled-brick element Solid186. For the MFC element, the electrode and Kapton layers are modelled similarly with the use of a Solid186 element, while the piezoelectric fiber layer with a homogenized material is modelled by using a Solid226-node coupled brick element. In addition, it has been assumed that modelling of the adhesive layer can be omitted due to its thickness which is less than 15 μm. Finally, this leads to determining the computational model of the considered structure shown in Figure 2 consists of 120 of finite elements Solid226 of the length of 10 mm for the cantilever beam, 102 elements of type Solid226 for each passive layer (electrode and Kapton) and 102 elements of type Solid226 for the active layer.

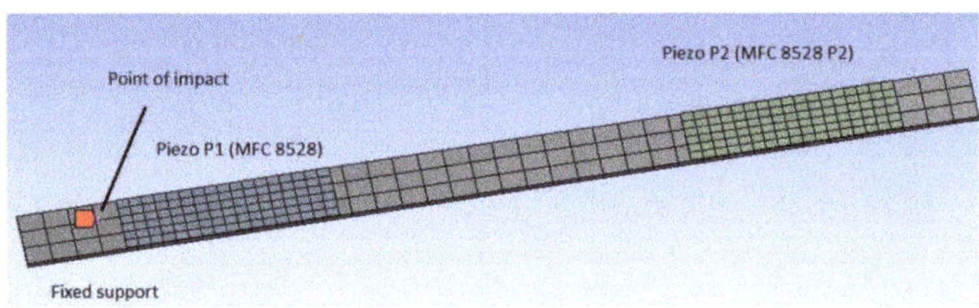

Figure 2. The numerical model of the smart intact structure with both piezo composites attached to the host structure.

Table 1. Material properties of homogenized MFC layer of MFC8528 P2.

Mechanical Parameters							
Young's Modulus (GPa)		Poisson's Ratio (-)		Shear Modulus (GPa)		Piezoelectric Charge Coefficient (pC/N)		Relative Permittivity (-)
E_x	31.6	v_{xy}	0.4	G_{xy}	4.9	d_{31}	−173	ε_r^T	2253
E_y	17.1	v_{yz}	0.2	G_{yz}	2.5	d_{32}	−150		
E_z	9.5	v_{xz}	0.4	G_{xz}	2.4	d_{33}	325		

Geometrical parameters						
overall length [mm]	overall width [mm]	active length [mm]	active width [mm]	thickness of fiber layer [μm]	thickness of electrode layer [μm]	thickness of Kaption layer [μm]
103	31	85	28	180	25	30

In the first step of numerical calculations, the eigenvalue problem of such a modelled structure is solved by using the Ansys software. For this purpose, the behavior of three different structures (one intact and two damage structures with one hole and two holes, respectively, of the diameter of 8 mm) are compared in the selected frequency range 1–400 Hz. The example of the damage structure with two holes is shown in Figure 3. The obtained eigenvalues for each structure are listed in Table 2.

Table 2. Values of natural frequencies of the intact structure and damage structures.

Mode of Vibration	Eigenvalues [Hz]		
	Intact Structure	Damage with One Hole	Damage with Two Holes
First	10.27	10.23	10.20
Second	63.74	63.52	63.25
Third	183.04	182.84	182.85
Fourth	369.79	369.12	368.13

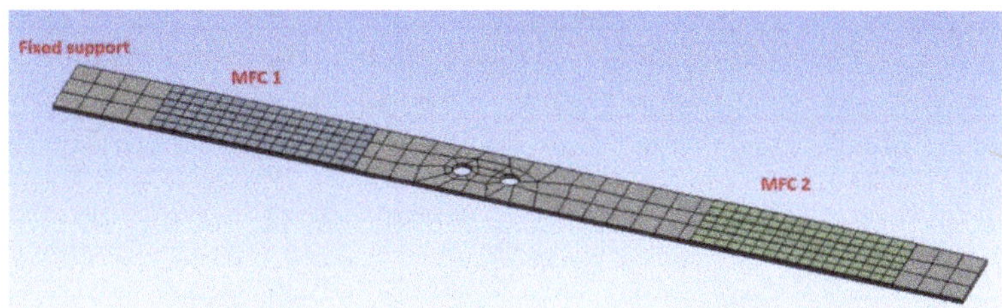

Figure 3. The numerical model of the damage structure with two holes drilling in region close to the end of the piezo-patch MFC1.

Taking into account the obtained result shown in Table 2, it can be noticed that the values of the first four lowest natural frequencies of the damage structures for the increasing number of holes in the structure decreases slightly. The obtained effect is insignificant from the point of view of considering an SHM system since the decrease is of about 0.5% versus the values of eigenvalues of the intact structure. This is thus can be omitted in further analysis.

Next, a harmonic analysis for each structure is performed. To do this, the computational models of intact and damage structures with shell models of the MFC elements were excited by an impulse load of 48N magnitude, while the vertical displacement was taken from specific nodes of brick models of piezo composites MFC1 and MFC2, respectively. Taking into account the results presented in Figure 4, it can be observed that the accuracy between the frequency responses of intact and damage structures is high, especially in resonance regions, where the vibration amplitudes of the damage structure are lower than that of the intact structure (see Figure 5). This leads to difficulties related to proper identification and interpretation of damage in the structure and the calculation of damage identification metrics as damage indicators. This issue is taken up in detail in the next section.

Further analysis of the harmonic responses performed for the FEM models of the intact and damage structures indicates that the distance between the location of the measurement and excitation is the cause of obtaining two different kinds of systems from a control strategy view point, linked to collocated for MFC1 sensor and non-collocated for MFC2 sensor, respectively. An example of the frequency response for the collocated system is the upper diagram shown in Figure 4, where resonance and anti-resonance frequencies are occurring. An example of the frequency response of the non-collocated system is presented in the lower diagram in Figure 4 where the increase of the distance between the point of measurement related to the location of the piezo MFC2 and the point of impact lead to omitting the first antiresonance frequency. Consequently, a proper location of the piezo-sensor on the structure should be also considered for SHM deployment.

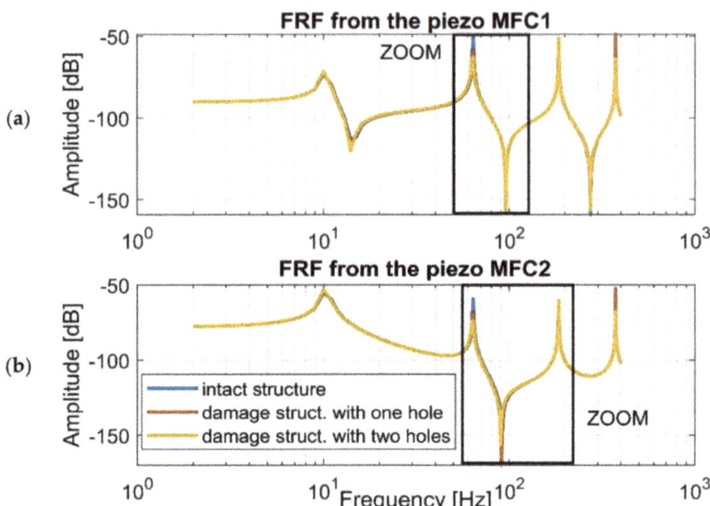

Figure 4. The frequency response function for the intact and damaged structures calculated in the selected frequency range of 1–400 Hz in terms of (**a**) the piezo MFC1 location, (**b**) the piezo MFC2 location.

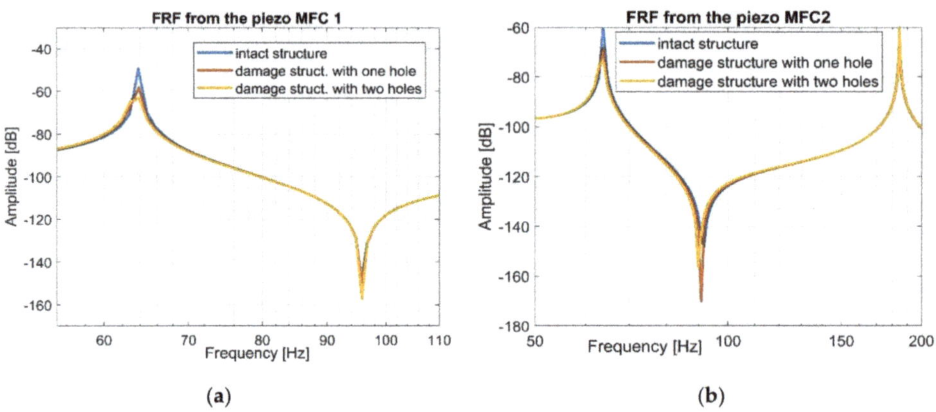

Figure 5. The frequency response function for the intact and damaged structures calculated in the selected frequency range in terms of (**a**) the piezo MFC1 location, (**b**) the piezo MFC2 location.

4. Experimental Verification

The process of frequency response function (FRF) verification of the intact and damage structures are further used to compute damage identification metrics. Two piezo-patch sensors, MFC 8528 P2, were attached to the host aluminium beam structure with the help of an adhesive UHU Plus (Figure 6). The first piezo MFC1 is located 40 mm from the fixed end, while the second one in the distance of 15 mm from the free end of the beam. The laboratory stand is retrofitted into the modal hammer developed by Bruel and Kjaer used to excite the considered structures to vibrations. The data acquisition module PXI 4496 is used to measure and record vibrations from the beam.

Taking numerical investigations of Section 3 into account, experiments on a real structure were carried out. An impulse load with a magnitude of 48 N is applied to the structure to a chosen point located 10 mm before the piezo MFC1 and 30 mm from the fixed end of the beam while structural vibrations are measured with both laminate composites, MFC1 and MFC2, respectively.

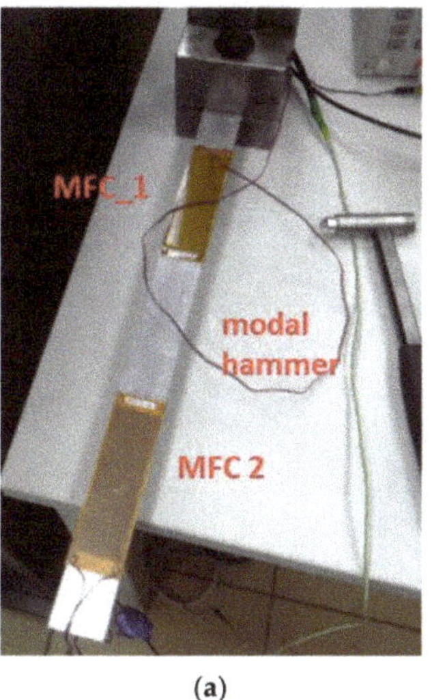

(a)

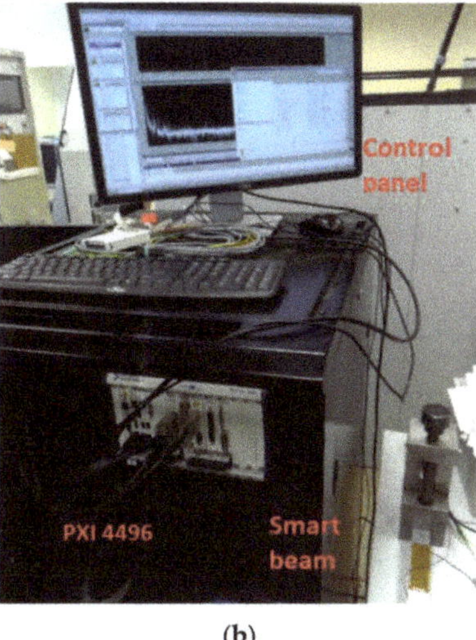

(b)

Figure 6. The view of a real laboratory stand during a lab test (**a**) the smart beam with both piezo-composites MFC 8528 P2, (**b**) the data acquisition module PXI 4496a.

First, the intact structure is investigated by applying the aforementioned impulse load to the beam at 4.2 s of measurement. Taking into account the recorded voltage from both piezo elements shown in Figure 7, it can be seen that the transient response measured with MFC2 is longer than that measured with MFC1. This is due to the fact that the piezo MFC1 is located closer to the fixed end of the beam where the damping is higher. Further analysis of the intact structure behavior requires transformation of input/output signals to the frequency domain and determining two frequency responses functions (FRFs) at MFC1 and MFC2 locations, respectively, as presented in Figure 8.

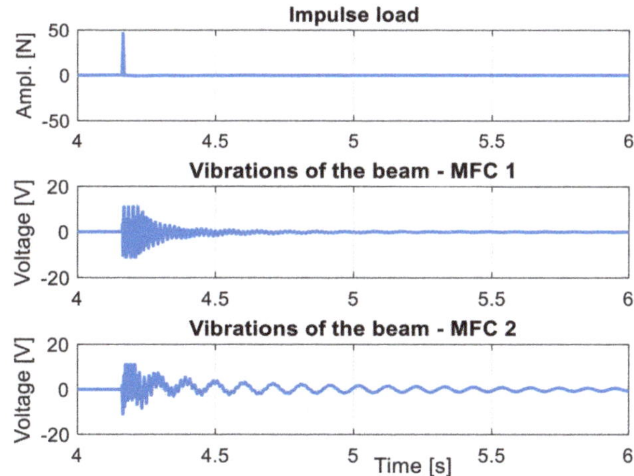

Figure 7. The excitation signal and measurement signals measured by MFC elements during analysis of the intact structure.

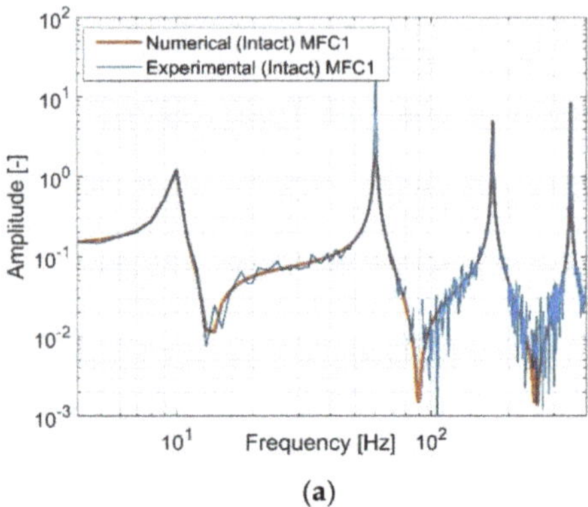 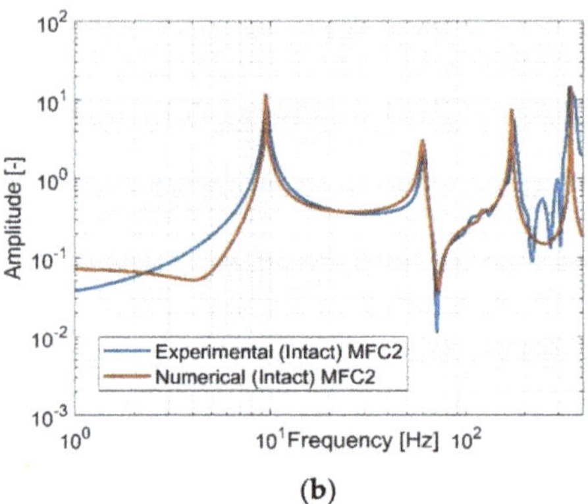

(a)   (b)

Figure 8. The comparison of the amplitude plots of the intact structure measured by using (**a**) piezo-patch MFC1, (**b**) piezo-patch MFC2.

In Figure 8, collocated piezo MFC1 and non-collocated piezo MFC2 aspects are observed and this allows to verify FRFs determined from computational models. The frequency response for piezo MFC1 has interchangeable character of the resonance and anti-resonance frequencies, while in the case of MFC2, the generated frequency response is without the first anti-resonance frequency. Thus, the distance between the sensor and the actuator is crucial to describe the behaviour of the structure.

Further analysis of the recorded frequency responses from both piezo-sensors showed also a high convergence between them, especially in the resonance areas, where the amplitude of vibrations from tests is close to the amplitude from the numerical model. In other areas it can observe mismatch between both FRFs that are due to a heterogenous adhesion between the piezo elements and the host structure and nonlinearity of the MFC, especially in the strain-displacement relation. In the effect, the real frequency response generated on the basis of the noisy measurement signal contains additional slight amplitude peaks especially in higher frequencies. However, despite these discrepancies, from monitoring point of view, it can be still conclude that those responses properly verify the numerical responses.

In the next step, an experimental test was carried out for the damage structure with one hole drilled in the distance of 25 mm from the end of the MFC1 piezo-patch sensor and 150 mm from the place of impact. Similarly to the previous case, this structure was excited to vibration again by using a modal hammer and applying the impulse load with the magnitude of 48 N at the same point. As a result, two measurement signals from both piezo-composites were measured by a PXI module that next allow to generate two separate frequency response functions showed in Figure 9.

Observing diagrams in Figure 9, it can be noticed again that the experimental frequency response functions are close to the frequency responses obtained based on the numerical model, especially in the resonance areas of the first four natural frequencies. In this case, it can be seen that the natural frequencies of the structure measured with the help of both piezo-composites, MFC1 and MFC2, are identical with those calculated from the numerical model, while their amplitudes, especially for those measured by using the MFC1, are less convergence. This behaviour results from higher damping in the real structure than it was assumed for the numerical model. Moreover, as it was just mentioned, it is caused by heterogenous adhesion between the sensor and the host structure, nonlinearity of the piezo sensors, as well as the noisy measurement signal that is used to generate real frequency response. The mismatch is also representative of typical tests.

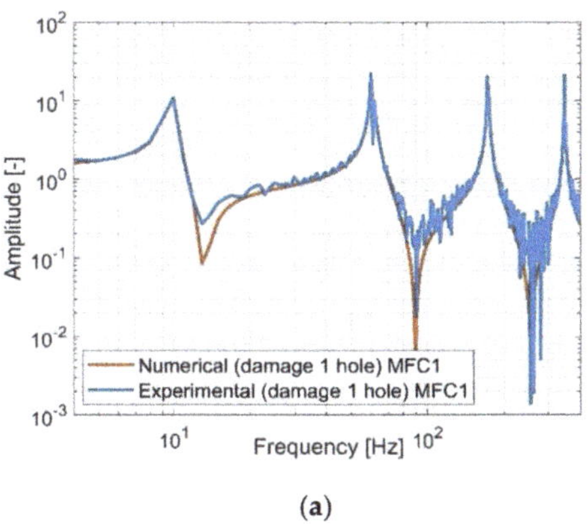
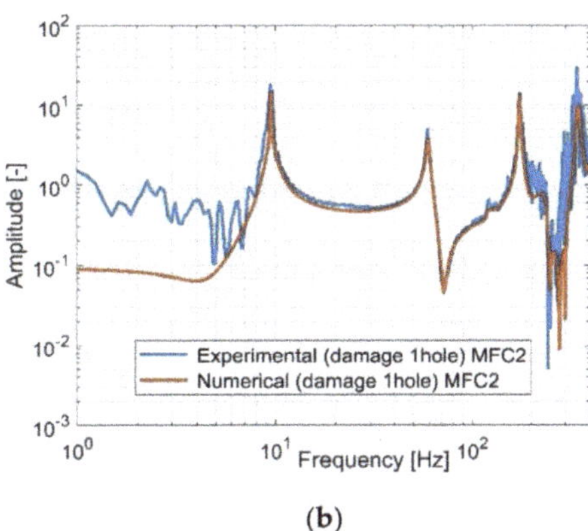

(a)   **(b)**

Figure 9. The comparison of the amplitude diagram of the damage structure with one hole measured by using (**a**) piezo-patch MFC1, (**b**) piezo-patch MFC2.

In the same way, the structure with two holes located very close to the MFC1 sensor was investigated. In this case, the second hole was located 12 mm from the first one and 37 mm from the end of the piezo MFC1. Taking into account Figure 10, it can be observed that the amplitudes of the structure vibration on the generated FRF from the piezo MFC1 are close to the amplitudes vibrations calculated based on the numerical model. Another behaviour that can be seen in the case of the FRF analysis from the piezo MFC2 is the high convergence only in resonant areas. The main reasons of this mismatch can be attributed to a heterogenous adhesion between the bottom surface of the piezo MFC2 and the top surface of the aluminium beam, nonlinearity of the piezo-composite and noisy measurement signal. Again, despite some discrepancies between them located outside the resonance areas, the FRF generated from the lab stand can be assumed as correct.

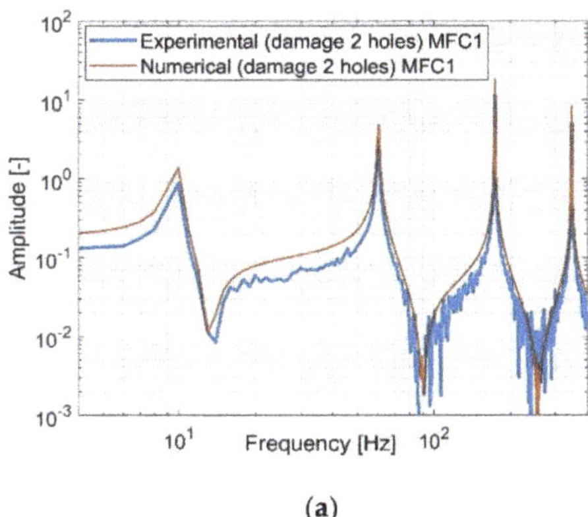
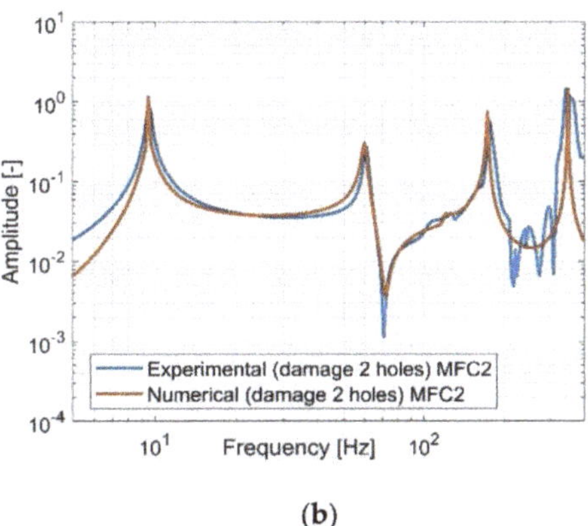

(a)   **(b)**

Figure 10. The comparison of the amplitude plots of the damage structure with two drilled holes measured by using (**a**) piezo-patch MFC1, (**b**) piezo-patch MFC2.

The last step of the experimental test was collecting all the generated FRF from both piezo sensors—MFC1 and MFC2—to perform their analysis and assess the real value of the damage identification metrics. Taking into account Figure 11, it can be observed that the increasing number of holes in the damage structure and the decreasing stiffness of the structure in chosen areas of the beam do not lead to a change in the natural frequencies but affects only the amplitude of the structure in the resonance areas. It this way, the conclusion from the analysis of the computational model has been verified. Further analysis of these diagrams indicates also that the decrease of the beam stiffness resulting from drilling the holes in the areas located very close the MFC1 sensor leads also to the amplitude increase of vibrations measured by the MFC1 sensor but only for the first lowest natural frequencies. In the case of the piezo MFC2, it can be observed that drilling one hole leads firstly to the increase of the vibration amplitude while drilling another hole leads to its decrease. A similar effect is also presented in Figure 12 where the power spectrum of measurement signals from both piezo-sensors is analyzed. Finally, taking into account the generated FRF's and the power spectrum of signals from the piezo composites, it can be concluded that those diagrams are insufficient to identify the damage in the structure properly. For this reason, the damage identification metrics should be calculated.

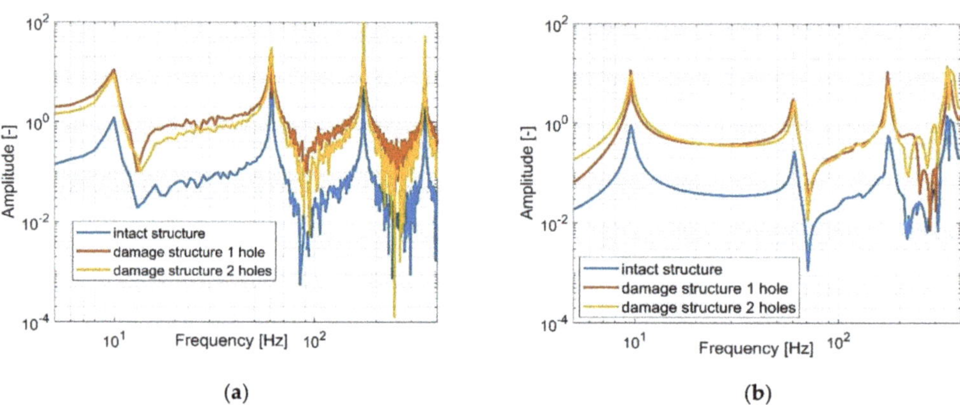

(a)   (b)

Figure 11. The comparison of the FRF of the intact beam and damage structures with one hole and two holes measured by (a) piezo-patch MFC1, (b) the piezo-patch MFC2.

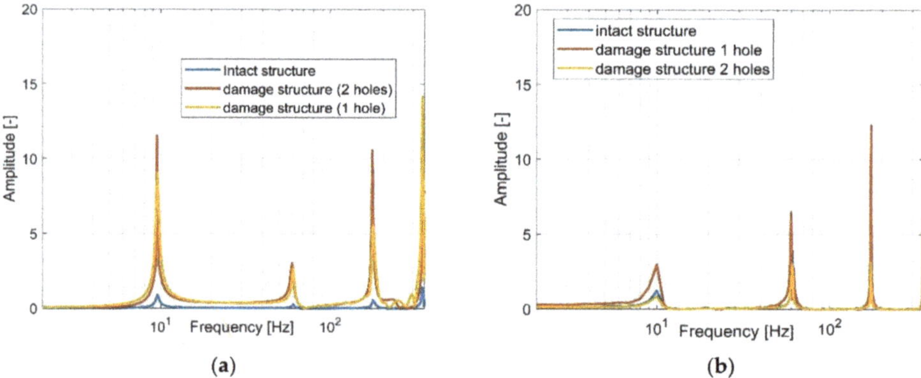

(a)   (b)

Figure 12. The comparison of power spectrum of the intact beam and damage structures with one hole and two holes measured by (a) piezo-patch MFC1, (b) the piezo-patch MFC2.

5. Damage Identification Metrics and Discussion

The numerical and experimental investigations of the intact structure and damage structures presented in the previous sections showed problems with a proper identification of damage in the structure only in terms of FRFs, because the dynamics of these structures is scattered when the frequency increases. Therefore, in order to assess the precision of the damage type and the damage localization, the damage identification metrics should be calculated. Taking this fact into account, five different damage identification metrics M1–M5 taken from [41–45] were calculated for each considered damage structures, and the results of the calculations are presented in this section. In addition, in order to perform better analysis, the metric M2 for the intact structure is calculated as a reference value as well as the damage indicators M1–M5 on the basis of the computational model with three holes. The calculation of damage indices M1, M3, M4, and M5 cannot be done for a healthy baseline because they represent a relationship between the damaged and the intact structure, and consequently a reference M2 marker can link the performances together.

In the first step, their values were calculated based on the frequency responses from the numerical models and then from the FRF generated (Figures 8–10) from the laboratory setup.

The metrices considered in this paper can be divided into two groups: quantitative indicators M1–M3 given by Equations (25)–(27)and qualitative indicators M4–M5 given by Equations (28) and (29). The calculation of these metrics for the first group were performed in terms of a specific frequency value to assess the damage localization in the structure, while in the case of the second group, in terms of the selected frequency range 1–400 Hz, to assess the level of damage. Finally, the results obtained from the computational model considering the damage structure with three holes were collected in Table 3, while results for the experimental response without the structure with the most number of holes are presented in in Table 4. In addition, in order to easier analyze, the damage metric M2 for the intact and damage structures with one and two holes is also presented in the form of a diagram in Figure 13.

$$M_1 = \max\left(\left.\frac{H_I(f_i)}{H_d(f_i)}\right|_{P1}, \left.\frac{H_I(f_i)}{H_d(f_i)}\right|_{P2} \right), \tag{25}$$

where:

$H_I(f), H_d(f)$ denotes frequency response of the intact and damage structure, respectively.

$$M_2 = \max\left(\frac{H_d(f_i)|_{P2}}{H_d(f_i)|_{P1}} \right), \tag{26}$$

$$M_3 = \frac{\log(H_D(f_i) - H_I(f_i))}{\log(H_I(f_i))} * 100\%. \tag{27}$$

Table 3. Results of the damage metrics M1, M2, and M3 calculated in terms of numerical models.

Frequency Response Function	Damage Metrics				
	Metric M1		Metric M2	Metric M3 [%]	
Frequency [Hz]	10.1	60.8	13.5	9.97	60.8
Sensor	MFC_1	MFC_2	-	MFC_1	MFC_2
Intact structure	-	-	**66.18**	-	-
Damage structure (1 hole)	5.874	1.340	26.23	234.1	203.2
Damage structure (2 holes)	7.403	1.682	9.44	253.0	237.5
Damage structures (3 holes)	9.508	1.956	3.37	286.3	265.3

Table 4. Results of the damage metrics M1, M2, and M3 calculated in terms of the generated FRF from the lab stand.

Frequency Response Function	Damage Metrics				
	Metric M1		Metric M2	Metric M3 [%]	
Frequency [Hz]	10.1	60.8	13.9	10.1	60.5
Sensor	MFC_1	MFC_2	-	MFC_1	MFC_2

Table 4. *Cont.*

Frequency Response Function	Damage Metrics				
	Metric M1		**Metric M2**		**Metric M3 [%]**
Intact structure	-	-	**65.10**	-	-
Damage structure (1 hole)	5.399	1.278	26.41	230.1	209.1
Damage structure (2 holes)	7.620	1.569	10.07	257.0	241.5

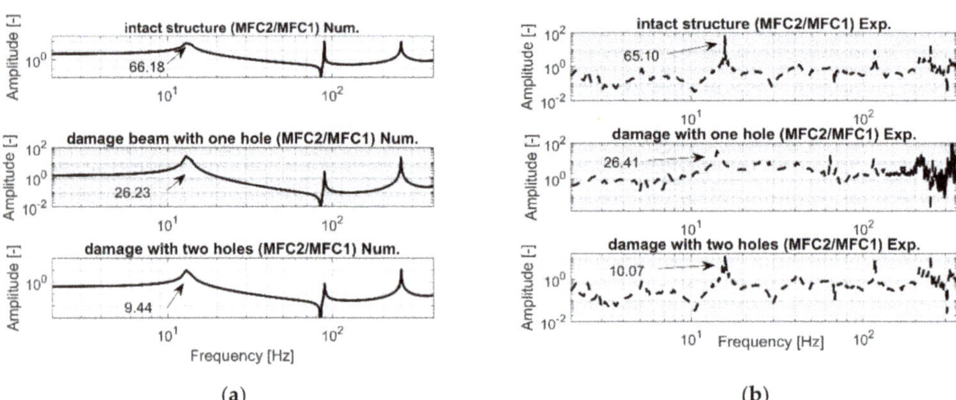

Figure 13. The comparison of damage identification metric M2 of the damage structures with one hole and two holes calculated based on (**a**) numerical approach, (**b**) experimental approach.

Taking into account the results collected in Tables 3 and 4 and Figure 13, it can be noticed that the experimental tests and the obtained values of the damage metrics M1, M2, and M3 verify in their numerical results. The analysis of the values collected in Tables 3 and 4 show that a gradual decrease of the structure stiffness in a chosen area of the structure leads to the increase of the values of M1 and M3. The inverse effect can be obtained in the case of the analysis of the damage metric M2 (Figure 13), where the increasing number of holes in the structure relative to the intact structure leads to a decrease of its maximum value. It is, however, important to note that all metrics exhibit a monotonicity of calibration against damage, which is important for SHM. Further analysis of these tables also shows that the calculated maximum values of the damage metrics M1 and M3 for the piezo-composite for MFC1 cases are higher than their values calculated for the piezo-patch MFC2. This leads to a conclusion that the damage in the structure is located closer the piezo MFC1 and the piezo MFC2. With adequate sensors, this can lead to the localization of damages better. The actual demand of the spacing of sensors will eventually depend on the demands of detection of the feature of interest in terms of extent and resolution, noise in the measured signal, and the excitation that generates the responses.

Next, the values of the qualitative indicators M4 and M5, given in Equations (28) and (29), were calculated to assess the level of damage in the structure in the selected frequency range of 1–400 Hz, with a frequency increment Δf of 0.00024 Hz (reciprocal of sampling frequency f_s = 4096 Hz). Similar to previous cases, the first frequency responses functions from the numerical model (Figures 7–9) were taken to calculate these values, and next, the FRF from the laboratory experiment to verify them. The obtained results for the damage metric M4 are shown in Figure 14, while the results for the damage metric M5 in Figure 15.

$$M_4 = \frac{\Delta f}{f_{high} - f_{low}} \sum_{i=1}^{n} \left| \frac{H_d(f_i) - H_I(f_i)}{H_I(f_i)} \right|, \tag{28}$$

where:

Δf—the frequency increment;
f_{high}—the upper frequency;
f_{low}—the lower frequency.

$$M_5 = \frac{\sum\limits_{i=1}^{n} |H_d(f_i) - H_I(f_i)|}{\sum\limits_{i=1}^{n} |H_I(f_i)|}. \tag{29}$$

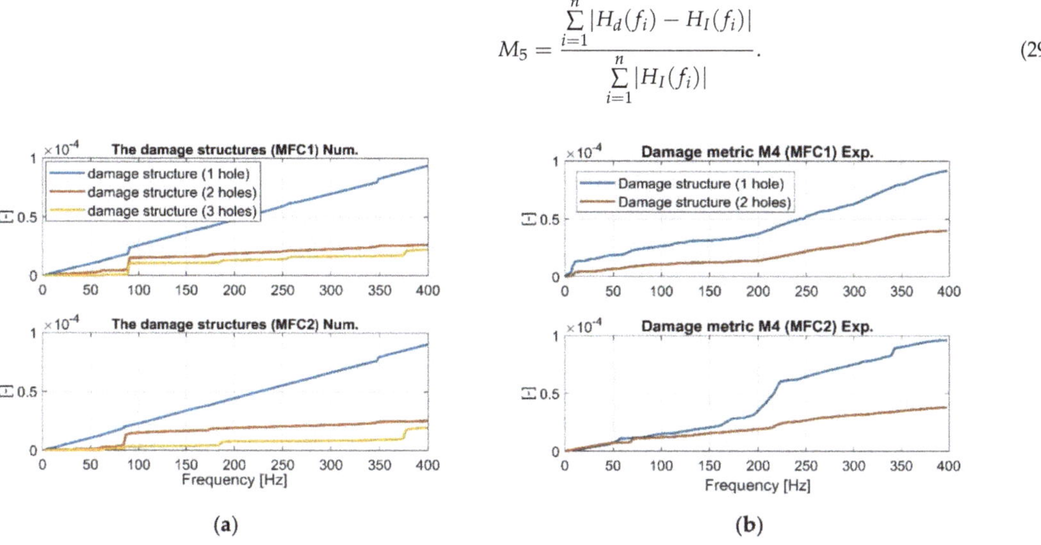

Figure 14. The comparison of damage identification metric M4 of the damage structures with one hole and two holes calculated based on (**a**) numerical approach, (**b**) experimental approach.

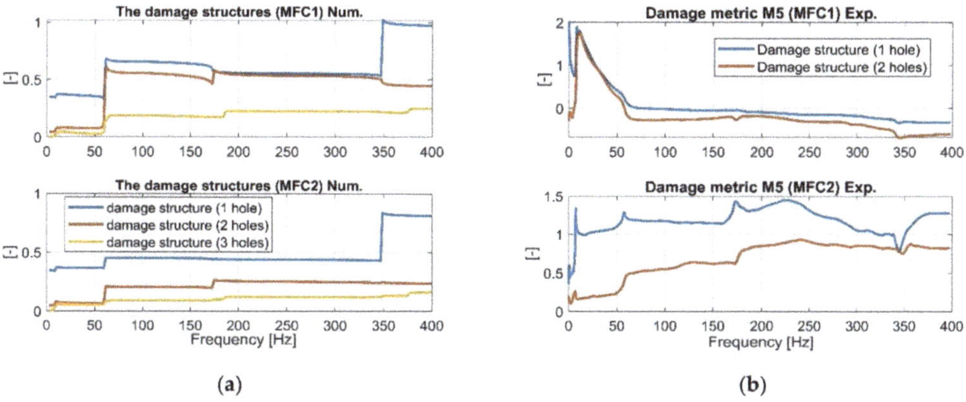

Figure 15. The comparison of damage identification metric M5 of the damage structures with one hole and two holes calculated based on (**a**) numerical approach, (**b**) experimental approach.

As observed in Figures 14b and 15b, the experimental results of the damage metrics M4 and M5 carried out for only two damage structures with one and two holes allow verifying the values obtained from the computational model. Analysis of M4 indicates a decrease of its values for a gradually increasing number of the drilled holes in a chosen area of the beam even for more damaged structure (see orange line in Figure 14a). A similar observation was noted during the analysis of M5 where its value for the structure with three holes were the lowest. Overall, damage metrics M1–M5 are useful to identify damage and its localization, and can support SHM for beam-like structures. Moreover, these results can be useful to build equivalent damage model and also create a fundamental, low-fidelity system which can lend itself to further studies.

6. Summary and Conclusions

The use of piezoelectric patches in SHM has expanded the possibilities of use of energy harvesting in recent times. Nonlinearity in the piezo patches with a potential application for SHM has led to investigations in this paper on structures composed of thin piezo-stripes by creating computational models for them. With the current focus on using traditional and modern sensors to aid digital twinning and model updating, such a focus on the behavior of the sensors becomes even more important. Composite structures are making new inroads into a range of sectors, including renewable energy, and so this example is also relevant for future expansion in terms of sustainability of such solutions.

Taking this into account, the stress–strain effect in the laminate structure was analyzed first in this paper to create a fundamental background model. Next, a modal analysis of the chosen structures (intact and damaged) with piezo patch MFCs were carried out using FEM, establishing a homogenized model of MFC elements. Results presented in Figures 4 and 5 indicate that a gradual increase of the number of the holes drilled in the beam in a chosen area slightly affects the values of the resonance and antiresonance frequencies and leads to a decrease of the vibration amplitude due to changes in local stiffness. Further analysis of response to harmonic loading, performed on the FEM models for the intact and damage structures indicates that the distance between the measurement location and excitation leads to collocated (sensor MFC1) and non-collocated (sensor MFC2) systems. Taking the FRF of the collocated system into account (Figures 8a, 9a and 10a) the inversion of resonance and anti-resonance frequencies is observed. For non-collocated system (Figures 8b, 9b and 10b), the increase of distance between the sensor and the location of excitation leads to the estimation of FRFs while omitting the first antiresonance frequency.

Next, FRFs of the FEM models of the intact and damaged structures are used to assess the damage location. Five different damage indexes were calculated in this regard, comprising of three quantitative and two qualitative indicators, estimated as a function of locations of the sensors on the structure. Estimated damage indices in Table 3 show that a gradual decrease of the beam stiffness leads to consistent and monotonic change of the indexes (26% increase of M1, 20–30% increase of M3 and decrease of M2) with respect to the reference value. This consistency of the damage indicators in the presence of realistic conditions is desirable and is indicative of their robustness.

Additionally, taking into account the values in Tables 3 and 4, it can be noticed that the damage of the structure is located closer to the MFC1 sensor and consequently, higher values of damage indicators closer to the harvesting sensors can also be used to identify the approximate location of the damages. With reasonably spaced sensors in the context of damage detection resolution requirements, this can provide information in terms of detection of the damage location. The qualitative metrics also show a decrease in values for increasing number of holes in the structure, as observed from numerical simulations. This consistency of multiple metrics to describe same damage changes also opens up the possibility of using combined metrics to have a more robust detection scheme.

Experimental investigations carried out in the laboratory (one intact and two damage) with an attached PEH allowed to verify the numerical results indicate that estimated FRFs from piezo-sensors MFC1 and MFC2 are consistent with changes due to damage. Subsequent analysis (Figures 11 and 12) also confirms that the beam stiffness decreasing in a chosen region by drilling an increasing number of holes slightly affects the values of resonances and antiresonances, but significantly affects their amplitude within the range of low frequencies. This is especially illustrated in the vicinity of the first natural frequency where drilling subsequent holes leads to the increase of their values (Figure 12a—MFC1), while in the case of the measurement using MFC2 sensor, the amplitude of vibrations increases and then decreases. Heterogenous adhesion between the harvesting elements and the host structure can lead to such a situation. Summarizing the experimental tests, it can be said that the damage location on the basis only on the analysis of FRF is difficult for energy harvesting and it must be processed further to create relevant markers of damage detection. The proposed damage metrics in this paper illustrate how such markers can be developed and combined, especially in an output-only context. The results of Table 4 indicate similar trends of the proposed metrics as compared to what was observed through numerical simulations. Higher values of damage metrics were observed for sensor closer to the damage (Figures 14b and 15b) along with distinctive and consistent difference over the entire testing range of 1–400 Hz.

This work can act as a reference point for modelling and the expectation of performance of such energy harvesting based SHM sensors for applications in civil/ mechanical systems.

Author Contributions: A.K.: concept, supervision, methodology, numerical calculation, experimental investigations, writing of original manuscript; V.P.: writing and editing of the reviewing manuscript; K.L.: determine of computational model for the damage structure with three holes, numerical calculation of damage metrics M1-M5 for the damage structure with three holes. All authors have read and agreed to the published version of the manuscript.

Funding: Project financing through the program of the Minister of Science and Higher Education of Poland named "Regional Initiative of Excellence" in 2019–2022 project number 011/RID/2018/19 amount of financing 12,000,000 PLN".

Institutional Review Board Statement: Not applicable.

Informed Consent Statement: Not applicable.

Data Availability Statement: Data for the experiments are available from the authors on request.

Acknowledgments: Vikram Pakrashi would like to acknowledge the Science Foundation Ireland MaREI Centre 12/RC/2302_2 and the UCD Energy Institute. Andrzej Koszewnik would like to acknowledge for participation in the program of the Minister of Science and Higher Education of Poland named "Regional Initiative of Excellence" in 2019–2022 project number 011/RID/2018/19 amount of financing 12,000,000 PLN that is realize on the Faculty of Mechanical Engineering of Bialystok University of Technology.

Conflicts of Interest: The authors declare no conflict of interest. The funders had no role in the design of the study; in the collection, analyses, or interpretation of data; in the writing of the manuscript, or in the decision to publish the results.

References

1. Premount, A. *Twelve Lectures of Structural Dynamics*; Springer Publisher: Berlin, Germany, 2013.
2. Mystkowski, A.; Koszewnik, A. Mu-synthesis robust control of 3D bar structure vibration using piezo-stacks. *Mech. Syst. Signal Process.* **2016**, *78*, 18–27. [CrossRef]
3. Ambrożkiewicz, B.; Wolszczak, P.; Litak, G. Modelling of Electromagnetic Energy Harvester with Rotational Pendulum Using Mechanical Vibrations to Scavenge Electrical Energy. *Appl. Sci.* **2020**, *10*, 671. [CrossRef]
4. Muller, F.; Krack, M. Explanation of the self-adaptive dynamics of a harmonically forced beam with a sliding mass. *Arch. Appl. Mech.* **2020**, *90*, 1569–1582. [CrossRef]
5. Qiu, Z.; Li, C.; Zhang, X. Experimental study of active vibration control for a kind of two-link flexible manipulator. *Mech. Syst. Signal Process.* **2019**, *118*, 623–644. [CrossRef]
6. Sante, D.; Fibre, R. Optic Sensors for Structural Health Monitoring of Aircraft Composite Structures: Recent Advances and Applications. *Sensors* **2015**, *18*, 18666–18713. [CrossRef]
7. Ruqiang, Y.; Xuefeng, C.; Subhas, C. *Mukhopadhyay, Structural Health Monitoring, An Advances Signal Processing Perspective*; Springer Publisher: Berlin, Germany, 2017.
8. Ganguli, R. *Structural Health Monitoring: A Non-Deterministic*; Springer Publisher: Berlin, Germany, 2020.
9. Capineri, L.; Bulleti, A. Ultrasonic Guided-Waves Sensors and Integrated Structural Health Monitoring Systems for Impact Detection and Localization: A Review. *Sensors* **2021**, *21*, 2929. [CrossRef]
10. Micknes, T.; Schulz, M.; Sundaresan, M.; Ghoshal, A. Structural Health Monitoring of Aircraft joint. *Mech. Syst. Signal Process.* **2003**, *17*, 285–303. [CrossRef]
11. Chang, C.; Chen, L.-W. Detection of the location and size of cracks in the multiple cracked beam by spatial wavelet based approach. *Mech. Syst. Signal Process.* **2005**, *19*, 139–155. [CrossRef]

12. Staaf, L.G.H.; Smith, A.D.; Laundgren, P. Effective piezoelectric energy harvesting with bandwidth enhancement by asymmetry augmented self-tuning of conjoined cantilevers. *Int. J. Mech. Sci.* **2019**, *150*, 1–11. [CrossRef]

13. Koszewnik, A.; Oldziej, D. Performance assessment of an energy harvesting system located on a copter. *Eur. Phys. J. Spéc. Top.* **2019**, *228*, 1677–1692. [CrossRef]

14. Koszewnik, A. Analytical Modeling and Experimental Validation of an Energy Harvesting System for the Smart Plate with an Integrated Piezo-Harvester. *Sensors* **2019**, *19*, 812. [CrossRef]

15. Bo, L.D.; Gardonio, P.; Casagrande, D.; Saggini, S. Smart panel with sweeping and switching piezoelectric patch vibration absorbers: Experimental results. *Mech. Syst. Signal Process.* **2018**, *120*, 308–325.

16. Koszewnik, A. The influence of a slider gap in the beam–slider structure with an MFC element on energy harvesting from the system: Experimental case. *Acta Mech.* **2020**, *232*, 819–833. [CrossRef]

17. Zhao, X.; Gao, H.; Zhang, G.; Ayhan, B.; Yan, F.; Kwan, C.; Rose, J.L. Active health monitoring of an aircraft wing with embedded piezoelectric sensor/actuator network: I. Defect detection, localization and growth monitoring. *Smart Mater. Struct.* **2007**, *16*, 1208–1217. [CrossRef]

18. Gao, Z.; Zhu, X.; Fang, Y.; Zhang, H. Active monitoring and vibration control of smart structure aircraft based on FBG sensors and PZT actuators. *Aerosp. Sci. Technol.* **2017**, *63*, 101–109. [CrossRef]

19. Na, W.S.; Baek, J. Piezoelectric Impedance-Based Non-Destructive Testing Method for Possible Identification of Composite Debonding Depth. *Micromachines* **2019**, *10*, 621. [CrossRef]

20. Kisa, M. Free vibration analysis of a cantilever composite beam with multiple cracks. *Compos. Sci. Technol.* **2004**, *64*, 1391–1402. [CrossRef]

21. 'Leary, K.; Pakrashi, V.; Kelliher, D. Optimization of composite material tower for offshore wind turbine structures. *Renew. Energy* **2019**, *140*, 928–942. [CrossRef]

22. Okosun, F.; Cahill, P.; Hazra, B.; Pakrashi, V. Vibration-based leak detection and monitoring of water pipes using output-only piezoelectric sensors. *Eur. Phys. J. Spéc. Top.* **2019**, *228*, 1659–1675. [CrossRef]

23. Okosun, F.; Guerin, S.; Celikin, M.; Pakrashi, V. Flexible amino acid-based energy harvesting for structural health monitoring of water pipes. *Cell Rep. Phys. Sci.* **2021**, *2*, 100434.

24. Cahill, P.; Hazra, B.; Karoumi, R.; Mathewson, A.; Pakrashi, V. Vibration energy harvesting based monitoring of an operational bridge undergoing forced vibration and train passage. *Mech. Syst. Signal Process.* **2018**, *106*, 265–283. [CrossRef]

25. Cahill, P.; Ni Nuallain, N.A.; Jackson, N.; Mathewson, A.; Karoumi, R.; Pakrashi, V. Energy Harvesting from Train-Induced Response in Bridges. *J. Bridg. Eng.* **2014**, *19*, 04014034. [CrossRef]

26. Liao, W.-H.; Wang, D.H.; Huang, S.L. Wireless Monitoring of Cable Tension of Cable-Stayed Bridges Using PVDF Piezoelectric Films. *J. Intell. Mater. Syst. Struct.* **2001**, *12*, 331–339. [CrossRef]

27. Zelenika, S.; Hadas, Z.; Bader, S.; Becker, T.; Gljuščić, P.; Hlinka, J.; Janak, L.; Kamenar, E.; Ksica, F.; Kyratsi, T.; et al. Energy harvesting Technologies for Structural Health Monitoring of Airplane Components—A review. *Sensors* **2020**, *20*, 2020. [CrossRef]

28. Ksica, F.; Hadas, Z.; Hlinka, J. Integration and test of piezocomposite sensors for structure health monitoring in aerospace. *Measurement* **2019**, *147*, 106861. [CrossRef]

29. Rabinovitch, O. Geometrically nonlinear behavior of piezoelectric laminated plates. *Smart Mater. Struct.* **2005**, *14*, 785–798. [CrossRef]

30. Marinkovic, D.; Köppe, H.; Gabbert, U. Degenerated shell element for geometrically nonlinear analysis of thin-walled piezoelectric active structures. *Smart Mater. Struct.* **2008**, *17*, 015030. [CrossRef]

31. Neto, M.A.; Leal, R.P.; Yu, W. A triangular finite element with drilling degrees of freedom for static and dynamic analysis of smart laminated structures. *Comput. Struct.* **2012**, *108–109*, 61–74. [CrossRef]

32. Nestorović, T.; Marinkovic, D.; Chandrashekar, G.; Marinković, Z.; Trajkov, M. Implementation of a user defined piezoelectric shell element for analysis of active structures. *Finite Elem. Anal. Des.* **2012**, *52*, 11–22. [CrossRef]

33. Nath, J.K.; Kapuria, S. Coupled efficient layer wise and smeared third order theories for vibration of smart piezo-laminated cylindrical shell. *Compos. Struct.* **2012**, *94*, 1886–1899. [CrossRef]

34. Bai, Y.; Tofel, P.; Palosaari, J.; Jantunen, H.; Juuti, J. A Game Changer: A Multifunctional Perovskite Exhibiting Giant Ferroelectricity and Narrow Bandgap with Potential Application in a Truly Monolithic Multienergy Harvester or Sensor. *Adv. Mater.* **2017**, *29*, 1700767. [CrossRef]

35. Jaksic, V.; Kennedy, C.R.; Grogan, D.M.; Leen, S.B.; Brádaigh, C. Influence of Composite Fatigue Properties on Marine Tidal Turbine Blade Design. In *Durability of Composites in a Marine Environment 2. Solid Mechanics and Its Applications*; Davies, P., Rajapakse, Y., Eds.; Springer: Cham, Switzerland, 2017; Volume 244, pp. 195–223.

36. Vathakkattil, J.; Hao, G.; Pakrashi, V. Extreme value estimates using vibration energy harvesting. *J. Sound Vib.* **2018**, *437*, 29–39. [CrossRef]

37. Vathakkattil, J.; Hao, G.; Pakrashi, V. Fragility analysis using vibration energy harvesters. *Eur. Phys. J. Spec. Top.* **2019**, *228*, 1625–1633. [CrossRef]

38. Sartorato, M.; Medeiros, R.; Tita, V. A finite element formulation for smart piezoelectric composites shells: Mathematical formulation, computational analysis and experimental evaluation. *Compos. Struct.* **2015**, *127*, 185–198. [CrossRef]

39. Saratoro, M.; Medeiros, R.; Vandepitte, D.; Tita, V. Porcusupporting SHM system design: Damage identification via numerical analysis. *Mech. Syst. Signal Process.* **2017**, *84*, 445–461. [CrossRef]

40. Rubes, O.; Tofel, P.; Macku, R.; Skarvada, P.; Ksica, F.; Hadas, Z. Piezoelectric Micro-fiber Composite Structure for Sensing and Energy Harvesting Applications. In Proceedings of the 18th International Conference on Mechatronics-Mechatronika (ME), Brno, Czech Republic, 5–7 December 2018; pp. 1–6.
41. Prada, M.A.; Toivola, J.; Kulla, J.; Hollmen, J. Three way analysis of structural health monitoring data. *Neurocomputing* **2012**, *80*, 119–128. [CrossRef]
42. Rahmatalla, S.; Eun, H.C.; Lee, E.T. Damage detection from the variation of parameter matrices estimated by incomplete FRF data. *Smart Struct. Syst.* **2012**, *9*, 55–70. [CrossRef]
43. Monaco, E.; Franco, R.; Lecce, L. Experimental and numerical activities on damage detection using magnetostrictive actuators and statistical analysis. *J. Intell. Mater. Struct.* **2000**, *11*, 567–578. [CrossRef]
44. Salawu, O. Detection of structural damage through changes in frequency: A review. *Eng. Struct.* **1997**, *19*, 718–723. [CrossRef]
45. Chen, H.P. *Strucutral Health Monitoring of Large Civil Engineering Structures*; Willey Black Well Publisher: Hoboken, NJ, USA, 2018.

sensors

MDPI

Article

Experimentally Verified Analytical Models of Piezoelectric Cantilevers in Different Design Configurations

Zdenek Machu [1,2,*], Ondrej Rubes [1], Oldrich Sevecek [1] and Zdenek Hadas [1]

[1] Laboratory of Energy Harvesting, Institute of Solid Mechanics, Mechatronics and Biomechanics, Faculty of Mechanical Engineering, Brno University of Technology, 616 69 Brno, Czech Republic; Ondrej.Rubes@vut.cz (O.R.); sevecek@fme.vutbr.cz (O.S.); hadas@fme.vutbr.cz (Z.H.)

[2] Institute of Physics, Czech Academy of Sciences, 182 21 Prague, Czech Republic

* Correspondence: Zdenek.Machu@vutbr.cz

Abstract: This paper deals with analytical modelling of piezoelectric energy harvesting systems for generating useful electricity from ambient vibrations and comparing the usefulness of materials commonly used in designing such harvesters for energy harvesting applications. The kinetic energy harvesters have the potential to be used as an autonomous source of energy for wireless applications. Here in this paper, the considered energy harvesting device is designed as a piezoelectric cantilever beam with different piezoelectric materials in both bimorph and unimorph configurations. For both these configurations a single degree-of-freedom model of a kinematically excited cantilever with a full and partial electrode length respecting the dimensions of added tip mass is derived. The analytical model is based on Euler-Bernoulli beam theory and its output is successfully verified with available experimental results of piezoelectric energy harvesters in three different configurations. The electrical output of the derived model for the three different materials (PZT-5A, PZZN-PLZT and PVDF) and design configurations is in accordance with lab measurements which are presented in the paper. Therefore, this model can be used for predicting the amount of harvested power in a particular vibratory environment. Finally, the derived analytical model was used to compare the energy harvesting effectiveness of the three considered materials for both simple harmonic excitation and random vibrations of the corresponding harvesters. The comparison revealed that both PZT-5A and PZZN-PLZT are an excellent choice for energy harvesting purposes thanks to high electrical power output, whereas PVDF should be used only for sensing applications due to low harvested electrical power output.

Keywords: energy harvesting; vibrations; piezoelectric; analytical model; beam model; equivalent model; power prediction

Citation: Machu, Z.; Rubes, O.; Sevecek, O.; Hadas, Z. Experimentally Verified Analytical Models of Piezoelectric Cantilevers in Different Design Configurations. *Sensors* **2021**, *21*, 6759. https://doi.org/10.3390/s21206759

Academic Editor: Salvatore Salamone

Received: 17 September 2021
Accepted: 9 October 2021
Published: 12 October 2021

Publisher's Note: MDPI stays neutral with regard to jurisdictional claims in published maps and institutional affiliations.

1. Introduction

Energy harvesting is more than 20 years a hot topic in the field of wireless sensing [1] since it allows for converting various energy types from ambient sources into an electrical one. Although the amount of such harvested energy is usually small (tens of μW up to several mW), it can be used as a source of electrical power for modern, low power-consuming sensors that are typically used in wearable electronics and industrial applications [2] where powering using cables is not feasible (either due to a hazardous environment or complex setup). Piezoelectric kinetic energy harvesters in the form of a vibrating multilayer structure with piezoelectric layers [3] are commonly used in vibration energy harvesting applications, where the structure is excited by an ambient source of vibrations. The main task of kinetic energy harvesters is then to transform the mechanical energy of ambient vibrations, mainly those of machine frames or human body movement, into useful electrical energy by means of the direct piezoelectric phenomenon.

The main goal in the field of energy harvesting is to design a kinetic energy harvester which is capable to generate a sufficient amount of electrical energy in a particular vibratory environment [4] in order to power some other electronic equipment. However, each application has its requirements or limits for dimensions and weight of the harvester; the principle of energy harvesting can be used practically everywhere, for example in the field of medicine [5], wearables [6], portables [7], aircrafts [8], structural health monitoring of railways [9] or bridges [10].

It has been proved many times that for harvesting energy from ambient vibrations the kinematically excited cantilever beam is one of the most effective designs of a piezoelectric energy harvester. The fundamental and also the most important issue of this solution is the choice of a suitable piezoelectric material for effective electromechanical conversion. The review of commonly used piezoelectric materials and structures for energy harvesting purposes is summarized in publication [11], where it is shown that not only the material itself but also the intended operational mode significantly affects the amount of harvested power due to a great variation in piezoelectric coefficients. The highest piezoelectric coefficients (generally, the higher the coefficients, the higher the amount of harvested power) are provided by piezoceramic materials [12], especially those based on lead (PZT). As a non-toxic alternative, new lead-free piezoceramic materials have been developed which are based on multifunctional Perovskite [13] or structured layers made of Barium and Titanate [14]. Besides these piezoceramic materials which are inherently very brittle and stiff, there are also more flexible materials such as macro-fiber composites which are very promising in the area of strain energy harvesting [15] and piezopolymers which are summarized in review paper [16]. An example of a cantilever harvesting device based on a piezoelectric polymer (PVDF) is presented in paper [17] and the effectivity of PVDF in energy harvesting applications is nowadays widely discussed [6].

In conclusion, the two most important factors that determine the effectiveness of a vibrational energy harvesting device are the used piezoelectric material and the harvester's geometry. Many recent works were concerned about the optimal harvester's geometry for selected piezoelectric material, e.g., [18], but the effectivity of various piezoelectric materials has not been widely discussed yet. Both the selection of efficient piezoelectric material and suitable geometry of the harvester can be solved with an appropriate model of the piezoelectric resonator. For this reason, the presented paper is organized as follows. First, derivation of an analytical beam model of a kinematically excited piezoelectric cantilever in both bimorph/unimorph configurations which also respects the dimensions of used tip mass. This beam model is subsequently reduced to a single degree-of-freedom (DOF) system using the first mode shape function. Although, the derivation of a coupled electromechanical model was published several times, e.g., [19–23], here, we also show the effect of chosen mode shape function which is used in reducing the beam model into single DOF model. Then, the model is verified with 3 different experimental results. Finally, the main aim of this paper is to provide a methodology based on a verified model that can be used to compare the effectivity of materials commonly used in energy harvesting applications.

2. Model of Piezoelectric Vibration Energy Harvester

In order to harvest as much energy from vibrations as possible, it is paramount to properly design dimensions of the harvester and optimize its electrical impedance. This goal can easily be achieved with an analytical model which is able to predict electromechanical response of piezoelectric energy harvesters. Therefore, here in this paper a single DOF model of a kinematically excited cantilever in both bimorph/unimorph configurations is derived. This analytical model is based on Euler-Bernoulli (thin) beam theory and its output is compared with results from experiments conducted with three different piezoelectric harvesters described further in the following section. Then, the derived model is used in a comparative study to compare the piezoelectric materials used in the experiments in terms of energy harvesting efficiency.

2.1. Bimorph Cantilever Beam with Piezoelectric Layers in Series

A piezoelectric cantilever beam harvester in a bimorph configuration is shown in Figure 1. This beam model with dimensions $L \times B \times H$, where $H = 2\,h_p + h_s$, is used for obtaining a single DOF analytical model. The clamped end is kinematically excited with a time-harmonic base acceleration $a(t)$ from an external source of vibrations. Piezoelectric layers are in operational mode 31 (in-thickness polarization of piezoelectric layers and axial bending deformation of the harvester) whose polarization is denoted by arrow symbols in the figure. These layers have electrodes present over a region of dimensions $L_E \times B$ which is mentioned further in the text as section V_E; the remaining portion of piezoelectric layers is not polarized (mentioned as section V_R), and thus is not affected by the piezoelectric effect. A tip mass M_t of negligible rotary inertia is attached to the free end of the beam spanning over the length L_{Mt}—this section is denoted as $V_{R,\,Mt}$. The bimorph model is reduced to a single DOF model which describes the movement of the bimorph's free end q relative to the moving clamped end.

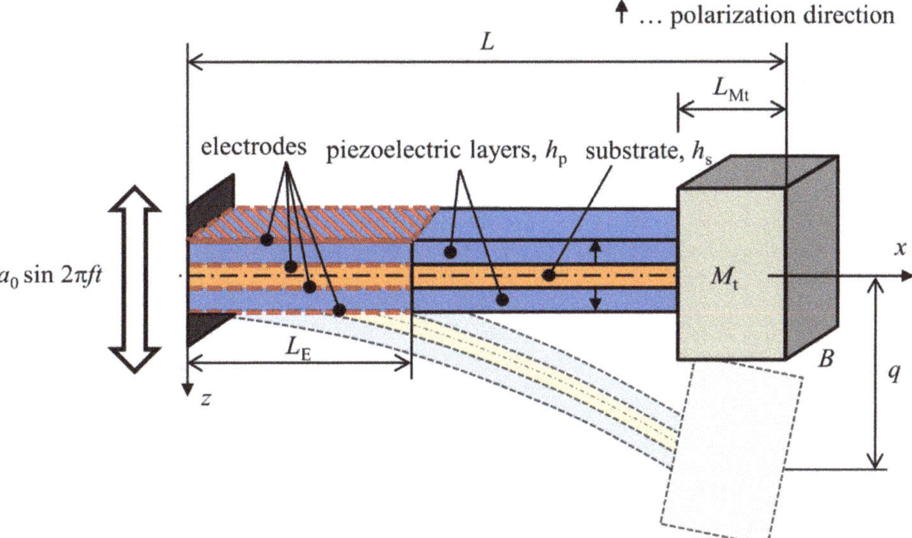

Figure 1. Geometric model of a piezoelectric bimorph in operational mode 31.

The assumptions of Euler-Bernoulli beam theory combined with those of the classical laminate theory concerning continuous strain throughout the layers of a multilayer structure imply that the strain ε_x within bimorph's layers can be expressed as

$$\varepsilon_x = -\frac{\partial^2 w}{\partial x^2} z, \tag{1}$$

where w is transverse displacement of the beam's centerline relative to the movement of the excited clamped end. Since the beam is split into a section V_E which is affected by the piezoelectric effect, section V_R which is not affected by the piezoelectric effect and section $V_{R,\,Mt}$ with distributed tip mass attached, the displacement needs to be a piecewise function defined as

$$w(x, t) = \begin{cases} w_1(x, t) \text{ for } x \in [0,\, L_E] \\ w_2(x, t) \text{ for } x \in (L_E,\, L - L_{Mt}] \\ w_3(x, t) \text{ for } x \in (L - L_{Mt},\, L] \end{cases}. \tag{2}$$

For these functions it must hold that they are continuous and smooth at intersection points, i.e., they must satisfy following conditions:

$$w_1(x = L_E, t) = w_2(x = L_E, t)$$
$$w_2(x = L - L_{Mt}, t) = w_3(x = L - L_{Mt}, t)$$
$$\frac{\partial w_1}{\partial x}(x = L_E, t) = \frac{\partial w_2}{\partial x}(x = L_E, t)$$
$$\frac{\partial w_2}{\partial x}(x = L - L_{Mt}, t) = \frac{\partial w_3}{\partial x}(x = L - L_{Mt}, t)$$

(3)

The piezoelectric constitutive relations, which are required further in the derivation of the analytical model, take the following form for the uniaxial stress state [23]:

$$\varepsilon_x = s_{11}^E \sigma_x + d_{31} E_z,$$

(4)

$$D_z = d_{31} \sigma_x + \epsilon_{33}^T E_z,$$

(5)

where ε_x represents normal strain, s_{11}^E is mechanical compliance measured at constant electric field, σ_x is normal stress, d_{31} is the mode-31 component of the piezoelectric charge coefficient matrix, E_z is the z-component of electric field intensity, D_z is the z-component of electric flux density and ϵ_{33}^T is permittivity in direction of the polarization axis measured at constant mechanical stress. From Equation (4) the expression for the stress σ_x is extracted as

$$\sigma_x = \left(s_{11}^E\right)^{-1}\varepsilon_x - \left(s_{11}^E\right)^{-1}d_{31}E_z .$$

(6)

Note that the reciprocal value of s_{11}^E equals to the elastic modulus Y_p of used piezoelectric material. Equation (6) can then be rewritten as

$$\sigma_x = Y_p\varepsilon_x - \underbrace{Y_p d_{31}E_z}_{e_{31}} = Y_p\varepsilon_x - e_{31}E_z ,$$

(7)

where e_{31} is the piezoelectric modulus. Substituting (7) into (5) yields

$$D_z = \underbrace{d_{31}Y_p\varepsilon_x}_{e_{31}} + \underbrace{\left(\epsilon_{33}^T - d_{31}e_{31}\right)}_{\epsilon_{33}^S}E_z = e_{31}\varepsilon_x + \epsilon_{33}^S E_z,$$

(8)

where ϵ_{33}^S is the permittivity of used piezoelectric material measured at constant strain.

Piezoelectric materials are dielectrics and, as a consequence of the Gauss' law, it holds that $\partial D_z/\partial z = 0$ [24]. This implies that D_z = const. Since the strain term in (8) changes linearly with the z-coordinate, we can use its mean value at the center of the n-th piezoelectric layer; the z-coordinate of the center of n-th layer is denoted as z_{Tpn}. Next, the fundamentals of electricity, see e.g., [25], state that integration of E_z over the thickness of a particular layer yields the voltage drop for the given layer. In order to make D_z independent of the z-coordinate, E_z must be a linear function of this coordinate with its mean value at the center of n-th layer given by

$$E_z(z_{Tpn}) = -\frac{U_n}{h_p} = -\frac{U}{2h_p}$$

(9)

where U is the magnitude of generated voltage drop and h_p is the thickness of piezoelectric layers. By using Equations (1) and (9) in Equation (8) and by assuming that D_z is a layer-wise function, one receives:

$$D_z = -e_{31}z_{Tpn}\frac{\partial^2 w}{\partial x^2} - \frac{\epsilon_{33}^S}{2h_p}U.$$

(10)

By combining Equations (8) and (10), the following expression for E_z is obtained:

$$E_z = \frac{e_{31}}{\epsilon_{33}^S}\frac{\partial^2 w}{\partial x^2}\left(z - z_{\mathrm{T}pn}\right) - \frac{U}{2h_p}. \tag{11}$$

According to [26], the electric current I generated by two in-series connected piezo-electric layers can be expressed as

$$I = \frac{U}{R_1} = \frac{\mathrm{d}}{\mathrm{d}t}\left(\iint_{A_E} D_z \mathrm{d}x\mathrm{d}y\right), \tag{12}$$

where the subscript A_E denotes the area of electrodes and R_1 represents the connected resistance. Inserting (10) into Equation (12) yields

$$\frac{U}{R_1} = \frac{\mathrm{d}}{\mathrm{d}t}\left[\iint_{A_E}\left(-e_{31}z_{\mathrm{T}pn}\frac{\partial^2 w}{\partial x^2} - \frac{\epsilon_{33}^S}{2h_p}U\right)\mathrm{d}x\mathrm{d}y\right]. \tag{13}$$

Integrating terms in (13) leads to a following PDE which governs the electrical behavior of the considered bimorph

$$C_{\mathrm{eq}}\frac{\mathrm{d}U}{\mathrm{d}t} + \frac{1}{R_l}U = \kappa\left.\frac{\partial^2 w}{\partial x \partial t}\right|_{x=L_E}, \tag{14}$$

where C_{eq} denotes the bimorph's equivalent capacitance defined as

$$C_{\mathrm{eq}} = \epsilon_{33}^S\frac{BL_E}{2h_p}, \tag{15}$$

and κ represents generic electromechanical coupling defined as

$$\kappa = -Be_{31}z_{\mathrm{T}p}. \tag{16}$$

As a next step, Equations (11) and (1) can be inserted into Equation (7) to obtain an expression for stress σ_x within the polarized piezoelectric layers:

$$\sigma_x = -Y_{\mathrm{p}}\frac{\partial^2 w}{\partial x^2}z + \frac{e_{31}^2}{\epsilon_{33}^S}\frac{\partial^2 w}{\partial x^2}\left(z_{\mathrm{T}pn} - z\right) + \frac{e_{31}}{2h_p}U. \tag{17}$$

Within the other layers (non-polarized piezoelectric layers and the substrate) the stress obeys the Hooke's law:

$$\sigma_x = -Y_{\mathrm{n}}\frac{\partial^2 w}{\partial x^2}z, \tag{18}$$

where Y_{n} denotes elastic modulus of the used piezoelectric material Y_{p} or the substrate Y_{s}.

Total energy stored in the considered bimorph upon vibrations consists of kinetic energy E_{k}, strain energy E_{p} and the work done by inertial forces due to kinematic excitation W_{ext}. Kinetic energy of the considered beam can be written as

$$\begin{aligned}
E_{\mathrm{k}} &= \frac{1}{2}\iiint_{V_E}\left[\rho_n\left(\frac{\partial w_1}{\partial t}\right)^2\right]\mathrm{d}V + \frac{1}{2}\iiint_{V_R}\left[\rho_n\left(\frac{\partial w_2}{\partial t}\right)^2\right]\mathrm{d}V + \frac{1}{2}\iiint_{V_{R,\,Mt}}\left[\left(\rho_n + \frac{M_t}{L_{Mt}BH}\right)\left(\frac{\partial w_3}{\partial t}\right)^2\right]\mathrm{d}V \\
&= \frac{1}{2}\int_0^{L_E}\left[m^*\left(\frac{\partial w_1}{\partial t}\right)^2\right]\mathrm{d}x + \frac{1}{2}\int_{L_E}^{L-L_{Mt}}\left[m^*\left(\frac{\partial w_2}{\partial t}\right)^2\right]\mathrm{d}x + \frac{1}{2}\int_{L-L_{Mt}}^{L}\left[\left(m^* + \frac{M_t}{L_{Mt}}\right)\left(\frac{\partial w_3}{\partial t}\right)^2\right]\mathrm{d}x,
\end{aligned} \tag{19}$$

where ρ_n is density of the n-th layer, M_t is the attached tip mass and m^* is the bimorph's mass per unit of its length defined as

$$m^* = B\left(\rho_s h_s + 2\rho_p h_p\right). \tag{20}$$

Strain energy stored in the bimorph can be expressed as

$$E_p = \tfrac{1}{2} \iiint\limits_{V_E} [\varepsilon_x \sigma_x]\,\mathrm{d}V + \tfrac{1}{2} \iiint\limits_{V_R} [Y_n \varepsilon_x^2]\,\mathrm{d}V + \tfrac{1}{2} \iiint\limits_{V_{R,\,Mt}} [Y_n \varepsilon_x^2]\,\mathrm{d}V$$

$$= \tfrac{1}{2} \int\limits_0^{L_E} \left[J_{\text{piezo}}^* \left(\frac{\partial^2 w_1}{\partial x^2}\right)^2 - \kappa w_1 \frac{\mathrm{d}\delta}{\mathrm{d}x}(x - L_E) U \right]\mathrm{d}x + \tfrac{1}{2} \int\limits_{L_E}^{L - L_{Mt}} \left[J^* \left(\frac{\partial^2 w_2}{\partial x^2}\right)^2 \right]\mathrm{d}x + \tfrac{1}{2} \int\limits_{L - L_{Mt}}^{L} \left[J^* \left(\frac{\partial^2 w_3}{\partial x^2}\right)^2 \right]\mathrm{d}x\,, \tag{21}$$

where $\mathrm{d}\delta/\mathrm{d}x$ is the first derivative of Dirac's delta function, J_{piezo}^* denotes bending stiffness of the beam section where the polarization of a piezoelectric material is considered (over the length L_E) defined as

$$J_{\text{piezo}}^* = \frac{1}{12} Y_s B h_s^3 + 2 Y_p \left[\frac{1}{12} B h_p^3 + \left(\frac{h_p}{2} + \frac{h_s}{2}\right)^2 B h_p \right] + v, \tag{22}$$

where

$$v = 2 \times \frac{1}{12} \frac{e_{31}^2}{\epsilon_{33}^S} B h_p^3\,, \tag{23}$$

and J^* is bending stiffness of the non-polarized section of the beam (the rest of the beam outside the length L_E) defined as

$$J^* = \frac{1}{12} Y_s B h_s^3 + 2 Y_p \left[\frac{1}{12} B h_p^3 + \left(\frac{h_p}{2} + \frac{h_s}{2}\right)^2 B h_p \right]. \tag{24}$$

The work done by inertia forces due to kinematic excitation is defined as

$$W_{\text{ext}} = - \iiint\limits_{V_E} [\rho_n a_0 w_1]\,\mathrm{d}V - \iiint\limits_{V_R} [\rho_n a_0 w_2]\,\mathrm{d}V - \iiint\limits_{V_{R,\,Mt}} \left[\left(\rho_n + \frac{M_t}{L_{Mt} B H}\right) a_0 w_3 \right]\mathrm{d}V$$

$$= - \int\limits_0^{L_E} [m^* a_0 w_1]\,\mathrm{d}x - \int\limits_{L_E}^{L - L_{Mt}} [m^* a_0 w_2]\,\mathrm{d}x - \int\limits_{L - L_{Mt}}^{L} \left[\left(m^* + \frac{M_t}{L_{Mt}}\right) a_0 w_3 \right]\mathrm{d}x. \tag{25}$$

Subsequently, Hamilton's variational principle [27] is used to obtain equations of motion in the form of PDEs with a nonzero right-hand side. The equations of motion for the polarized portion of the beam and for the non-polarized portions of the beam take the following form:

$$J_{\text{piezo}}^* \frac{\partial^4 w_1}{\partial x^4} + m^* \frac{\partial^2 w_1}{\partial t^2} - \kappa \frac{\mathrm{d}\delta}{\mathrm{d}x}(x - L_E) U = -m^* a_0, \quad x \in [0, L_E] \tag{26}$$

$$J^* \frac{\partial^4 w_2}{\partial x^4} + m^* \frac{\partial^2 w_2}{\partial t^2} = -m^* a_0, \quad x \in (L_E, L - L_{Mt}] \tag{27}$$

$$J^* \frac{\partial^4 w_3}{\partial x^4} + \left(m^* + \frac{M_t}{L_{Mt}}\right) \frac{\partial^2 w_3}{\partial t^2} = -\left(m^* + \frac{M_t}{L_{Mt}}\right) a_0, \quad x \in (L - L_{Mt}, L] \tag{28}$$

Equations above, however, do not account for damping; therefore, they have to be extended with a damping term. Here, we shall consider the stiffness damping term from Rayleigh's Damping theorem:

$$J_{\text{piezo}}^* \frac{\partial^4 w_1}{\partial x^4} + \frac{2 b_r}{\Omega_1} J_{\text{piezo}}^* \frac{\partial^5 w_1}{\partial x^4 \partial t} + m^* \frac{\partial^2 w_1}{\partial t^2} - \kappa \frac{\mathrm{d}\delta}{\mathrm{d}x}(x - L_E) U = -m^* a_0, \quad x \in [0, L_E] \tag{29}$$

$$J^*\frac{\partial^4 w_2}{\partial x^4} + \frac{2b_r}{\Omega_1}J^*\frac{\partial^5 w_2}{\partial x^4 \partial t} + m^*\frac{\partial^2 w_2}{\partial t^2} = -m^* a_0, \ x \in (L_E, \ L - L_{Mt}] \tag{30}$$

$$J^*\frac{\partial^4 w_3}{\partial x^4} + \frac{2b_r}{\Omega_1}J^*\frac{\partial^5 w_3}{\partial x^4 \partial t} + \left(m^* + \frac{M_t}{L_{Mt}}\right)\frac{\partial^2 w_3}{\partial t^2} = -\left(m^* + \frac{M_t}{L_{Mt}}\right)a_0, \ x \in (L - L_{Mt}, \ L] \tag{31}$$

where b_r is the considered damping ratio and Ω_1 is the value of the beam's first eigenfrequency. Equations (29)–(31) together with (14) form a complete equation system which describes the electromechanical response of the considered bimorph.

However, this system of PDEs is actually not very effective to be used in modelling of energy harvesting devices because of its complexity, thus its transformation into a much simpler single DOF model is necessary. In the scope of vibrational energy harvesting applications, the beam is kinematically excited with frequencies very close or equal to the harvester's first resonant frequency $f_{1,r}$. This fact means that beam vibrations are composed mostly of the first vibrational mode and, as a consequence, the beam's displacement relative to the base movement in all sections (V_E, V_R and $V_{R,Mt}$) can be written as

$$w(x, t) \approx \phi_1(x)\eta_1(t), \tag{32}$$

where ϕ_1 is the mode shape function of the first mode and η_1 is its modal coordinate. The shape function ϕ_1 can be approximated with an arbitrary function that resembles the shape of the first bending mode. Although Erturk in [26] recommends using an approximative function which accounts for a tip mass at the beam's free end, such a function is not appropriate for tip masses spanning over a finite length of the beam. Therefore, the following expression was chosen to simplify his approximative function into:

$$\phi_1(x) = C_1 \cdot \left[\cos\frac{\lambda_1}{L}x - \cos h\frac{\lambda_1}{L}x + \varsigma_1\left(\sin\frac{\lambda_1}{L}x - \sin h\frac{\lambda_1}{L}x\right)\right], \tag{33}$$

where

$$\varsigma_1 = \frac{\sin\lambda_1 - \sin h\lambda_1}{\cos\lambda_1 + \cos h\lambda_1}. \tag{34}$$

The eigenvalue λ_1 is obtained as the first positive root of the following transcendental equation [26]

$$1 + \cos\lambda_1\cos h\lambda_1 = 0. \tag{35}$$

Equation (33) accurately describes the first mode shape of a beam without a tip mass. Further in the paper it will be shown that simpler functions which deviate from the actual shape overestimate the beam's stiffness and influence the calculated results.

The constant C_1 in (33) should be evaluated so that ϕ_1 is mass-normalized to prevent numerical errors in further calculations of the model's parameters, i.e., ϕ_1 satisfies the following condition

$$\int_0^{L - L_{Mt}} \phi_1(x)m^*\phi_1(x)\mathrm{d}x + \int_{L - L_{Mt}}^{L} \phi_1(x)\left(m^* + \frac{M_t}{L_{Mt}}\right)\phi_1(x)\mathrm{d}x = 1 \tag{36}$$

Then, approximation (32) can be inserted into the equation system (14), (29)–(31) which can now be solved effectively using the Galerkin method [27], resulting into a much simpler equation system:

$$M\frac{\mathrm{d}^2\eta}{\mathrm{d}t^2} + B\frac{\mathrm{d}\eta}{\mathrm{d}t} + K\eta + \theta U = F, \tag{37}$$

$$C_{eq}\frac{\mathrm{d}U}{\mathrm{d}t} + \frac{1}{R_l}U = \theta\frac{\mathrm{d}\eta}{\mathrm{d}t}, \tag{38}$$

where

$$M = \int\limits_{0}^{L-L_{Mt}} \phi_1(x)m^*\phi_1(x)\mathrm{d}x + \int\limits_{L-L_{Mt}}^{L} \phi_1(x)\left(m^* + \frac{M_t}{L_{Mt}}\right)\phi_1(x)\mathrm{d}x = 1$$

$$K = J^*_{\mathrm{piezo}}\int\limits_{0}^{L_E}\left[\frac{\mathrm{d}^2\phi_1(x)}{\mathrm{d}x^2}\right]^2\mathrm{d}x + J^*\int\limits_{L_E}^{L}\left[\frac{\mathrm{d}^2\phi_1(x)}{\mathrm{d}x^2}\right]^2\mathrm{d}x = \Omega_1^2$$

$$B = 2b_r\Omega_1 \tag{39}$$

$$\theta = \kappa\frac{\mathrm{d}\phi_1}{\mathrm{d}x}\bigg|_{x=L_E}$$

$$F = -\left[m^*a_0\int\limits_{0}^{L-L_{Mt}}\phi_1(x)\mathrm{d}x + \left(m^* + \frac{M_t}{L_{Mt}}\right)a_0\int\limits_{L-L_{Mt}}^{L}\phi_1(x)\mathrm{d}x\right]$$

2.1.1. Effect of Chosen Mode Shape Function on Model Output

This section addresses the effect of the chosen approximative mode shape function ϕ_1 on the model's behavior. To this purpose, a reference configuration of a piezoelectric harvester with a significant tip mass is needed. This requirement is satisfied by a PZT-5A bimorph from a well-known work of Erturk and Inman [26].

To analyze the influence of the chosen approximative function ϕ_1 on the model's output, the actual mode shape of the reference harvester is needed. To obtain the actual mode shape, a 3D numerical model of the reference harvester was created in commercial FE software ANSYS APDL made of approx. 1,000 SOLID186 higher-order elements. The actual mode shape denoted as $\phi_{1,\mathrm{true}}$ was obtained from a modal analysis of the model using the path post-processing tool. Both the actual mode shape $\phi_{1,\mathrm{true}}$ and the approximation $\phi_{1,\mathrm{approx}}$ defined by (33) normed to unity are plotted in Figure 2a. While mode shapes in the graph look almost identical, a much clearer distinction can be seen in Figure 2b by plotting their first derivatives with respect to x (the slope of the mode shape). Here, the actual mode shape $\phi_{1,\mathrm{true}}$ shows a much higher degree of compliance (higher value of $\mathrm{d}\phi_1/\mathrm{d}x$) at the beam's free end, which is crucial for high power output. This increase in compliance at the beam's free end is caused by the presence of the tip mass. Therefore, the presence of heavy tip masses at the beam's free end causes an increase in beam's compliance near its free end which cannot be accounted for using simpler approximative functions, such as polynomials. Using simpler approximative functions will lead to stiffer behavior of the beam model and result in higher resonant frequencies and underestimation of generated electrical power. Nevertheless, the errors in the model's output by using (33) are not significant as demonstrated further in the paper.

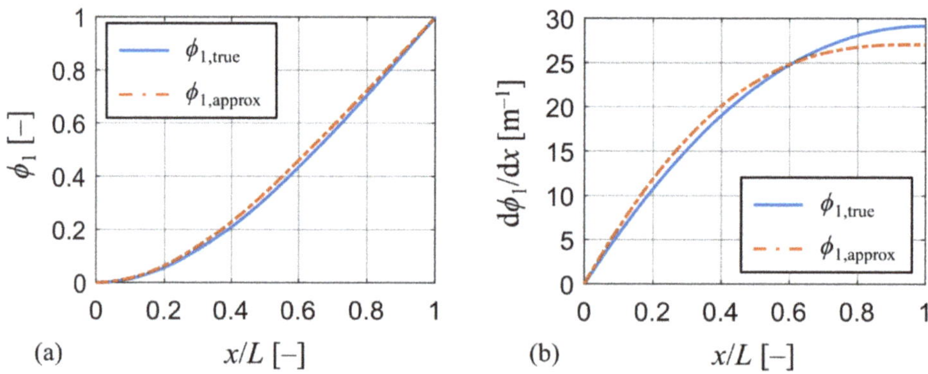

Figure 2. (**a**) Comparison between approximation and true mode shape; (**b**) comparison of slopes between approximation and true mode shape.

2.1.2. Single DOF Model of Bimorph Configuration

The system of Equations (37) and (38) is still not suitable for prediction of harvested power due to using the modal coordinate. For this reason, these equations are transformed from the modal coordinate into direct calculation of the relative movement of the bimorph's free end q defined by (32) as

$$q(t) = \phi_1(x = L)\eta_1(t), \tag{40}$$

Inserting (40) into (37) and (38) transforms the equation system into the sought single DOF model:

$$M_{\text{eff}}\frac{\mathrm{d}^2 q}{\mathrm{d}t^2} + B_{\text{eff}}\frac{\mathrm{d}q}{\mathrm{d}t} + K_{\text{eff}}q + \theta_{\text{eff}}U = F_{\text{eff}} \tag{41}$$

$$C_{\text{eq}}\frac{\mathrm{d}U}{\mathrm{d}t} + \frac{1}{R_l}U = \theta_{\text{eff}}\frac{\mathrm{d}q}{\mathrm{d}t} \tag{42}$$

where

$$\begin{aligned}
M_{\text{eff}} &= \frac{M}{\phi_1^2(x=L)} = \frac{1}{\phi_1^2(x=L)} \\
K_{\text{eff}} &= \frac{K}{\phi_1^2(x=L)} = \frac{\Omega_1^2}{\phi_1^2(x=L)} \\
B_{\text{eff}} &= \frac{B}{\phi_1^2(x=L)} = \frac{2b_r\Omega_1}{\phi_1^2(x=L)} \\
\theta_{\text{eff}} &= \frac{\theta}{\phi_1(x=L)} = \frac{\kappa}{\phi_1(x=L)}\frac{\mathrm{d}\phi_1}{\mathrm{d}x}\bigg|_{x=L_E} \\
F_{\text{eff}} &= \frac{F}{\phi_1(x=L)}
\end{aligned} \tag{43}$$

where capacitance C_{eq} is defined through Equation (15) and the connected resistive load R_l represents the useful electrical load.

2.2. Modification of Single DOF Model for Unimorph Configuration

The model of a unimorph configuration of a piezoelectric harvester which considers only one piezoelectric layer is commonly used with piezoelectric polymers. The unimorph geometric model shares the same parameters to that of a bimorph shown in Figure 1. The single DOF model for the unimorph uses exactly the same equations that were derived for the bimorph, i.e., (41) and (42). Contrary to the bimorph, however, the major difference lies in fact that the bimorph's neutral axis is coincident with its geometrical midplane, whereas this is not true in case of a unimorph. Therefore, the coefficients in (41) and (42) have to be re-defined to respect this fact.

First, the neutral axis of the unimorph z_N is calculated as [28]

$$z_N = \frac{Y_s z_{Ts}h_s + Y_p z_{Tp}h_p}{Y_s h_s + Y_p h_p} = \frac{1}{2}h_s h_p \frac{Y_p - Y_s}{Y_s h_s + Y_p h_p}. \tag{44}$$

Then, the coefficients in (41) and (42) are re-calculated with respect to the unimorph's neutral axis z_N using the same shape function ϕ_1 as in (33). First, the mass coefficient M_{eff} is defined as

$$M_{\text{eff}} = \frac{\int_0^{L - L_{\text{Mt}}} \phi_1(x)m^*\phi_1(x)\mathrm{d}x + \int_{L - L_{\text{Mt}}}^{L} \phi_1(x)\left(m^* + \frac{M_t}{L_{\text{Mt}}}\right)\phi_1(x)\mathrm{d}x}{\phi_1^2(x = L)}, \tag{45}$$

where m^* changes to

$$m^* = B\left(\rho_s h_s + \rho_p h_p\right). \tag{46}$$

Then, the stiffness coefficient K_{eff} changes to

$$K_{\text{eff}} = \frac{J_{\text{piezo}}^* \int_0^{L_E} \left(\frac{\mathrm{d}^2\phi_1(x)}{\mathrm{d}x^2}\right)^2 \mathrm{d}x + J^* \int_{L_E}^{L} \left(\frac{\mathrm{d}^2\phi_1(x)}{\mathrm{d}x^2}\right)^2 \mathrm{d}x}{\phi_1^2(x = L)}, \tag{47}$$

for which the term J^*_{piezo} is defined as

$$J^*_{\text{piezo}} = Y_s B\left[\frac{1}{3}\left(\frac{h_s - h_p}{2}\right)^3 + \frac{1}{3}\left(\frac{h_s + h_p}{2}\right)^3 - z_N^2 h_s\right] + Y_p B\left[\frac{1}{3}\left(\frac{h_s + h_p}{2}\right)^3 - \frac{1}{3}\left(\frac{h_s - h_p}{2}\right)^3 - z_N^2 h_p\right] + v,\tag{48}$$

where

$$v = \frac{e_{31}^2}{\epsilon_{33}^S} B\left[\frac{1}{3}\left(\frac{h_s + h_p}{2}\right)^3 - \frac{1}{3}\left(\frac{h_s - h_p}{2}\right)^3 - z_{Tp}^2 h_p\right],\tag{49}$$

and J^* is defined as

$$J^* = Y_s B\left[\frac{1}{3}\left(\frac{h_s - h_p}{2}\right)^3 + \frac{1}{3}\left(\frac{h_s + h_p}{2}\right)^3 - z_N^2 h_s\right] + Y_p B\left[\frac{1}{3}\left(\frac{h_s + h_p}{2}\right)^3 - \frac{1}{3}\left(\frac{h_s - h_p}{2}\right)^3 - z_N^2 h_p\right].\tag{50}$$

Next, the damping coefficient B_{eff} is defined as

$$B_{\text{eff}} = \frac{2b_r}{\Omega_1} K_{\text{eff}},\tag{51}$$

and the electromechanical coupling coefficient θ_{eff} changes to

$$\theta_{\text{eff}} = \frac{\kappa}{\phi_1(x = L)} \left.\frac{\mathrm{d}\phi_1}{\mathrm{d}x}\right|_{x = L_E},\tag{52}$$

where κ is re-calculated with respect to the neutral axis z_N as

$$\kappa = -e_{31}\left(z_{Tp} - z_N\right).\tag{53}$$

Then, the effective load F_{eff} is defined as

$$F_{\text{eff}} = -\frac{m^* a_0 \int_0^{L - L_{Mt}} \phi_1(x)\mathrm{d}x + \left(m^* + \frac{M_t}{L_{Mt}}\right) a_0 \int_{L - L_{Mt}}^{L} \phi_1(x)\mathrm{d}x}{\phi_1(x = L)},\tag{54}$$

and the equivalent capacity C_{eq} is defined as

$$C_{\text{eq}} = \epsilon_{33}^S \frac{B L_E}{h_p}.\tag{55}$$

3. Verification of Analytical Model Based on Experimental Results

In this chapter, the derived single DOF model is verified for a time-harmonic kinematic excitation. Three different piezoelectric energy harvesters with known geometry and materials are analyzed and their measured responses are compared with the simulations of the derived single DOF model.

The first experiment is a well-known published work of Erturk and Inman [26] where the authors used a bimorph with PZT-5A piezoelectric material and electrodes spanning over the whole bimorph's length, providing a linear dynamic response. The other two experiments included both bimorph and unimorph configurations of piezoelectric harvesters with a partial electrode length. Both these experiments were conducted in laboratories of Brno University of Technology. The first of these experiments used a bimorph made of PZZN-PLZT piezoceramic [6] which exerted a weak non-linear response in the frequency domain. The second experiment used a simple unimorph configuration for wearables with a thin PVDF layer. Geometrical parameters and material data of individual harvesters for both the piezoelectric layer and the substrate are summarized in Tables 1 and 2, respectively. Data for the PZT-5A bimorph is extracted from [26], the PZZN-PLZT from [6] and the PVDF from [16].

Table 1. Parameters of individual harvesters used in experiments.

Harvester Type (Configuration)	L [mm]	L_E [mm]	L_{Mt} [mm]	B [mm]	h_s [mm]	h_p [mm]	M_t [g]
PZT-5A (bimorph)	50.8	50.8	–	31.8	0.14	0.26	12
PZZN-PLZT (bimorph)	40	25	15	10	0.1	0.2	10
PVDF (unimorph)	71.9	49.2	4	10	0.3	0.13	2.6

Table 2. Material properties of piezoelectric layers and substrates for each harvester used in experiments.

Harvester Type (Configuration)	Material	P [kg/m^3]	Y [GPa]	d_{31} [C/N]	$\epsilon_{33}^S/\epsilon_0$ [$-$]
PZT-5A (bimorph)	PZT-5A	7800	66	-190×10^{-12}	1500
	Brass shim	9000	105	–	–
PZZN-PLZT (bimorph)	PZNN-PLZT	7800	62.5	-195×10^{-12}	1850
	Steel shim	7850	210	–	–
PVDF (unimorph)	PVDF	1760	2	-19×10^{-12}	12
	Steel shim	7850	210	–	–

3.1. PZT-5A Bimorph with a Full Electrode Length and a Linear Response

This experiment was described and published in detail in paper [26]. This experimental work has a very high impact and for this reason it was used in our analysis as an etalon for the other piezoelectric harvesters. The geometric model of this piezoelectric harvester is in accordance with the model in Figure 1. The bimorph's piezoelectric layers were made of PZT-5A and the substrate was made of brass. It included electrodes covering the whole bimorph's length for harvesting the generated charge. Geometric parameters of the bimorph and properties of used materials are summarized above in Tables 1 and 2, respectively. This bimorph had an experimentally determined damping ratio $b_r = 0.027$. Since the authors did not state a full description of the tip mass' position and dimensions, it is assumed that the tip mass is located exactly at the bimorph's free end with dimensions allowing for considering the tip mass as a point particle.

This harvester was subjected to a time-harmonic kinematic excitation with a varying forcing frequency f. The experiment mapped how amplitudes of generated electrical power and amplitudes of velocity of the bimorph's free end change with a varying forcing frequency upon different values of connected resistive load. Furthermore, the experiment mapped how the peak values of generated electrical power vary with connected resistive load at a specific forcing frequency.

A comparison of published and measured results with the output of our analytical model is presented in Figure 3. The experiment tracked how the peak values of generated electrical power and the velocity amplitude at the beam's free end dq_0/dt change with a varying forcing frequency. The results are displayed for three different values of used resistive load: 1 kΩ, 33 kΩ and 470 kΩ. The graphs show a good match between the output of the analytical model and the obtained experimental data for all three used resistive loads. Note that some discrepancies exist upon the first resonant frequency of the bimorph; this is mainly due to a steep gradient of calculated results near the first resonant frequency. The reader should also note that for resistive loads of 1 kΩ and 33 kΩ there is a slight difference in resonant frequencies between the real bimorph and the analytical single DOF model. This deviation is caused by the used approximative function ϕ_1 which does not account for a concentrated tip mass at the beam's free end. Therefore, as mentioned earlier, the used approximative function forces the beam to behave slightly stiffer and lowers the amount of generated electrical power.

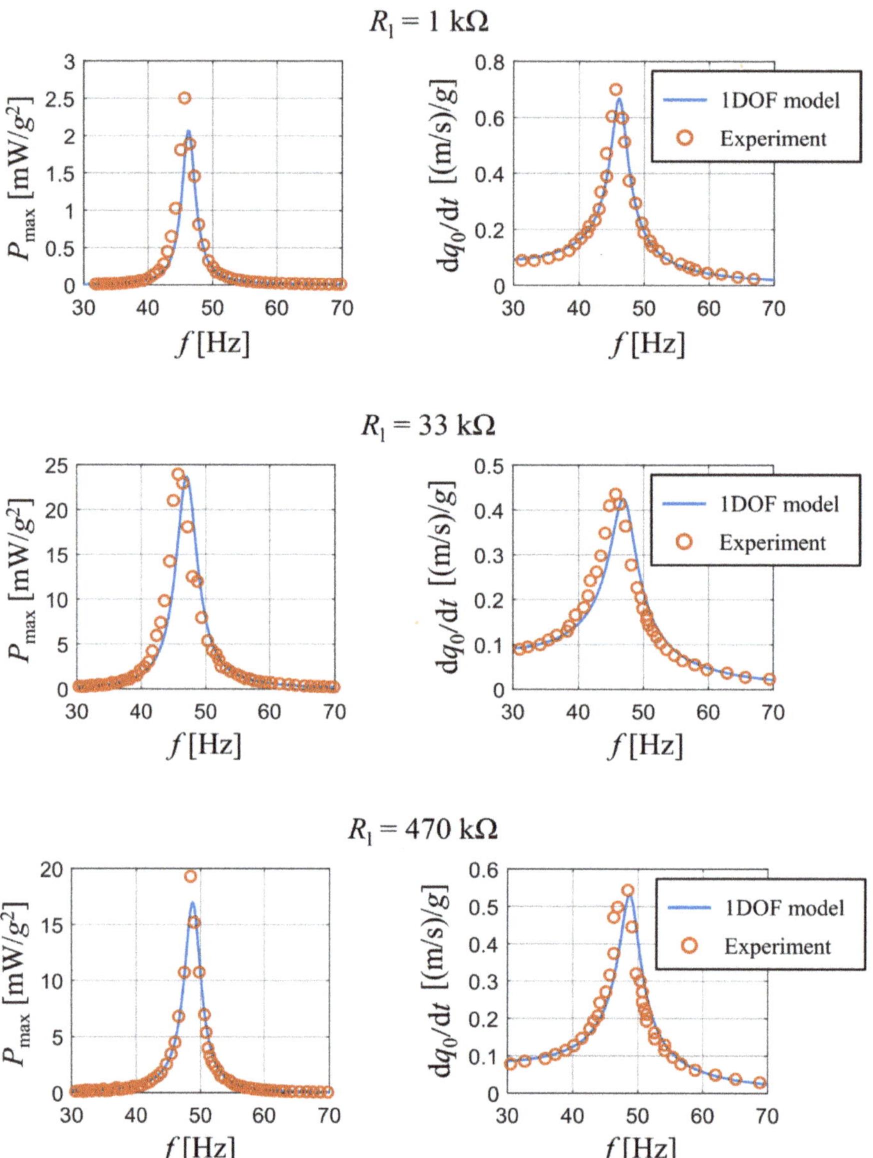

Figure 3. Comparison of electrical power and velocity of tip mass for both experimental results [26] and analytical model.

The output of analytical model matches perfectly with the experimental results for both the short-circuit frequency f_{SC} and the open-circuit frequency f_{OC} of this coupled electromechanical system. The short-circuit and open-circuit frequency are the first resonant frequencies in case of $R_1 = 0$ and $R_1 \rightarrow \infty$, respectively. The match of simulation results with the measured ones for various values of resistive load R_1 and kinematic excitation at both the short-circuit frequency f_{SC} and the open-circuit frequency f_{OC} is shown in Figure 4. Both states correspond with operations slightly below and above the resonance excitation for various values of resistive load, which determine the value of actual resonance frequency.

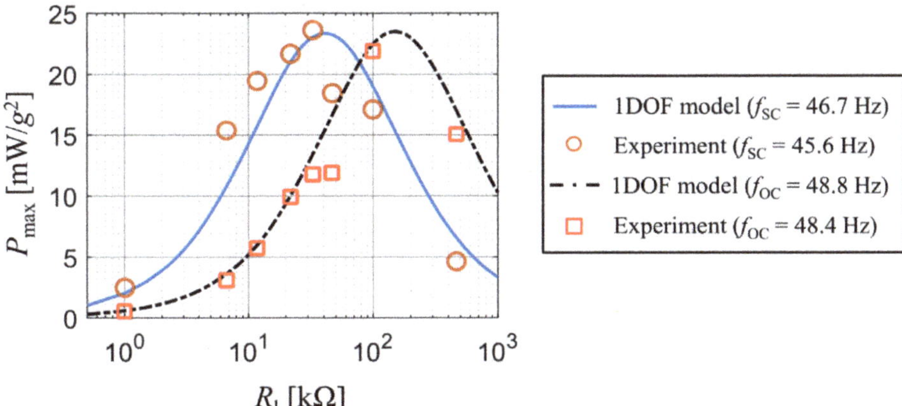

Figure 4. Peak power values as a function of resistive load upon excitation at short-circuit resonance frequency and the open-circuit resonance frequency.

While in case of the open-circuit forcing frequency the results of the analytical single DOF model agree with the measured values, for the short-circuit case the calculated values from the single DOF model are slightly shifted towards higher values of resistive load. Also note that there are differences in both frequencies between the real bimorph and the single DOF model. While in case of the open-circuit frequency this difference is very small, for the short-circuit frequency this difference is notably larger and affects the value of optimal resistive load for which the generated electrical power reaches its peak value. This is caused by the used approximative function ϕ_1 which causes the beam to behave stiffer. Nevertheless, this inaccuracy is negligible in terms of using analytical models for a rough prediction of the generated power when the system is excited by real vibrations.

3.2. PZNN-PLZT Bimorph with Partial Electrode Length and Weak Non-Linear Response

The experiment with PZNN-PLZT bimorph (see Figure 5) was conducted in a laboratory at Brno University of Technology with a partial electrode length (there are no electrodes under the tip mass). Geometrical parameters of this bimorph are summarized above in Table 1. The bimorph's piezoelectric layers were made of PZNN-PLZT, which is in detail described in [6], and the substrate was made of a common steel shim. Properties of these materials are listed above in Table 2. Electrodes were made using a thin silver tape casting. Since the silver electrodes were substantially thinner than other layers, they were not accounted for in the calculation of single DOF model parameters due to their negligible effect on the net mass and the beam's stiffness. The bimorph had an experimentally determined damping ratio $b_r = 0.025$ via an impulse response in the short-circuit state.

The clamping of the used bimorph was kinematically excited at several forcing frequencies near the bimorph's first resonant frequency with a constant acceleration amplitude $a_0 = 0.1$ g. The aim of this experiment was to track results, namely the RMS of generated voltage and RMS of velocity of the tip mass, at different excitation frequencies close to the bimorph's first natural frequency. Also, the optimal resistive load was sought at which the bimorph generates maximal electrical power at its current first resonant frequency which slightly varies with changes in R_l.

A comparison between the measured data and the calculated output of the analytical model is shown in Figure 6, namely the RMS values of output voltage and those of velocity of the bimorph's free end as a function of forcing frequency. The results are displayed for two values of used resistive load: 1 MΩ and 10 MΩ. The measured data shows a weak non-linear softening dynamic behavior; however, the analytical single DOF model with linearized parameters still shows a very good degree of accuracy for both used resistive loads in terms of achieved amplitudes.

Figure 5. PZNN-PLZT bimorph used in experiment.

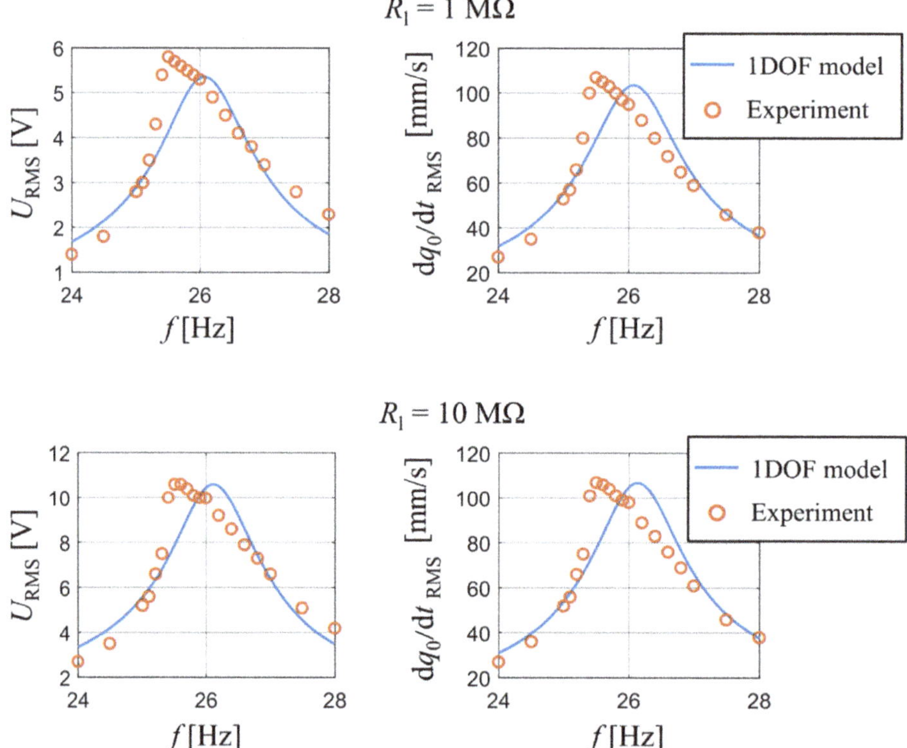

Figure 6. Comparison of generated voltage and velocity of harvester's tip mass obtained from measurement and developed analytical model for R_l are 1 MΩ and 10 MΩ.

Our experiment also tracked the values of generated electrical power as a function of used resistive load. During the measurement, the forcing frequency was adjusted for each value of resistive load so that it matched the bimorph's actual first resonant frequency. Both the analytical model and the experiment show (Figure 7) that the optimal resistive load is approx. 1.5 MΩ and, at the same time, also the maximal values of generated electrical power calculated with the single DOF model agree with experimental data at all used values of resistive load. The reader should note here that the curve from the analytical model is slightly shifted to higher values of R_1 which is, similarly as in the previous experiment, due to the used approximative function ϕ_1, which makes the beam model behave slightly stiffer.

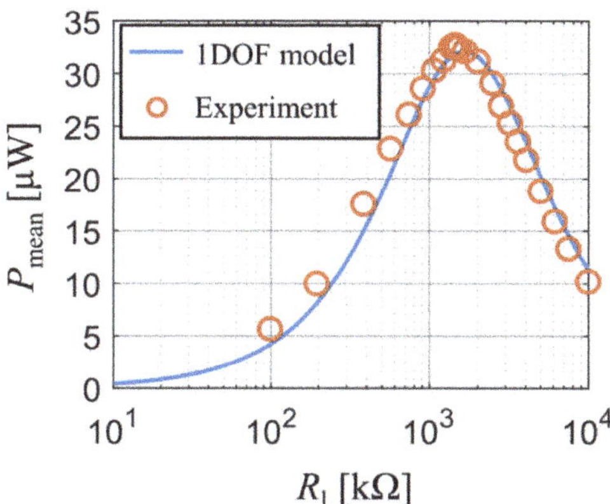

Figure 7. Power as function of resistive load upon excitation at actual resonant frequencies.

3.3. PVDF Unimorph with a Partial Electrode Length and a Linear Response

PVDF piezoelectric energy harvesters are very often presented as a suitable kinetic energy harvester [29] and for this reason the PVDF material was chosen for the last experiment, which was also conducted in a laboratory at Brno University of Technology. The PVDF foil is used in a unimorph configuration of a clamped cantilever with a partial electrode length shown in Figure 8. Parameters of this unimorph are listed above in Table 1. The unimorph's piezoelectric layer is a PVDF foil and the substrate is a steel shim [30]. Properties of these materials are summarized above in Table 2. Electrodes were made using a thin silver tape casting. The silver electrodes were not accounted in the calculation of single DOF model parameters as in the previous model due to their negligible effect on the net mass and beam's stiffness. The clamping of the used unimorph was kinematically excited at several forcing frequencies near the unimorph's first natural frequency ($f_{1,r}$ = 18.7 Hz) with a constant acceleration amplitude a_0 = 0.035 g. The unimorph had an experimentally determined damping ratio b_r = 0.0065 via an analysis of impulse response in the short-circuit state.

This experiment measured the RMS of output voltage and the amplitude of velocity of the tip mass at different forcing frequencies close to the unimorph's first natural frequency. A comparison between the measured data and the calculated output of the analytical model is shown in Figure 9, namely the RMS values of output voltage U and amplitudes of velocity of the tip mass as a function of a forcing frequency for R_1 = 10 MΩ. One can see that the first resonant frequency of the analytical model is again slightly higher due to used approximative function ϕ_1; nevertheless, the calculated values from the analytical model agree with the measured ones.

3.4. Single DOF Model Parameters of Considered Harvesters

The calculated parameters of individual harvesters which were used as input in analytical models are summarized in Table 3. The values of effective load F_{eff} were normalized with respect to 1 g of base acceleration.

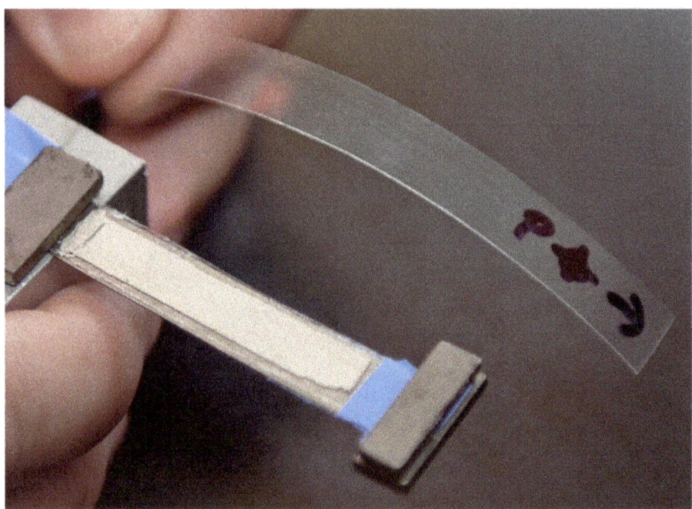

Figure 8. PVDF unimorph used in the experiment and a strip of PVDF foil used as the piezoelectric layer.

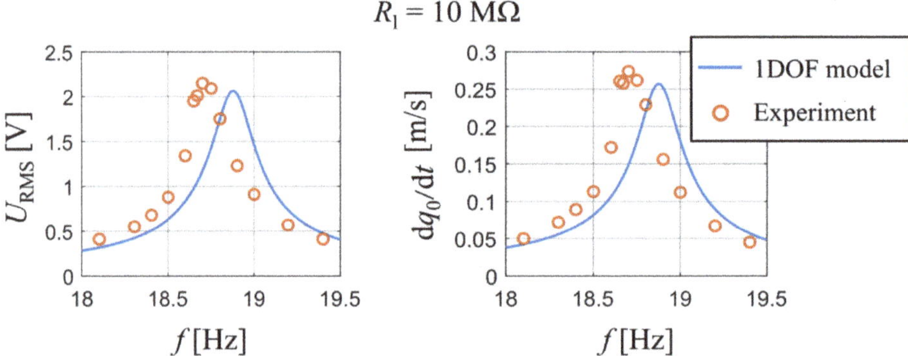

Figure 9. Comparison of output voltage and velocity of tip mass obtained from the measurement and by using the developed analytical model for $R_1 = 10$ MΩ.

Table 3. Parameters of each piezoelectric harvester used in the analytical model.

Harvester Type	M_{eff} [g]	B_{eff} [Ns/m]	K_{eff} [N/m]	F_{eff} [N/g]	θ_{eff} [N/V]	C_{eq} [F]
PZT-5A	14.1	2.24×10^{-1}	1218.10	1.51×10^{-1}	2.20×10^{-3}	4.12×10^{-8}
PZZN-PLZT	6.10	5.13×10^{-2}	164.56	7.90×10^{-2}	6.03×10^{-5}	3.65×10^{-9}
PVDF	2.90	4.40×10^{-3}	40.38	3.27×10^{-2}	1.21×10^{-6}	3.08×10^{-10}

The steady-state results calculated with the developed analytical single DOF model and parameters given in Table 3 showed an excellent agreement with all presented experiments. Combined with low usage of computer resources, the developed analytical model presents a simple and very effective tool for proper designing of piezoelectric harvesters. Moreover, the model can be used to simulate transient responses of the considered harvester (represented by its geometry and materials) to arbitrary time-dependent loads as demonstrated further in the text. For increased accuracy outside the vicinity of the harvester's first resonant frequency, additional mode shapes (ϕ_2, ϕ_3, etc.) can be supplemented. Moreover, the single DOF model can easily be extended to support calculations of strain and stress levels within the beam's layers to determine a maximal allowable load as shown in [31].

4. Comparison of Piezoelectric Materials for Kinetic Energy Harvesting Purposes

Since the analytical single DOF model of a piezoelectric harvester was successfully validated for both unimorph and bimorph configurations using the data from three different piezoelectric materials and experiments, the verified models of three harvesters will enable to determine the effectivity of used piezoelectric materials in different energy harvesting applications. Here, the three materials considered in the scope of this work (PZT-5A, PZZN-PLZT and PVDF) are compared in terms of harvested electrical power when subjected to harmonic vibrations (lab shaker) and in terms of harvested electrical energy when subjected to random vibrations (human body movement).

4.1. Harmonic Vibrations Case

To compare the output of harmonically excited piezoelectric harvesters made of different piezoelectric materials, their dynamic parameters must be similar, that is, their effective mass M_{eff} and eigenfrequency f_1. This can be done by changing dimensions of the considered harvesters; however, doing this will also lead to changes in the piezoelectric coupling coefficient and ultimately making the comparison invalid. To overcome this issue and maintain comparability, the volume of polarized piezoelectric materials was kept constant. The dimensions of polarized piezoelectric material in case of piezoceramic bimorphs were fixed at values $L_E \times B \times h_p = 40 \times 10 \times 0.26$ mm and in case of PVDF unimorph at $L_E \times B \times h_p = 40 \times 40 \times 0.13$ mm due to manufacturing limits of PVDF foils (these must be thin but can span over a large area). Then, to reduce the complexity of this optimizing task, the value of L_{Mt} was fixed at 5 mm and the thickness of the substrate h_s was fixed at 0.15 mm (0.3 mm) in case of bimorphs (PVDF unimorph). Thus, the only parameters left for optimizing were the total length of the harvester L and and the tip mass M_t. These parameters were then tuned (see Table 4) to achieve values of M_{eff} and f_1 common to all three harvesters. Upon the study, the harvesters were forced with a time-harmonic base acceleration for various values of resistive load R_l at their actual resonant frequencies.

Table 4. Tuned dimensions of harvesters and their equivalent single DOF model parameters used in the comparison.

Harvester Type (Configuration)	L [mm]	L_E [mm]	L_{Mt} [mm]	B [mm]	h_s [mm]	h_p [mm]	M_t [g]
PZT-5A (bimorph)	68.8	40	5	10	0.15	0.26	3.67
PZZN-PLZT (bimorph)	54.3	40	5	10	0.15	0.26	3.99
PVDF (unimorph)	71.9	40	5	40	0.3	0.13	2.60

Harvester Type (Configuration)	M_{eff} [g]	B_{eff} [Ns/m]	K_{eff} [N/m]	f_1 [Hz]	F_{eff}/g [N/1g]	θ_{eff} [N/V]	C_{eq} [F]
PZT-5A (bimorph)	4.21	4.45×10^{-2}	161.25	31.13	5.02×10^{-2}	5.14×10^{-4}	1.02×10^{-8}
PZZN-PLZT (bimorph)	4.21	4.20×10^{-2}	161.05	31.13	4.75×10^{-2}	6.36×10^{-5}	4.50×10^{-9}
PVDF (unimorph)	4.21	1.07×10^{-2}	161.25	31.13	5.40×10^{-2}	5.54×10^{-6}	1.30×10^{-9}

The comparison (see Figure 10) revealed that PZT-5A is the best suitable material of the considered ones for energy harvesting purposes thanks to its high power output (several mW per 1 g of base acceleration) and a broad range of optimal resistive load (approx. 150 kΩ to 1.1 MΩ) due to its strong piezoelectric coupling. PZZN-PLZT is also suitable for energy harvesting applications since it offers high power output of about 1 mW per 1 g of base acceleration for resistive loads close to 1 MΩ. On the other hand, PVDF generates the least amount of power of the three materials and due to its very high optimal resistive load it is not sufficient for energy harvesting applications. Note that the strong piezoelectric coupling in case of PZT-5A harvester significantly damps the power output between the two optimal resistive loads which results in a local minimum surrounded by two local maxima.

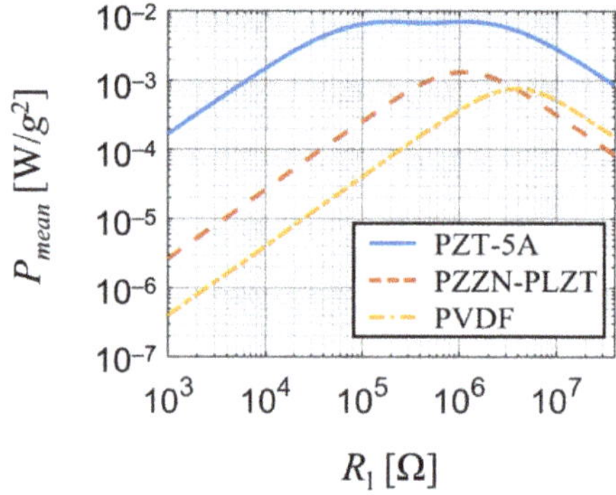

Figure 10. Comparison of harvested electrical power among the considered materials for various resistive loads upon simple harmonic forcing at the first resonant frequency.

4.2. Random Vibrations Case

For this comparison, typical mechanical vibrations of a human forearm which are encountered in wearables applications were measured [32] and analyzed using the developed model. The measured time-course of acceleration $a(t)$, see Figure 11a, is generated by a random movement of the forearm and can be thought of as a representative of random vibrations as can be seen from its spectrogram in Figure 11b. Therefore, it is perfect for the comparison of energy harvesting devices since a steady-state response, whose magnitude depends on how close the forcing frequency is to the harvester's resonant frequency, will not occur.

The measured acceleration $a(t)$ is used as input for a transient analysis of the derived single DOF analytical model, where the applied force is a function of acceleration data. This model of coupled electro-mechanical system is realized in Matlab Simulink simulation environment [33] and its aim is to track the amount of harvested electrical energy for various values of resistive load R_1. Simulation results of predicted harvested energy for this wearable operation for the harvesters used in the experiments and the tuned harvesters from Section 4.1 are shown in Figure 11c,d, respectively.

In case of unmodified harvesters used in the experiments (Figure 11c), the PZT-5A harvester is able to convert the most of mechanical energy (~0.17 mJ) among the three compared harvesters and has a low value of optimal resistive load (~118 kΩ). The energy output of PZZN-PLZT harvester (~0.07 mJ) is of the same order as the one of the PZT-5A harvester, however its much higher optimal resistive load (~1.82 MΩ) makes it less suitable for energy harvesting applications. On the contrary, the PVDF harvester is not suitable for energy harvesting applications at all due to very low energy output (~0.07 μJ), but it will find its use in sensing applications due to very high value of optimal resistive load (~28 MΩ). The same also applies to results in case of tuned harvesters (Figure 11d), where the tuned PZT-5A harvester once again shows that the energy harvesting properties of PZT-5A are far superior to those of PZZN-PLZT and PVDF. The increase in harvested electrical energy for the tuned PZT-5A harvester compared to the unmodified geometry is due to its lower resonant frequency which was reduced from 46.8 Hz to 31.1 Hz.

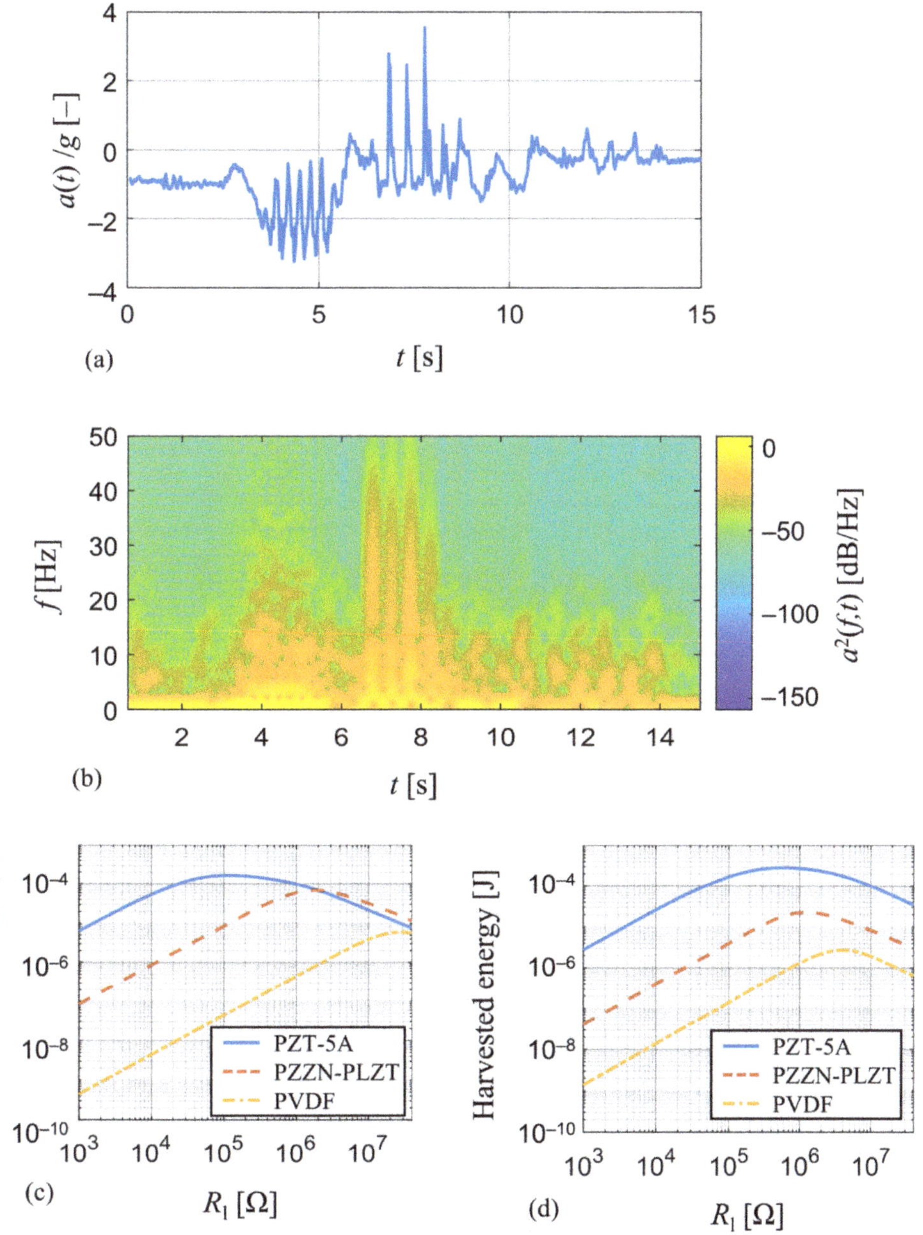

Figure 11. (a) Measured acceleration of a random movement of a human wearable and (b) its spectrogram; a comparison of harvested electrical energy from the human forearm movement among (c) the piezoelectric harvesters used in the experiments and (d) the tuned piezoelectric harvesters from Section 4.1 for different values of resistive load.

The results of both these comparisons showed that both PZT-5A and PZZN-PLZT piezoceramic harvesters are suitable for energy harvesting purposes, although operating at different values of resistive load, whereas the harvester based on piezoelectric polymer PVDF provides an insufficient energy harvesting system due to a very low amount of harvested energy. However this energy harvester should primarily be used in sensing applications [34].

5. Conclusions

The main aim of this paper was to compare the effectiveness of materials commonly used in energy harvesting operations using a single DOF model and at the same time analyze the effect of used mode shape function on simulation results. The single DOF model of a cantilever piezoelectric harvester in both bimorph and unimorph configurations was derived based on Euler-Bernoulli beam theory. Output of the model was confronted with available experimental data obtained from three different piezoelectric harvesters (PZT-5A bimorph, PZZN-PLZT bimorph and PVDF unimorph) and showed a good degree of accuracy. It is obvious that the presented model of an energy harvester can be used for various piezoelectric materials. Therefore, the developed single DOF analytical model represents a simple and very helpful tool for designing piezoceramic vibration energy harvesters. Moreover, it could easily be employed to check if a particular kinetic energy harvester provides sufficient output power for the intended application. Or inversely, the model could be used to design a piezoceramic harvester with optimized operational parameters and dimensions due to the model's ability to predict the amount of harvested energy in particular operational conditions. Moreover, the developed model can easily be extended to support calculations of strain and stress levels within harvester's layers for further assessments concerning strength and fatigue limits.

The model is primarily intended for operations of the harvester at frequencies where vibrations consist mostly of the first mode shape, since it offers the best operational conditions for energy harvesting (no strain nodes). If higher vibrational modes are of interest, the developed model can easily be extended by supplementing their respective shape functions and using the superposition principle. It was found that the quality of the used approximative function for the first mode shape affects the beam model's stiffness in such a way so that simpler (less accurate) approximative functions force the beam model to behave stiffer, i.e., its resonant frequency being shifted to higher values. Also, the way how device layers and electrodes are assembled can also affect the stiffness of the system and could potentially result in a weak nonlinearity, which was observed in one of our experiments. Nevertheless, a typical assembly of layers and the approximative shape function used in this work (the shape of the beam's first vibrational mode without a tip mass) still shows a good degree of accuracy.

The single DOF model itself poses as a very effective tool whose main advantages are low computer resources usage and the ability to calculate transient responses for arbitrary time-dependent loads. Both these features were employed in an energy harvesting effectivity comparison of the three materials used in the scope of this work. The materials comprised of PZT-5A, which is known nowadays to be one of the best materials for energy harvesting purposes, and PZZN-PLZT and PVDF which are used in our laboratory for designing vibration energy harvesting devices. The comparison was split into two parts with respect to forcing: case of simple harmonic vibrations and case of random vibrations. The case of simple harmonic vibrations was carried out so that the harvesters' dimensions were tuned in order to achieve common value of seismic mass and resonant frequency for all three harvesters and at the same time the harvesters had same volume of polarized piezoelectric material. In case of random vibrations the harvesters were subjected to a non-harmonic and non-periodic vibrations typical for wearables applications. Results from both comparison cases showed that the piezoceramic harvesters (PZT-5A and PZZN-PLZT) are a perfect choice for energy harvesting applications, though geometry and electrical load must be optimized. On the contrary, the PVDF harvester is not suitable for energy harvesting purposes due to very low values of harvested energy despite many recent papers reporting the otherwise, and its potential lies in sensing applications.

Author Contributions: Conceptualization, Z.H.; methodology, Z.M. and O.S.; software, Z.M. and O.S.; validation, O.R.; formal analysis, Z.M. and O.S.; investigation, Z.M. and O.R.; resources, Z.H.; data curation, O.R.; writing—original draft preparation, Z.M.; writing—review and editing, Z.H. and O.S.; visualization, Z.M.; supervision, Z.H. and O.S.; project administration, Z.H.; funding acquisition, Z.H. All authors have read and agreed to the published version of the manuscript.

Funding: This presented research and development was supported by the Czech Science Foundation project GA19-17457S "Manufacturing and analysis of flexible piezoelectric layers for smart engineerin", Czech Republic.

Institutional Review Board Statement: Not applicable.

Informed Consent Statement: Not applicable.

Conflicts of Interest: The authors declare no conflict of interest. The funders had no role in the design of the study; in the collection, analyses, or interpretation of data; in the writing of the manuscript, or in the decision to publish the results.

References

1. Roundy, S.; Wright, P.K.; Rabaey, J. A study of low level vibrations as a power source for wireless sensor nodes. *Comput. Commun.* **2003**, *26*, 1131–1144. [CrossRef]
2. Mitcheson, P.D.; Yeatman, E.; Rao, G.K.; Holmes, A.S.; Green, T. Energy Harvesting From Human and Machine Motion for Wireless Electronic Devices. *Proc. IEEE* **2008**, *96*, 1457–1486. [CrossRef]
3. Roundy, S.; Wright, P.K. A piezoelectric vibration based generator for wireless electronics. *Smart Mater. Struct.* **2004**, *13*, 1131–1142. [CrossRef]
4. Hadas, Z.; Smilek, J.; Rubes, O. Analyses of electromagnetic and piezoelectric systems for efficient vibration energy harvesting. In *Smart Sensors, Actuators, and MEMS VIII*; Fonseca, L., Prunnila, M., Peiner, E., Eds.; SPIE MICROTECHNOLOGIES: Barcelona, Spain, 2017.
5. Gljušćić, P.; Zelenika, S.; Blažević, D.; Kamenar, E. Kinetic energy harvesting for wearable medical sensors. *Sensors* **2019**, *19*, 4922. [CrossRef]
6. Bai, Y.; Tofel, P.; Hadas, Z.; Smilek, J.; Lošák, P.; Skarvada, P.; Macku, R. Investigation of a cantilever structured piezoelectric energy harvester used for wearable devices with random vibration input. *Mech. Syst. Signal Process.* **2018**, *106*, 303–318. [CrossRef]
7. Paulo, J.; Gaspar, P.D. Review and future trend of energy harvesting methods for portable medical devices. In *WCE 2010—World Congress on Engineering 2010*; Ao, S.I., Gelman, L., Hukins, D., Eds.; IAENG Society of Electrical Engineering: London, UK, 2010.
8. Zelenika, S.; Hadas, Z.; Bader, S.; Becker, T.; Gljušćić, P.; Hlinka, J.; Janak, L.; Kamenar, E.; Ksica, F.; Kyratsi, T.; et al. Energy Harvesting Technologies for Structural Health Monitoring of Airplane Components—A Review. *Sensors* **2020**, *20*, 6685. [CrossRef] [PubMed]
9. Duarte, F.; Ferreira, A. Energy harvesting on railway tracks: State-of-the-art. *Proc. Inst. Civ. Eng. Transp.* **2017**, *170*, 123–130. [CrossRef]
10. Cahill, P.; Hanley, C.; Jaksic, V.; Mathewson, A.; Pakrashi, V. Energy harvesting for monitoring bridges over their operational life. In Proceedings of the 8th European Workshop on Structural Health Monitoring, EWSHM 2016, Bilbao, Spain, 5–8 July 2016; Volume 4, pp. 1–11.
11. Bowen, C.R.; Kim, H.A.; Weaver, P.M.; Dunn, S. Piezoelectric and ferroelectric materials and structures for energy harvesting applications. *Energy Environ. Sci.* **2014**, *7*, 25–44. [CrossRef]
12. Panda, P.K.; Sahoo, B. PZT to lead free piezo ceramics: A review. *Ferroelectrics* **2015**, *474*, 128–143. [CrossRef]
13. Bai, Y.; Tofel, P.; Palosaari, J.; Jantunen, H.; Juuti, J. A Game Changer: A Multifunctional Perovskite Exhibiting Giant Ferroelectricity and Narrow Bandgap with Potential Application in a Truly Monolithic Multienergy Harvester or Sensor. *Adv. Mater.* **2017**, *29*, 1700767. [CrossRef] [PubMed]
14. Tofel, P.; Machu, Z.; Chlup, Z.; Hadraba, H.; Drdlik, D.; Sevecek, O.; Majer, Z.; Holcman, V.; Hadas, Z. Novel layered architecture based on $Al_2O_3/ZrO_2/BaTiO_3$ for SMART piezoceramic electromechanical converters. *Eur. Phys. J. Spec. Top.* **2019**, *228*, 1575–1588. [CrossRef]
15. Pozzi, M.; Canziani, A.; Durazo-Cardenas, I.; Zhu, M. Experimental characterisation of macro fibre composites and monolithic piezoelectric transducers for strain energy harvesting. In *Smart Structures (NDE)*; Kundu, T., Ed.; SPIE: Bellingham, DC, USA, 2012.
16. Sappati, K.K.; Bhadra, S. Piezoelectric polymer and paper substrates: A review. *Sensors* **2018**, *18*, 3605. [CrossRef] [PubMed]
17. Song, J.; Zhao, G.; Li, B.; Wang, J. Design optimization of PVDF-based piezoelectric energy harvesters. *Heliyon* **2017**, *3*, e00377. [CrossRef] [PubMed]
18. Kim, M.; Dugundji, J.; Wardle, B.L. Efficiency of piezoelectric mechanical vibration energy harvesting. *Smart Mater. Struct.* **2015**, *24*, 055006. [CrossRef]

19. Erturk, A.; Inman, D.J. Issues in mathematical modeling of piezoelectric energy harvesters. *Smart Mater. Struct. Smart Mater. Struct* **2008**, *17*, 065016. [CrossRef]
20. Liao, Y.; Liang, J. Maximum power, optimal load, and impedance analysis of piezoelectric vibration energy harvesters. *Smart Mater. Struct.* **2018**, *27*, 075053. [CrossRef]
21. Li, X.; Upadrashta, D.; Yu, K.; Yang, Y. Sandwich piezoelectric energy harvester: Analytical modeling and experimental validation. *Energy Convers. Manag.* **2018**, *176*, 69–85. [CrossRef]
22. Mendonca, L.S.; Martins, L.T.; Radecker, M.; Bisogno, F.; Killat, D. Normalized Modeling of Piezoelectric Energy Harvester Based on Equivalence Transformation and Unit-Less Parameters. *J. Microelectromech. Syst.* **2019**, *28*, 666–677. [CrossRef]
23. Machů, Z.; Ševeček, O.; Hadas, Z.; Kotoul, M. Modeling of electromechanical response and fracture resistance of multilayer piezoelectric energy harvester with residual stresses. *J. Intell. Mater. Syst. Struct.* **2020**, *31*, 2261–2287. [CrossRef]
24. Meitzler, A. *176-1987 IEEE Standard on Piezoelectricity*; IEEE: New York, NY, USA, 1988.
25. Halliday, D.; Resnick, R.; Walker, J. *Fundamentals of Physics Extended*, 9th ed.; Wiley: Hoboken, NJ, USA, 2010.
26. Erturk, A.; Inman, D.J. An experimentally validated bimorph cantilever model for piezoelectric energy harvesting from base excitations. *Smart Mater. Struct.* **2009**, *18*, 025009. [CrossRef]
27. Reddy, J.N. *Energy Principles and Variational Methods in Applied Mechanics*; John Wiley & Sons, Ltd.: Hoboken, NJ, USA, 2017.
28. Flores-Domínguez, M. Modeling of the Bending Stiffness of a Bimaterial Beam by the Approximation of One-Dimensional of Laminated Theory. *J. Eng. Res. Appl.* **2014**, *4*, 492–497.
29. Zhao, J.; You, Z. Models for 31-Mode PVDF Energy Harvester for Wearable Applications. *Sci. World J.* **2014**, *2014*, 893496. [CrossRef] [PubMed]
30. Hadas, Z.; Rubes, O.; Tofel, P.; Machu, Z.; Riha, D.; Sevecek, O.; Kastyl, J.; Sobola, D.; Castkova, K. Piezoelectric PVDF Elements and Systems for Mechanical Engineering Applications. In Proceedings of the 2020 19th International Conference on Mechatronics—Mechatronika (ME), Prague, Czech Republic, 2–4 December 2020.
31. Rubes, O.; Machu, Z.; Sevecek, O.; Hadas, Z. Crack Protective Layered Architecture of Lead-Free Piezoelectric Energy Harvester in Bistable Configuration. *Sensors* **2020**, *20*, 5808. [CrossRef]
32. Rubes, O.; Hadas, Z. Design and Simulation of Bistable Piezoceramic Cantilever for Energy Harvesting from Slow Swinging Movement. In Proceedings of the 2018 IEEE 18th International Conference on Power Electronics and Motion Control, PEMC 2018, Budapest, Hungary, 26–30 August 2018.
33. Rubes, O.; Brablc, M.; Hadas, Z. Nonlinear vibration energy harvester: Design and oscillating stability analyses. *Mech. Syst. Signal Process.* **2019**, *125*, 170–184. [CrossRef]
34. Fitzgerald, P.C.; Malekjafarian, A.; Bhowmik, B.; Prendergast, L.J.; Cahill, P.; Kim, C.-W.; Hazra, B.; Pakrashi, V.; Obrien, E.J. Scour Damage Detection and Structural Health Monitoring of a Laboratory-Scaled Bridge Using a Vibration Energy Harvesting Device. *Sensors* **2019**, *19*, 2572. [CrossRef]

sensors

MDPI

Article

A New Method to Perform Direct Efficiency Measurement and Power Flow Analysis in Vibration Energy Harvesters

Jan Kunz [1], Jiri Fialka [1,2], Stanislav Pikula [1], Petr Benes [1,2,*], Jakub Krejci [1], Stanislav Klusacek [1,2] and Zdenek Havranek [1,2]

1. Department of Control and Instrumentation, Faculty of Electrical Engineering and Communications, Brno University of Technology, Technicka 3082/12, 616 00 Brno, Czech Republic; xkunzj00@vutbr.cz (J.K.); jiri.fialka@ceitec.vutbr.cz (J.F.); pikula@vutbr.cz (S.P.); xkrejc44@vutbr.cz (J.K.); klusacek@vutbr.cz (S.K.); zdenek.havranek@ceitec.vutbr.cz (Z.H.)
2. CEITEC—Central European Institute of Technology, Brno University of Technology, Purkynova 656/123, 612 00 Brno, Czech Republic
* Correspondence: benesp@vutbr.cz

Citation: Kunz, J.; Fialka, J.; Pikula, S.; Benes, P.; Krejci, J.; Klusacek, S.; Havranek, Z. A New Method to Perform Direct Efficiency Measurement and Power Flow Analysis in Vibration Energy Harvesters. *Sensors* **2021**, *21*, 2388. https://doi.org/10.3390/s21072388

Academic Editor: Amir H. Alavi

Received: 5 March 2021
Accepted: 25 March 2021
Published: 30 March 2021

Abstract: Measuring the efficiency of piezo energy harvesters (PEHs) according to the definition constitutes a challenging task. The power consumption is often established in a simplified manner, by ignoring the mechanical losses and focusing exclusively on the mechanical power of the PEH. Generally, the input power is calculated from the PEH's parameters. To improve the procedure, we have designed a method exploiting a measurement system that can directly establish the definition-based efficiency for different vibration amplitudes, frequencies, and resistance loads. Importantly, the parameters of the PEH need not be known. The input power is determined from the vibration source; therefore, the method is suitable for comparing different types of PEHs. The novel system exhibits a combined absolute uncertainty of less than 0.5% and allows quantifying the losses. The approach was tested with two commercially available PEHs, namely, a lead zirconate titanate (PZT) MIDE PPA-1011 and a polyvinylidene fluoride (PVDF) TE LDTM-028K. To facilitate comparison with the proposed efficiency, we calculated and measured the quantity also by using one of the standard options (simplified efficiency). The standard concept yields higher values, especially in PVDFs. The difference arises from the device's low stiffness, which produces high displacement that is proportional to the losses. Simultaneously, the insufficient stiffness markedly reduces the PEH's mechanical power. This effect cannot be detected via the standard techniques. We identified the main sources of loss in the damping of the movement by the surrounding air and thermal losses. The latter source is caused by internal and interlayer friction.

Keywords: piezoelectric; piezoelectric ceramic; lead zirconate titanate (PZT); polyvinylidene fluoride (PVDF); energy harvesting; efficiency; efficiency measurement; power conversion; power flow

1. Introduction

The increasing demands on the monitoring and predictive maintenance of mechanical structures, intelligent building control, and other remote sensing applications generate the need to install electronics at isolated locations. Such sites are often characterized by difficult access and a lack of infrastructure. Self-powered electronic devices thus became an essential prerequisite. These instruments and apparatuses gather electricity from sources such as light, temperature difference, and diverse forms of kinetic energy, including ambient mechanical vibrations [1,2]. The vibrations can be converted to usable electrical energy by means of piezoelectric energy harvesters (PEHs).

This type of movement is present in machines and structures where collecting relevant operational information has a beneficial impact, namely in industrial equipment, cars, aircraft, and bridges [1,3]. Moreover, PEHs are also employed in harvesting from human movement [1]; substantial effort has therefore been invested in developing and improving PEHs to generate as much energy as possible from the oscillations.

The power output constitutes the key criterion to determine the applicability of a PEH; however, this factor does not provide sufficient information if individual PEHs are to be compared. The reason is that the maximum power output is defined mainly by the PEH's dimensions and vibration amplitude. In view of these facts, Beeby et al. [4] proposed normalized power density (NPD), which divides the power output by the vibration amplitude and the volume of the harvester. Although such metrics appear suitable for comparing PEHs, the authors also mention certain drawbacks. One of the disadvantages rests in ignoring important factors, including the bandwidth [4]. Generally, however, Beeby et al.'s approach is simple, straightforward, and easily applicable. Other researchers developed more sophisticated methodologies, involving the frequency parameters; these techniques were analyzed comprehensively by Hadas et al. [5], who outlined their benefits and drawbacks.

A portion of relevant papers interpret the PEH as an energy converter, meaning that the power output and related alternatives are not considered convenient for a comparison of PEHs. The authors thus employ efficiency as the criterion to examine the devices, as is common in other energy converters. The relevant theories and measurement procedures were summarized and discussed by Yang et al. [6]. In their paper [6] the efficiency not only embodies an essential factor in the development and optimization of PEHs but also finds use as a parameter allowing comparison between energy harvesting methods.

Efficiency

The efficiency is the ratio between the power output and input. In PEHs, the electrical power output is measurable without difficulty. Determining the mechanical power input, however, constitutes a more challenging task, as the process cannot be easily monitored [6]. To establish the efficiency, the power input of the harvester is often simplified to its mechanical power [6,7] because this quantity can be easily calculated via the PEH's parameters.

The calculation is performed by means of, for example, the widely applied formula proposed by Richards et al. [7], which defines the efficiency for the resonant frequency and optimal load [6]:

$$\eta = \frac{1}{2}\frac{k_{\mathrm{sys}}^2}{1-k_{\mathrm{sys}}^2} \left/ \left(\frac{1}{Q} + \frac{1}{2}\frac{k_{\mathrm{sys}}^2}{1-k_{\mathrm{sys}}^2} \right), \right. \tag{1}$$

where η is the efficiency, and k_{sys} and Q denote the system coupling coefficient and the quality factor of the PEH, respectively.

Some of the researchers [6,8,9] expanded the efficiency calculation to involve broader frequency and load ranges. Regrettably, neither the calculating process nor the measurement of the parameters are unified, resulting in that the determined efficiencies oscillate between less than 1% [10,11] and more than 80% [8,12] in lead zirconate titanate (PZT) PEHs.

The role of the system coupling coefficient (k_{sys}) is not identical with that of the electromechanical coupling coefficient (k_{eff}) defined in [13,14]. The latter embodies a piezomaterial coefficient, whereas the former (k_{sys}) describes the whole structure of the PEH. The material coupling coefficient can be obtained from the material resonance and anti-resonance frequencies [13]:

$$k_{\mathrm{sys}}^2 = \frac{f_{\mathrm{o}}^2 - f_{\mathrm{s}}^2}{f_{\mathrm{o}}^2}, \tag{2}$$

where f_{o} and f_{s} are the open and the short natural frequencies of the PEH, respectively.

Equation (2) has found wide use [6–8], in spite of the fact that it is valid only for undamped systems (damping coefficient, $\zeta = 0$) [15,16]. Utilizing Formula (2) in damped systems ($\zeta > 0$) produces inaccurate parameters [15] and, consequently, imprecise efficiency.

Regrettably, this method (Equation (2)) for calculating the k_{sys} exhibits significant uncertainty due to two large numbers being subtracted close to each other; therefore, the efficiency calculation (Equation (1)) will comprise a major error, too. This error, however, is not evaluated in the literature [6,7,10,11,17,18].

Some experts [6,12] also measured experimentally the mechanical power of PEHs. Although such an attitude finds application in validating the derived formulas, it still ignores the mechanical losses. By extension, the parameters of the devices have to be employed to calculate the power. This simplification assumes that the ambient vibration source has an "infinite power" compared to the consumed power of the PEH. The mechanical losses can therefore be ignored, as they are easily replenished from the source [19]. In some situations, such as harvesting from human motion, this assumption is invalid, and the mechanical losses have to be taken into account [19]. Furthermore, the mechanical losses also constitute an important parameter in developing the PEH.

Liao and Sodano [20] established that the approach neglecting the mechanical losses is oversimplified, and they proposed that the input power should not be considered the mechanical power of the PEH. The authors recommend using the power fed into the device from the ambient source to maintain the steady state vibrations. Although this claim is correct, the calculation of the power from the parameters of the PEH cannot be characterized in the same manner. The reason rests in that the procedure involves only the internal and electrical losses and ignores the mechanical ones (e.g., the internal friction or aerodynamic drag). In this context, Yang et al. [6], for example, found close similarity between their method and the PEH's damping coefficient.

Based on our review of the literature, we can conclude that the mechanical losses are disregarded in the current efficiency measurements, which exploit the mechanical power of PEHs as the input. Furthermore, although the harvesters' parameters are utilized for the calculations, the measurement method and the uncertainties are either characterized unclearly or remain ignored [6,8,10,11].

Such issues then justify the need for a new measurement system capable of evaluating the efficiency via the mechanical power supplied to a PEH by an ambient vibration source (proposed efficiency). This power is determined from the force and velocity of the source (in our case, a vibration shaker), eliminating the necessity to use the parameters of the PEH in the calculation. Furthermore, the same measurement assesses the harvester's mechanical power, too, and the efficiency can then be established in a simplified manner (simplified measured). Importantly, the device's parameters (k_{sys} and Q) are calculable from the results to facilitate computing the efficiency according to (1) (simplified calculated). In addition, the uncertainties of all of the aforementioned approaches are determined during the measuring procedure.

As indicated, the unique setup allows us to yield the proposed efficiency, the simplified efficiencies, and the related uncertainties from merely a single measurement. The methods are thus compared in identical conditions. Expectably, the simplified efficiencies, being comparable, should exhibit the same values. The proposed efficiency nevertheless shows lower values, as it includes the mechanical losses.

The paper is organized as follows: The introductory sections set out the problem, together with relevant processes, procedures, and options. Section 2 debates the methodology, characterizing the power flow in piezoelectric harvesters, outlining the calculation of the input power, and describing the measurement system and its uncertainties. In Section 3, the main discussion and results are presented, including the outcomes obtained from a measurement of the efficiency of two different commercially available harvesters. The results of the proposed method are correlated to common simplified efficiencies. Finally, the power flow is discussed, in conjunction with a quantitative analysis of the losses.

2. Methods

Prior to describing the efficiency measurement system, we will discuss the power flow in PEHs (Figure 1). During each cycle, the energy is transmitted back and forth due to the oscillations of the system. In this chapter, however, only the power flow average through several cycles is assumed.

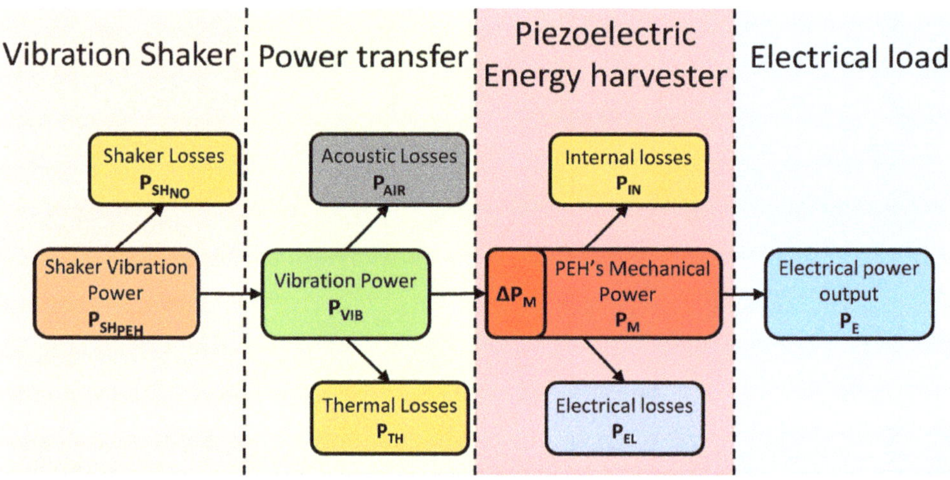

Figure 1. The power flow in a piezo energy harvester (PEH). The traditional approach to determine the efficiency considers the harvester's mechanical power (P_M) as the input power. Alternatively, according to the definition, the power fed into the PEH from the ambient source (P_{VIB}) has to be used as the input.

In a laboratory environment, the vibrations are generated by a vibration exciter (shaker), which produces a vibration power ($P_{SH_{PEH}}$) and has its own losses ($P_{SH_{NO}}$).

The power from the vibration source (P_{VIB}) is transferred to the vibration movement of the PEH, i.e., the harvester's mechanical power (P_M). Similarly to every other power transfer, this process generates some losses. In the presented case, the two main sources are the thermal losses (P_{TH}) generated by the internal friction in the PEH, and the acoustic losses (P_{AIR}), originating from the movement of the PEH in the air.

The mechanical power (P_M) is further transferred to the harvester's usable electrical power output (P_E). Here, the main sources of loss are the internal losses (P_{IL}), arising mainly from the electromechanical losses in the piezoelectric and electrical dissipation (P_{EL}) on the PEH's internal resistance.

In steady state vibrations, however, only a part of the PEH's mechanical power (ΔP_M) is dissipated in the form of the internal and electrical losses. The electrical power output is replenished from the vibration source, as proposed by [20]. Nevertheless, the amount of such replenished power is difficult to evaluate. For simplification we assume in the calculation that the entire mechanical power of the PEH is replenished from the vibration source. This approach also allows us to be consistent with other efficiency-related papers, except for [20]. Similarly to our procedure, however, the authors of this referenced source, too, were unable to measure the losses.

The applied simplification underestimates the thermal losses (P_{TH}) as these are calculated via the PEH's mechanical power (P_M). The efficiencies are not affected, because the traditional approach to determine the efficiency employs the mechanical power (P_M) as the input power [6]. However, according to the definition [20], the power supplied to the PEH from the ambient source, namely the P_{VIB} in our case, has to be used as the input. For clarification purposes, the methods to measure or calculate the powers expressed in Figure 1 are described in the sections below.

2.1. Shaker Vibration Power

The vibration power of the shaker (P_{SH}) is calculated out of the known force and velocity (Equation (3)) [6,18,21]. As both of the quantities are vectors, their phases have to be measured, too. In the steady state condition and at harmonic vibration, the power of the mechanical vibration is calculated as:

$$p_{SH} = \mathbf{F} \cdot \mathbf{v} = |\mathbf{F}||\mathbf{v}| \cos(\phi). \tag{3}$$

The velocity of the structure's base is either directly measured or, as in our case, calculated from the known acceleration. Since the vibrations are steady state and harmonic, the calculation simplifies into:

$$|\mathbf{v}| = \left| \int \mathbf{a}dt \right| = \int a_{ampl} \sin(2\pi ft + \phi_a)dt = \frac{a_{ampl}}{2\pi f} \cos(2\pi ft + \phi_a). \tag{4}$$

The electrodynamic shaker is, principally, a loudspeaker. To calculate the force generated by the shaker, we can then employ the formula of Ampere's law, commonly used for this purpose in acoustics [22]:

$$\mathbf{F} = l\mathbf{I} \times \mathbf{B}. \tag{5}$$

As the magnetic flux density B in the coil gap and the length of the wire l in the magnetic field remain constant, their product, k_{Bl}, can be easily determined [22], simplifying Equation (5) to read

$$|\mathbf{F}| = k_{Bl}|\mathbf{I}| = k_{Bl} I_{ampl} \sin(2\pi ft + \phi_I). \tag{6}$$

Substituting (4) and (6) into (3) and averaging the outcome through the period, we yield:

$$P_{SH} = \frac{1}{T} \int_T p_{SH}dt = \frac{1}{T} \int_T k_{Bl} I_{ampl} \sin(2\pi ft + \phi_I) \frac{a_{ampl}}{2\pi f} \cos(2\pi ft + \phi_a)dt$$
$$= k_{Bl} I_{rms} \frac{a_{rms}}{2\pi f} \cos\left(\phi - \frac{\pi}{2}\right), \tag{7}$$

where p_{SH} and P_{SH} are the shaker's instantaneous and average vibration power, respectively; F_{rms} and v_{rms} are the effective values of the vibration force and velocity, respectively; k_{Bl} denotes the shaker's Bl coefficient; I_{rms} and a_{rms} are the effective values of the drive current and acceleration, respectively; ϕ is the phase between those two vectors; and f represents the frequency of the vibrations.

2.2. Vibration Power

The measurement of the vibration power (P_{VIB}) delivered from the vibration source to the PEH is rather complicated [6]. In the case of steady state harmonic vibrations, however, the power can be determined from the vibration power of the shaker (P_{SH}).

The power generated by the vibration shaker (P_{SH}) depends on the vibration amplitude, frequency, and load on the vibration rod and fixture. Advantageously, the power to vibrate structures other than the PEH is measurable as well, by the same measurement but without the PEH mounted. Then, the vibration power (P_{VIB}) can be calculated simply by subtracting the vibration power without the PEH ($P_{SH_{NO}}$) from that with the device ($P_{SH_{PEH}}$). We have:

$$P_{VIB} = P_{SH_{PEH}} - P_{SH_{NO}}, \tag{8}$$

where P_{VIB} is the vibration power consumed by the PEH from the ambient vibration source, and $P_{SH_{PEH}}$ and $P_{SH_{NO}}$ stand for the shaker's vibration power with and without the PEH mounted, respectively.

2.3. Acoustic Losses

The motion of the piezoelectric cantilever is damped by the ambient air. The damping produces an acoustic power, calculable from the known deflection of the moving object. The acoustic power for a harmonic oscillation is given by [22], as follows:

$$P_{\text{air}} = AI = A\frac{1}{2}\rho c \omega^2 d^2, \tag{9}$$

where A is the area, I denotes the sound intensity, ρ represents the air density, c is the speed of sound in the air, ω stands for the angular frequency, and d is the oscillation amplitude.

Due to the nature of the cantilever deflection, the calculation can be materialized when the resonance mode is measured at n points evenly distributed on the surface of the cantilever. The measured deflections (d_i) are applicable in establishing the partial intensities (I_i), and we sum them to estimate the total acoustic power; we then have:

$$P_{air_{tot}} = \sum_{i=1}^{n} \frac{A}{n} I_n = \sum_{i=1}^{n} \frac{A}{n}\frac{1}{2}\rho c \omega^2 d_i^2. \tag{10}$$

When the deflections are measured on the edge of the structure, this arrangement has to be taken into account and then reflected in the calculation by an appropriate reduction of the corresponding area.

2.4. Thermal Losses

The power transfer from the vibration rig to the movement of the PEH, the PEH's mechanical power (P_{M}), involves losses, such as the internal friction between and in the material layers of the harvester. The losses dissipate some of the power to heat. By extension, in more general terms, losses also accompany the mechanical–electrical power conversion in the piezoelectric (Section 2.8) and the electrical losses (Section 2.7) on the internal resistance; both are also dissipated as heat. Thus, the powers of the individual loss components cannot be evaluated by measuring the heat.

Moreover, in PEHs, the overall power dissipated into heat is so small that it produces only an almost insignificant change in the harvester's surface temperature, for instance, in the PZT PEH at 1 g_{rms} the warming of the PEH caused by the thermal losses (14.93 mW; Table 1) is less than 0.02 K. For this reason, we were unable to measure the outcome.

Beneficially, the amount of thermal losses (P_{TH}) can be determined in an indirect manner, via:

$$P_{\text{TH}} = P_{\text{VIB}} - P_{\text{AIR}} - P_{\text{M}}, \tag{11}$$

where P_{TH} and P_{AIR} are the thermal and the acoustic losses, respectively, and P_{VIB} and P_{M} denote the vibration and the mechanical powers of the PEH, respectively.

2.5. PEH Mechanical Power

The mechanical power of a PEH (P_{M}) can be measured via Formula (12), derived by Yang et al. [6]:

$$P_{\text{M}} = \frac{1}{2}m_e a_{\text{X}} v_{\text{Z}} \sin(\phi_{\text{X}}), \tag{12}$$

where m_{e} is the PEH's effective mass, a_{x} denotes the amplitude of the relative acceleration of the PEH's tip, v_{Z} represents the amplitude of the base velocity, and ϕ_{X} is the phase difference between the base velocity and the PEH's tip acceleration.

2.6. Electrical Power Output

The electrical power output of the PEH (P_{E}) is measured on the resistive load; however, it alternates with the vibration frequency. Thus, the output has to be rectified to be able to power some types of electronics, and additional losses occur. For this reason, rectification was not involved in our experiments. Relevant power management circuits are also being intensively developed [5]; these are nevertheless not discussed in this paper. We have:

$$P_{\mathrm{E}} = \frac{U_{\mathrm{rms}}^2}{R_{\mathrm{L}}}, \tag{13}$$

where P_{E} is the PEH's electrical power output, and U_{rms} denotes the effective value of the voltage generated by the PEH on the preset resistive load, R_{L}.

2.7. Electrical Losses

To ensure the maximum flow of power from the PEH to the load, the optimum external load has to have the same value as the internal resistance of the PEH. Then, the power transfer is at a maximum but provides an efficiency of only 50%, meaning that half of the generated electrical power is the electrical power output (P_{E}) and the other half comprises the electrical losses (P_{EL}) dissipated on the internal resistance. Then, the electrical losses (P_{EL}) can be easily determined:

$$P_{\mathrm{EL}} = P_{\mathrm{E}}, \tag{14}$$

where P_{EL} denotes the electrical losses on the internal resistance, and P_{E} is the electrical power output.

2.8. Internal Losses

The mechanical–electrical power conversion is performed by the piezoelectric layer in the PEH. The portion of the mechanical power that is absorbed by the layer but remains unconverted into the electrical power (P_{E}) forms the internal losses (P_{IL}), which are dissipated into heat.

The efficiency of the conversion in the piezomaterial is defined by the electromechanical coupling coefficient (k_{eff}). The relevant value usually varies between 0.4 and 0.6 in the PZT ceramic but amounts to less than 0.3 in the polyvinylidene fluoride (PVDF) film. The conversion efficiency of the material enclosed in the PEH's structure is slightly different, and therefore the internal losses (P_{IL}) cannot be calculated directly by using this coefficient. However, indirect calculation is possible; we then have:

$$P_{\mathrm{IL}} = P_{\mathrm{M}} - P_{\mathrm{E}} - P_{\mathrm{EL}}, \tag{15}$$

where P_{IL} and P_{EL} are the internal and electrical losses, respectively, and P_{M} and P_{E} represent the mechanical and electrical powers, respectively.

2.9. Efficiency Calculation

To include the mechanical losses in the efficiency, we consider as an input power the vibration power supplied to the PEH from an ambient vibration source, P_{VIB}. We have:

$$\eta = \frac{P_{\mathrm{E}}}{P_{\mathrm{VIB}}} = \frac{\frac{U_{\mathrm{rms}}^2}{R_{\mathrm{L}}}}{P_{\mathrm{SH_{PEH}}} - P_{\mathrm{SH_{NO}}}}, \tag{16}$$

where η stands for the measured efficiency; P_{VIB} is the PEH's power input; P_{E} denotes the electrical power output, i.e., the voltage U generated across the load resistor, R_{L}; and $P_{\mathrm{SH_{PEH}}}$ and $P_{\mathrm{SH_{NO}}}$ represent the vibration power with and without the PEH mounted on the shaker, respectively.

2.10. Efficiency Measurement

To measure the efficiency, we improved the automated measurement system [23] to manage the vibration amplitudes at a lower error rate and to reduce the noise. Moreover, the data post-processing stage was redesigned to compute the power consumption of the PEH. The actual setup is described in detail within Section 2.12.

Because the efficiency results appear to be the most interesting near the PEH's mechanical resonance, this resonance is localized by using sweep sine vibration.

To obtain proper efficiency results, we need to ensure that the PEH vibrates in its first longitudinal mode without any torsional oscillation or clamping. Such a condition was verified via a mode measurement system utilizing a scanning laser vibrometer. This step is especially significant when the examined PEH has unknown parameters. Importantly, the measured mode can also be used for calculating the aerodynamic drag. The system is characterized in the following section.

2.11. Mode Measurement System

The applied scanning laser vibrometer system (Figure 2) consists of a Polytec OFV-5000 vibrometer controller and Polytec OFV-505 sensor head, two-axis ThorLabs GVS012 scanning mirrors, a National Instruments Compact DAQ-9174 chassis with an NI 9223 analog input card (to ensure the acquisition of the measured and the reference vibrations), and an NI 9263 voltage output card (to adjust the mirrors' positions).

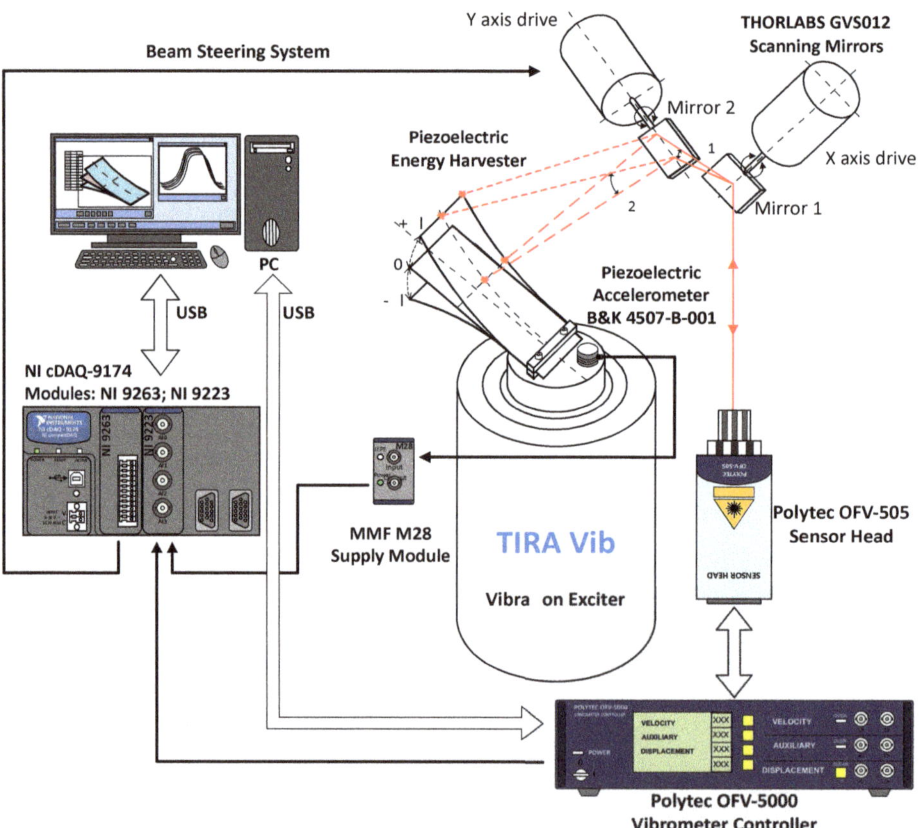

Figure 2. The scanning vibrometer setup to used verify that the first resonance mode has been reached.

The system is controlled by a custom-built, PC-based LabVIEW application, which defines and corrects the movement of the mirrors and the laser focus to measure at selected points on the PEH. In addition to these operations, the program acquires the vibrometer velocity data for every set point. The reference vibration was measured with a B&K 4507-B-001 accelerometer to enable the phase compensation of the discrete vibration measurements at the points selected on the PEH's surface. This step ensures the determination of the vibration mode and allows the measurement of the acoustic losses. As the laser vibrometer acquires the velocity of every measured point, it was integrated into the spectral domain to yield the displacement amplitude. The obtained data can be used to confirm the resonance mode and to calculate the acoustic power produced by the PEH.

2.12. Efficiency Measurement System

The system comprises two parts (Figure 3), one measuring the harvesters' parameters and the other controlling the vibrations. By definition, the measuring component sets the resistance via an Agilent 34970A data acquisition/switch unit and communicates the vibration parameters, frequency, and amplitude to its counterpart. Here, the vibrations are monitored and adjusted by a PI controller, and after they have settled down (error < 0.1% for 10 s, the system will measure the harvester's output voltage on the pre-set load, the displacement of the harvester's tip, and the shaker's current and voltage. The acceleration data are then transferred from the vibration control unit. The smaller displacement amplitudes are measured by a Polytec PDV 100 vibrometer, while in the higher ones a Micro-Epsilon optoNCDT 1401-20 optical sensor is employed. The measurement is performed with a 24-bit, ±5 V, NI 9234 analog input card or, if the measured harvester's output voltage reaches above 5 V, a 24-bit, ±30 V NI 9232 card. To minimize the noise (<40 μV and 127 μV, respectively) in the experiment, we selected sampling frequencies exhibiting the maximum decimation rate of 256. Moreover, the noise and possible scalloping loss were further reduced by extracting the effective value from the signal spectrum via an FFT with a flat top window [24].

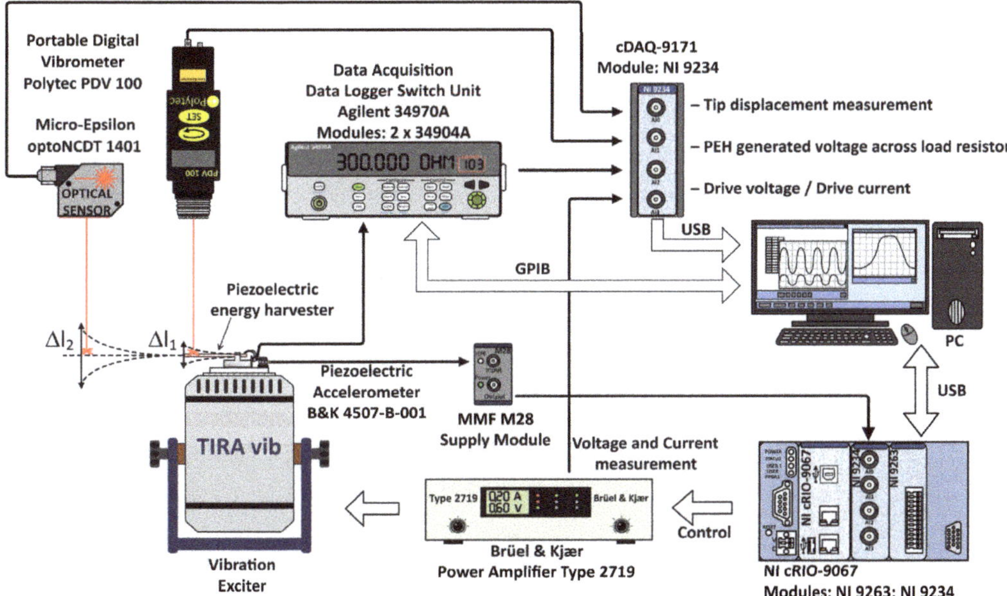

Figure 3. The system used to measure the power and efficiency.

The vibration control is ensured by a Compact RIO 9067 device equipped with NI 9234 analog input and NI 9263 analog output cards. The generated signal is transmitted to a B&K 2719 power amplifier, which drives a TIRA 52110 vibration shaker. The vibrations are measured by the B&K 4507-B-001 accelerometer, and the signal returns to the Compact RIO via an MFF M28 supply module. The accuracies of the accelerometer and the MMF supply module correspond to 0.4% (in the measured range of 20–300 Hz) and 0.5%, respectively;

both of the units are calibrated. To reduce the noise, we employed the method characterized above.

The resistive load is configured by a set of different parallel combinations of resistors on an Agilent 34904A matrix switch inserted into the Agilent 34970A measurement unit. Each of the two switches contains nine resistors, whose parallel combinations cover two decades of resistance (meaning that four decades in total are covered). Throughout the measurement process, the resistances are controlled by an Agilent 34401A multimeter, with the repeatability of the measurements amounting to less than 0.04%. When the parallel combination of the load and the measurement card input resistances produces a significant error, a voltage follower is employed. The entire system is located in an air-conditioned laboratory, where all the measurements take place.

2.13. Comparison with the Methodology Currently in Use

To facilitate comparison of the proposed efficiency measurement technique with the existing methodology, we calculated the efficiency values in the PEHs at the devices' resonant frequencies and optimum loads by using the standard approach. Even though novel computing formulas are available, such as that described in [25], these share the same simplified principle, which interprets the harvester's mechanical power as the input power. Thus, we decided to compare our new metrics with a well-established and widely used option (Equation (1)) [7] (simplified calculated). The procedure embodied in Equation (1) was experimentally verified by, for instance, the authors of reference [6], whose efficiency measurement concept is also used herein for comparison (simplified measured).

The data enabling the calculation of the aforementioned efficiencies were gathered during merely one measurement, meaning that the error that arises from diverse conditions cannot occur.

3. Results and Discussion

The measurement system was tested on two commercially available PEHs, one being a PZT PEH MIDE PPA-1011 (PZT PEH), which has a rigid structure with a FR4 base, and the other embodied in a PVDF TE LDTM-028K (PVDF PEH), which exhibits a flexible foil structure with an integrated 720 mg tip mass. The dimensions of these PEHs are shown in Figure 4; the capacitances equal 90.5 nF and 485 pF, respectively.

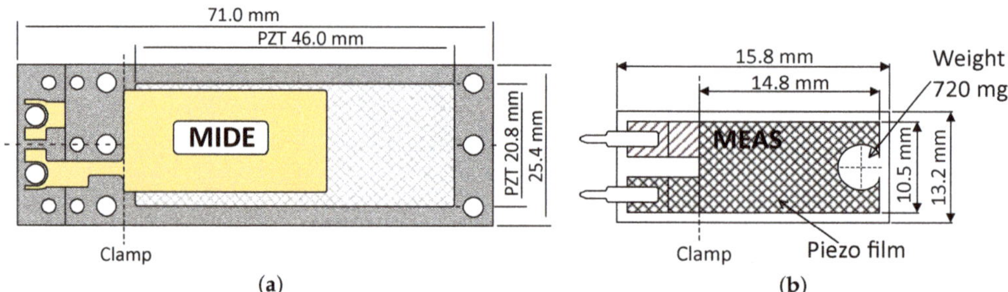

Figure 4. The tested PEHs: the MIDE PPA-1011 (**a**) and the TE LDTM-028K (**b**).

Initially, each resonance frequency was found via the sine sweep, and then the vibration mode (Figure 5) of the examined harvester was checked. The measured amplitudes enabled us to calculate the acoustic power corresponding to the aerodynamic drag. At the final stage, we carried out the same measurement cycles at different vibration amplitudes. The 3D proposed efficiency plots (side and top views) relating to selected vibration amplitudes are shown in Figure 6.

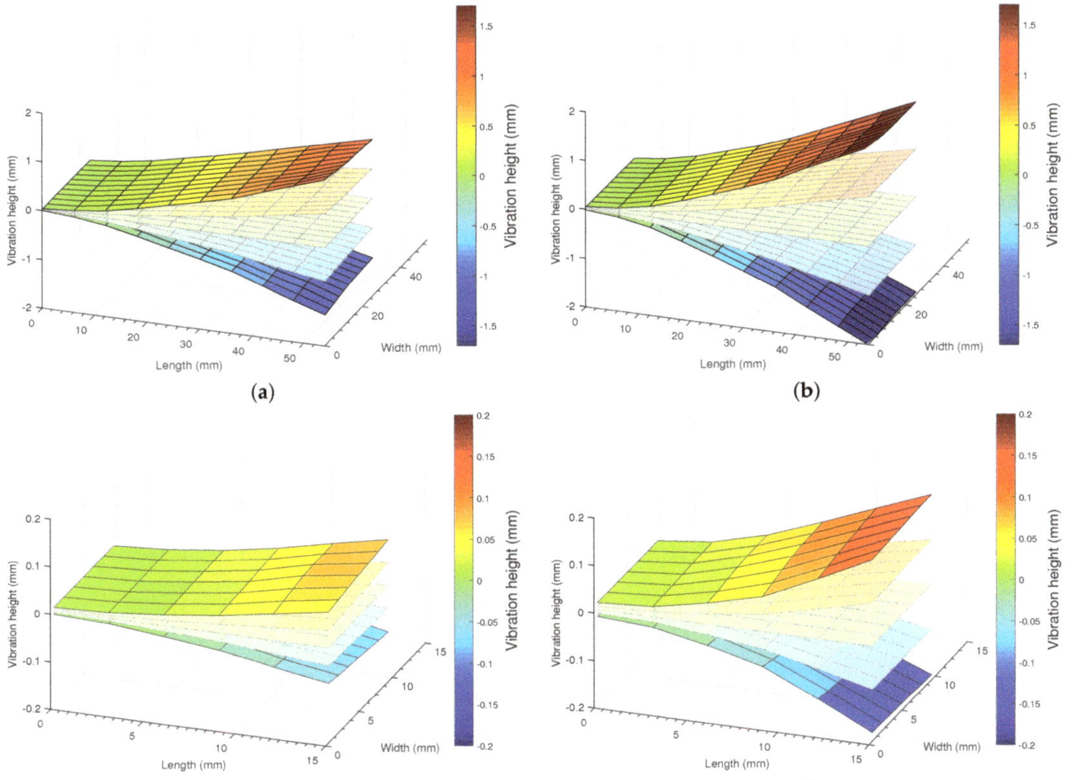

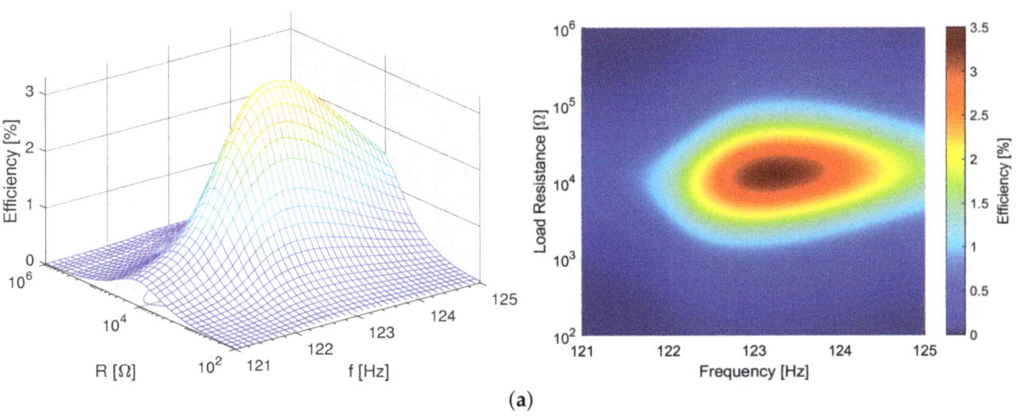

Figure 5. The confirmed oscillation: the first longitudinal mode in the MIDE PPA-1011 at 1 g_{rms} (**a**) and 2 g_{rms} (**b**); an in the TE LDTM-028K at 1 g_{rms} (**c**) and 2 g_{rms} (**d**).

Figure 6. *Cont.*

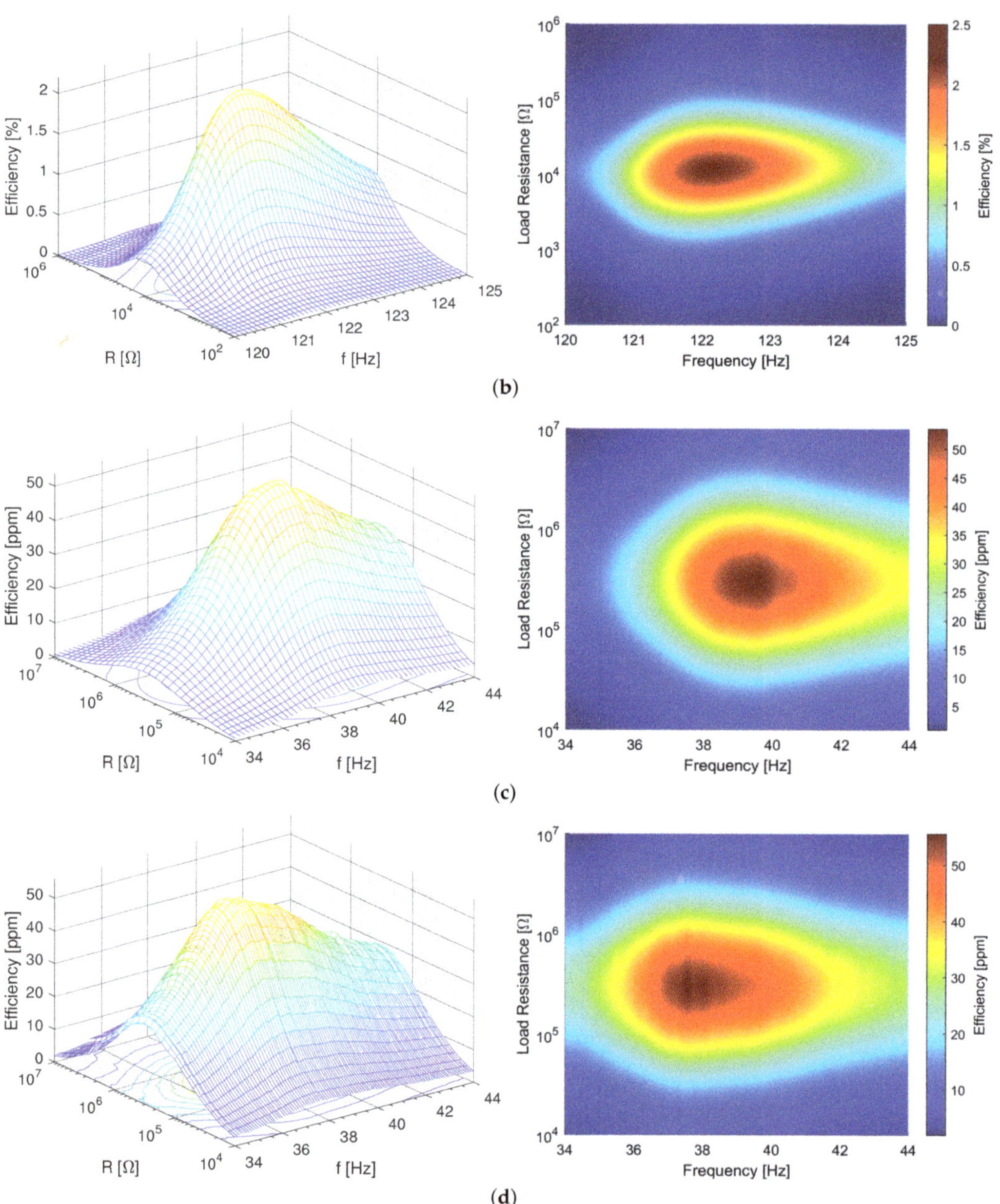

Figure 6. The proposed efficiency rates measured near the harvesters' resonance frequencies, at 1 g_{rms} and 2 g_{rms}: the zirconate titanate (PZT) PEH, (**a**,**b**, respectively); the polyvinylidene fluoride (PVDF) PEH, (**c**,**d**, respectively). The 3D plots and the top views are shown on the left-hand and the right-hand sides, respectively. Note the different efficiency scales for the two PEHs.

By common definition, the power generated by the PEHs increases with the vibration amplitude. In the PZT PEH, the power varied between 165 µW at 0.25 g_{rms} and 3.78 mW at 2 g_{rms}. The PVDF PEH, however, is smaller and exhibits a higher optimal load, and thus the power values dropped to between 11.6 nW at 0.25 g_{rms} and 793 nW at 2 g_{rms}.

3.1. Uncertainties of the Measurement System

The type B uncertainty was calculated analytically from the devices' specifications. In this uncertainty, the cause consisted in the errors of the NI measurement cards and the accelerometer chain. The error distribution in the individual devices was considered normal (Gaussian), allowing us to calculate the uncertainty. The type B uncertainty varied from 0.003 to $0.4\%_A$ for the interval of 1σ. The unit of efficiency is usually %, and the same unit normally expresses relative error. In this paper, for clarity, the symbol $\%_A$ thus denotes the measured efficiency or its absolute error, whereas $\%_R$ stands for the relative error.

The type A uncertainty was determined as the worst case scenario from all measurements at different frequencies and vibration amplitudes, with each of these uncertainties established from 300 repetitive measuring cycles. The result to express the uncertainty was computed from the post-processed values. In general terms, this method involves the random measurement errors and their filtering by means of an FFT with a flat top window. The relative type A uncertainty amounted to 0.05, 0.03, and $0.002\%_R$ in the current, acceleration, and phase, respectively. The absolute combined uncertainty relating to the proposed efficiency measurement ranged between $0.007\%_A$ and $0.5\%_A$ for the interval of 1σ.

3.2. Uncertainty in Simplified Efficiencies

In the simplified measured and proposed efficiencies, the uncertainties were calculated similarly, from the measurement system's uncertainties. The absolute combined uncertainty relating to the simplified measured efficiency varied between $0.015\%_A$ and $0.27\%_A$ for the interval of 1σ.

The uncertainty of the simplified calculated efficiency, however, arises from the uncertainty of the parameters k_{sys} and Q, which are calculated from relevant frequencies. For instance, in the PZT PEH, vibrating at 1 g_{rms}, the open natural frequency attained $f_o = 123.33$ Hz and the short one equaled $f_s = 123.00$ Hz, both at the relative uncertainty $0.12\%_R$. The two frequencies being close to each other, the relative uncertainty of k_{sys} amounted to $63.3\%_R$. This rate, above all, then caused the combined absolute uncertainty of the simplified calculated efficiency to be much higher than those in the measured efficiencies, namely, it ranged between 3.05 and $5.56\%_A$ for the interval of 1σ.

3.3. Efficiency of the PZT Ceramic Based Harvester

The diagram characterizing the PZT PEH (Figure 7) indicates that the simplified measured and simplified calculated efficiencies yielded similar results; the measured one, however, has a markedly lower uncertainty. These efficiencies are three to five times greater than the proposed efficiency because they did not take into account the mechanical losses.

Moreover, all types of efficiency exhibited a decreasing trend when the vibration amplitude rose. The proposed class nevertheless declined slightly faster due to the growth of the mechanical losses, which occurs with increasing vibration amplitude.

In the given context, Figure 8 shows the individual powers of the PEH. It is worth noting that the aerodynamic loss was higher than the PEH's mechanical power, the difference being approximately 50%. Furthermore, the amount of the vibration input power (P_{VIB}) transferred to the mechanical power of the PEH (P_M) decreased with increasing vibration amplitude from approx. 40% to 20%. This indicates that the mechanical losses increased with rising vibration amplitude.

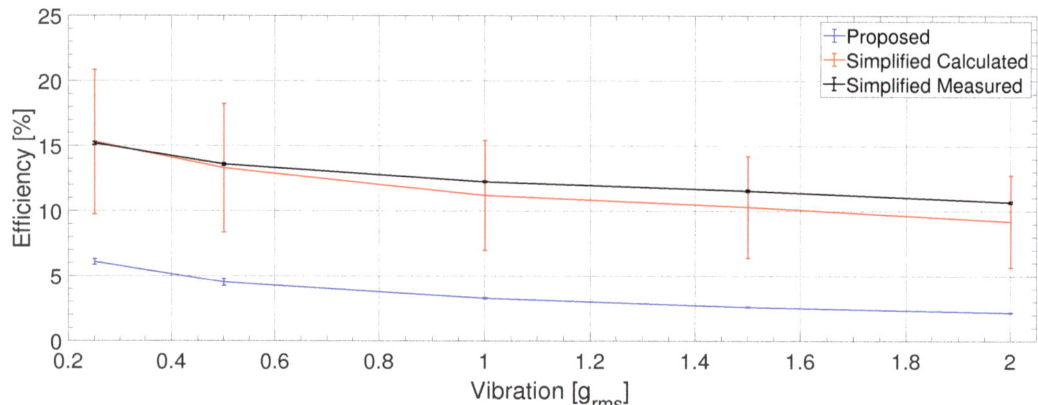

Figure 7. The PZT PEH: Comparing the efficiency values established via our method (proposed) with those calculated from the PEH's parameters for the optimum condition (simplified calculated) and measured from the PEH's mechanical power (simplified measured). The error bars represent the 1σ interval.

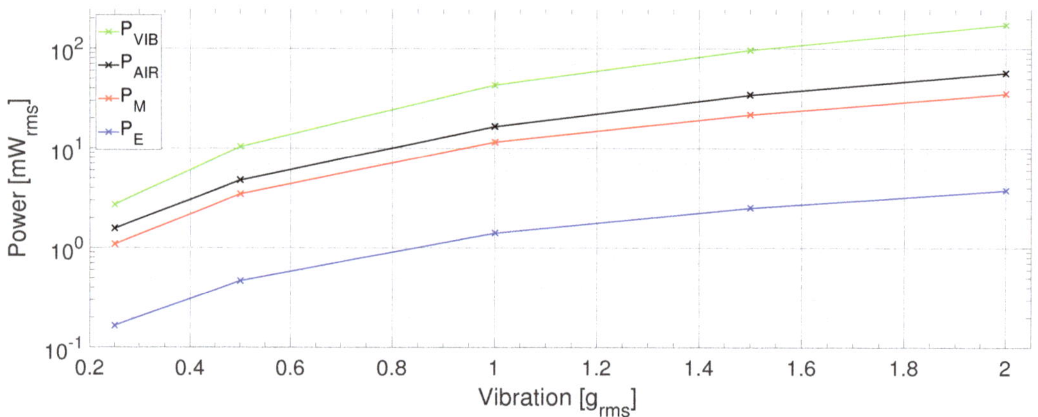

Figure 8. The PZT PEH: Comparing the power delivered by the shaker to the PEH, with the power dissipated in the air, the PEH mechanical vibration power, and the electrical power output.

3.4. Efficiency of the PVDF Foil Based Harvester

In the PVDF PEH, the simplified calculated and simplified measured efficiencies (Figure 9) amounted to approximately 1.5 %$_A$; the proposed efficiency was significantly smaller, varying between 0.005 and 0.015 %$_A$. Such a condition is caused mainly by the harvester's low stiffness, an effect stemming from the flexible structure. Thus, even though the displacement of the PEH's tip is notable, the vibration power remained very small, being less than 1%$_R$ of the input vibration power (Figure 10). Consequently, due the major displacement, the aerodynamic losses produced more than a half of the overall ones. Moreover, the rather high optimum load resistance (330 kΩ) means that the power output was small, ranging between 11.6 nW at 0.25 g$_{rms}$ and 793 nW at 2 g$_{rms}$.

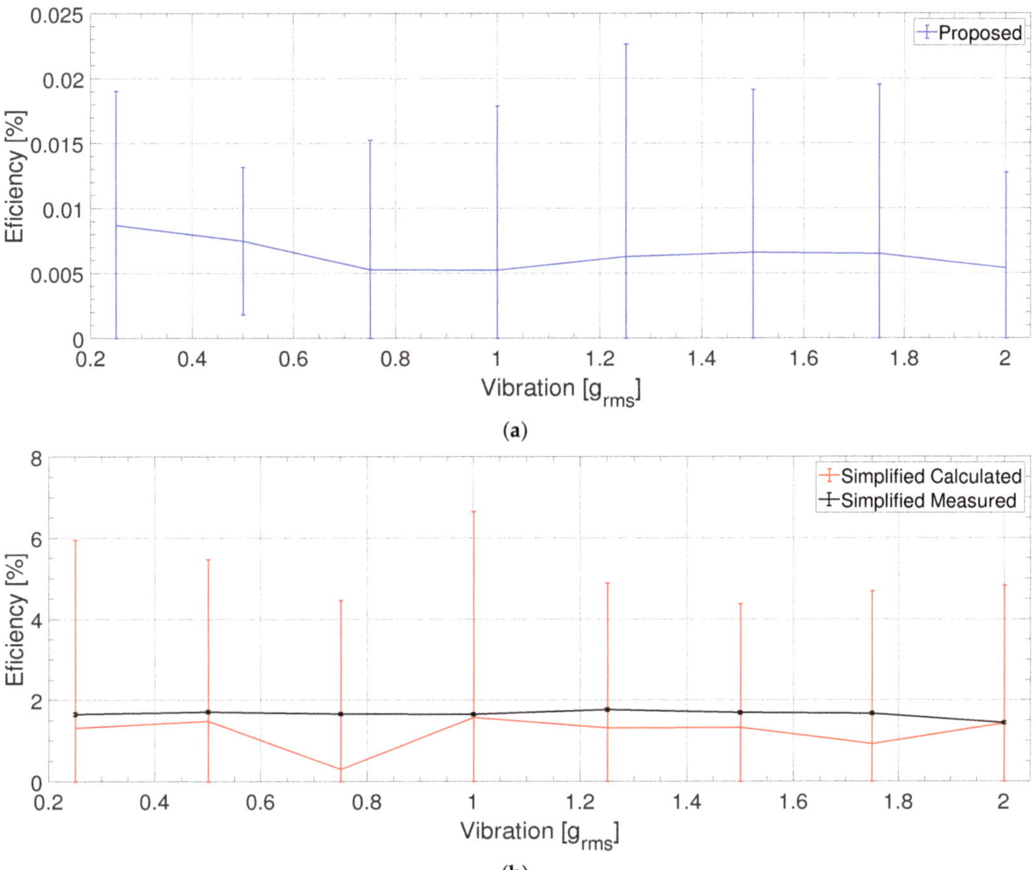

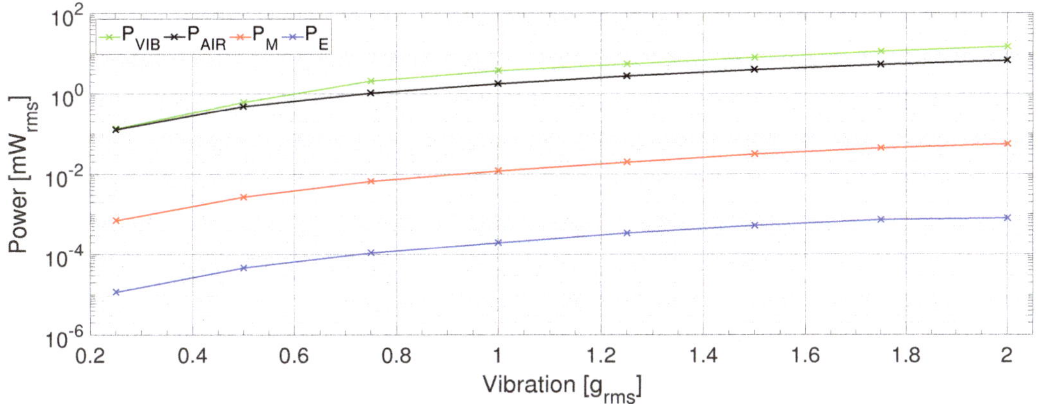

Figure 9. The PVDF PEH: Comparing the efficiency values established via our method (proposed) (**a**) with those calculated from the PEH's parameters for the optimum condition (simplified calculation) and measured from the PEH mechanical power (simplified measurement) (**b**). The error bars represent the 1σ interval.

Figure 10. The PVDF PEH: Comparing the power delivered by the shaker to the PEH with the power dissipated in the air, the PEH mechanical vibration power, and the electrical power output.

3.5. Comparing the Efficiencies

The simplified efficiencies (measured and calculated) exhibited the same results, similarly to the findings presented by Yang et al. [6].

The proposed efficiency showed lower values because the mechanical losses not only need to be taken into account but are also, as is obvious from the measured data, higher than the actual mechanical power of the PEH (P_M). In general terms, it is then possible to claim that the mechanical losses embody a significant factor affecting the performance of PEHs. Ignoring the losses during in the input power calculations involved in the computation of the simplified efficiency computations of the PEH can cause a significant error when different types of PEHs are compared for various purposes. Our novel method enables direct efficiency measurement and does not omit the mechanical losses of the PEH.

Even in situations where the mechanical losses can be neglected, the calculation of the efficiency from the parameters (1) is affected by a large uncertainty. This arises mainly from the uncertainty of the system coupling coefficient (k_{sys}). If the same efficiency is measured via the PEH's mechanical power (P_M), the uncertainty is significantly lower.

Moreover, excluding the harvester's parameters allows us to use the proposed method for comparing different principles of power generation from ambient vibrations. Our system is capable of measuring a wide range of efficiencies, from tens of percent to units of parts per million, and the shape of the data (Figure 6) appears to be credible even at low efficiency rates. The uncertainty of the measured efficiency is also provided.

3.6. Power Flow

Once the PEH has been characterized, we can determine the individual power components according to the classification in Figure 1. The individual power components in the absolute value and their contribution to the total power consumed by the PEH are shown in Table 1. These items of information allow us to localize the most significant losses in the power conversion.

Table 1. Comparing the individual power components in the tested PEHs at 1 g_{rms}.

Power Type		MIDE PPA-1011		TE LDTM-028K	
		[mW]	[%]	[μW]	[%]
Vibration power	P_{VIB}	43.04	100	3745	100
→ Acoustic losses	P_{AIR}	16.56	38.5	1770	47.3
→ Thermal losses	P_{TH}	14.93	34.7	1963	52.4
→ PEH mechanical power	P_M	11.55	26.8	11.94	0.318
→ Internal losses	P_{IL}	8.72	20.2	11.55	0.308
→ Electrical losses	P_{EL}	1.41	3.3	0.197	0.005
→ Electrical power output	P_E	1.41	3.3	0.197	0.005

The most significant types of loss were the acoustic and thermal ones, which together dissipated more than two thirds of the total power input. Regrettably, we were able to find only a few papers that focused on reducing such losses [26–28]. By contrast, many research teams are developing new piezomaterials with the aim to reduce the internal losses; these processes, however, dissipate only a small portion of the power input. Finally, the electrical losses are negligible when compared to the other categories. The intensive research and development of power conditioning circuits will nevertheless allow the electrical losses to be reduced even further in the future.

In conclusion, let us emphasize again that multiple efforts are being made to minimize the internal and electrical losses (which embody only a small portion of the total losses), whereas the type of loss that exhibits the maximum contribution to the total sum remains mostly ignored.

4. Conclusions

By reviewing the papers that discuss techniques for calculating the efficiency of PEHs, we concluded that researchers regularly compute the input power as the power stored in the harvester's movement. Such a procedure, however, causes the mechanical losses induced by the actual motion to be ignored. We designed a method to measure the efficiency of energy harvesters in such a manner that the mechanical losses generated by internal friction and aerodynamic drag are taken into account. Unlike the traditional approach, our technique relies on computing the input power out of the vibration power extracted from the shaker, with the losses preserved. In the experiments, a mode measurement system was employed to estimate the losses linked to the aerodynamic drag.

We developed a system capable of characterizing the power and measuring directly the overall efficiency of a harvester; the combined absolute uncertainty amounted to less than 0.5 $\%_A$. The measurement system was tested on commercially available PZT and PVDF PEHs: a MIDE PPA-1011 and a TE LDTM-028K. In the former, the efficiency at the resonant frequency and optimal load varied from 6.1 $\%_A$ at 0.25 g_{rms} to 2.2 $\%_A$ at 2 g_{rms}. In the latter, the same quantity reached approximately 0.007 $\%_A$ in all of the measured vibration amplitudes (0.25–2 g_{rms}).

The efficiency data relating to the optimum load and frequency were compared with the calculations based on the PEHs' coupling coefficients and quality factors. In the PZT harvester, the proposed measured efficiency rate was roughly three to five times lower than the efficiencies determined via the existing methodology. The PVDF device, however, exhibited rates markedly lower than those established in the PZT one. Generally, it is then possible to claim that the mechanical losses embody a significant factor affecting the performance of PEHs.

We localized the main sources of loss, identifying these with the thermal and acoustic losses caused by the movement of the PEH. The internal losses, arising from the piezoelectric efficiency, consist of only a small portion of the total dissipated power, and the electrical ones, generated by the internal resistance, are negligible. Regrettably, while efforts are being made to minimize those sources of loss that exert a minimum impact on the total losses, only a few research groups focus on the major sources.

As our metrics are independent of the PEHs' parameters, they can be applied in comparing the efficiencies of diverse vibration harvesters. The proposed method may accelerate the development of new vibration harvesters in general because the partial procedures comprised within the technique allow easy and accurate measurement of the devices' performances. By extension, as suggested above, an interesting benefit lies in the possibility of comparing the outcomes delivered by different research teams.

In terms of future research, we intend to improve the measurement system to better measure and model the individual types of loss. Such an aim should then produce an increase in the efficiency of piezo energy harvesters.

Author Contributions: Conceptualization, J.K. (Jan Kunz), J.F., S.P. and P.B.; methodology, J.K. (Jan Kunz), S.P. and P.B.; software, J.K. (Jan Kunz) and S.P.; validation, J.F., P.B., J.K. (Jakub Krejci) and Z.H.; formal analysis, J.K. (Jan Kunz) and S.P.; investigation, J.K. (Jan Kunz), J.F., S.P. and P.B.; resources, J.F., S.P., P.B. and Z.H.; data curation, J.K. (Jan Kunz), S.P. and J.K. (Jakub Krejci); writing—original draft preparation, J.K. (Jan Kunz) and S.P.; writing—review and editing, J.K. (Jan Kunz), S.P., J.F., P.B. and Z.H.; visualization, J.K. (Jan Kunz), J.F., S.P. and J.K. (Jakub Krejci); supervision, P.B., S.K. and Z.H.; project administration, S.K. and Z.H.; funding acquisition, S.K. and Z.H. All authors have read and agreed to the published version of the manuscript.

Funding: The completion of this paper was made possible by the grant No. FEKT-S-20-6205— "Research in Automation, Cybernetics and Artificial Intelligence within Industry 4.0" financially supported by the Internal science fund of Brno University of Technology. The work was supported by the infrastructure of RICAIP that has received funding from the European Union's Horizon 2020 research and innovation programme under grant agreement No. 857306 and from Ministry of Education, Youth and Sports under OP RDE grant agreement No CZ.02.1.01/0.0/0.0/17_043/0010085.

Institutional Review Board Statement: Not applicable.

Informed Consent Statement: Not applicable.

Data Availability Statement: The data used for the manuscript are available for researchers on request.

Acknowledgments: The authors gratefully acknowledge the contribution of Premysl Dohnal who edited and proofread this paper.

Conflicts of Interest: The authors declare no conflict of interest.

Abbreviations

The following abbreviations are used in this manuscript:

PEH	Piezoelectric Energy Harvester
PZT	Lead Zirconate Titanate
PVDF	Polyvinylidene Fluoride
k_{sys}	System Coupling Coefficient
Q	Quality Factor
f_o	Open Circuit Natural Frequency
f_s	Short Circuit Natural Frequency
P_{SH}	Shaker Vibration Power
$P_{SH_{PEH}}$	Shaker Vibration Power with Mounted PEH
$P_{SH_{NO}}$	Shaker Vibration Power without PEH
P_{VIB}	Vibration Power
P_{AIR}	Acoustic Losses
P_{TH}	Thermal Losses
P_M	PEH Mechanical Power
P_{IL}	Internal Losses
P_{EL}	Electrical Losses
P_E	Electrical Power Output
F	Vibration Force
v	Vibration Velocity
k_{Bl}	Shaker Bl Coefficient
a	Vibration Amplitude
f	Frequency of Vibration
$\%_A$	Percent as Absolute Quantity (used for the efficiency value and its absolute error)
$\%_R$	Percent as Relative Quantity (used for the relative error)

References

1. Beeby, S.P.; White, N.M. *Energy Harvesting for Autonomous Systems*; Artech House: Norwood, MA, USA, 2010.
2. Teso-Fz-Betoño, D.; Aramendia, I.; Martinez-Rico, J.; Fernandez-Gamiz, U.; Zulueta, E. Piezoelectric Energy Harvesting Controlled with an IGBT H-Bridge and Bidirectional Buck–Boost for Low-Cost 4G Devices. *Sensors* **2020**, *20*, 7039. [CrossRef]
3. Fitzgerald, P.C.; Malekjafarian, A.; Bhowmik, B.; Prendergast, L.J.; Cahill, P.; Kim, C.W.; Hazra, B.; Pakrashi, V.; OBrien, E.J. Scour Damage Detection and Structural Health Monitoring of a Laboratory-Scaled Bridge Using a Vibration Energy Harvesting Device. *Sensors* **2019**, *19*, 2572. [CrossRef]
4. Beeby, S.P.; Torah, R.; Tudor, M.; Glynne-Jones, P.; O'donnell, T.; Saha, C.; Roy, S. A micro electromagnetic generator for vibration energy harvesting. *J. Micromechanics Microengineering* **2007**, *17*, 1257. [CrossRef]
5. Hadas, Z.; Smilek, J.; Kanoun, O. (Eds.) *Energy Harvesting for Wireless Sensor Networks: Technology, Components and System Design*; Walter de Gruyter GmbH & Co KG: Berlin, Germany, 2018; pp. 45–63.
6. Yang, Z.; Erturk, A.; Zu, J. On the efficiency of piezoelectric energy harvesters. *Extrem. Mech. Lett.* **2017**, *15*, 26–37. [CrossRef]
7. Richards, C.D.; Anderson, M.J.; Bahr, D.F.; Richards, R.F. Efficiency of energy conversion for devices containing a piezoelectric component. *J. Micromech. Microeng.* **2004**, *14*, 717. [CrossRef]
8. Kim, M.; Dugundji, J.; Wardle, B.L. Efficiency of piezoelectric mechanical vibration energy harvesting. *Smart Mater. Struct.* **2015**, *24*, 055006. [CrossRef]
9. Shu, Y.; Lien, I. Analysis of power output for piezoelectric energy harvesting systems. *Smart Mater. Struct.* **2006**, *15*, 1499. [CrossRef]
10. Akaydin, H.D.; Elvin, N.; Andreopoulos, Y. The performance of a self-excited fluidic energy harvester. *Smart Mater. Struct.* **2012**, *21*, 025007. [CrossRef]

11. Janphuang, P.; Lockhart, R.; Henein, S.; Briand, D.; de Rooij, N.F. On the experimental determination of the efficiency of piezoelectric impact-type energy harvesters using a rotational flywheel. *J. Phys. Conf. Ser.* **2013**, *476*, 012137. [CrossRef]
12. Cho, J.; Richards, R.; Bahr, D.; Richards, C.; Anderson, M. Efficiency of energy conversion by piezoelectrics. *Appl. Phys. Lett.* **2006**, *89*, 104107. [CrossRef]
13. EN:50324-2. *Piezoelectric Properties of Ceramic Materials and Components Methods of Measurement—Low Power*, 3rd ed.; CENELEC: Brussels, Belgium, 2002.
14. IEC Standard. *Guide to Dynamic Measurements of Piezoelectric Ceramics with High Electromechanical Coupling*; IEC: Geneva, Switzerland, 1976.
15. Renno, J.M.; Daqaq, M.F.; Inman, D.J. On the optimal energy harvesting from a vibration source. *J. Sound Vib.* **2009**, *320*, 386–405. [CrossRef]
16. Erturk, A.; Inman, D.J. *Piezoelectric Energy Harvesting*; John Wiley & Sons: Hoboken, NJ, USA, 2011.
17. Ruan, J.J.; Lockhart, R.A.; Janphuang, P.; Quintero, A.V.; Briand, D.; de Rooij, N. An automatic test bench for complete characterization of vibration-energy harvesters. *IEEE Trans. Instrum. Meas.* **2013**, *62*, 2966–2973. [CrossRef]
18. Wang, X. A study of harvested power and energy harvesting efficiency using frequency response analyses of power variables. *Mech. Syst. Signal Process.* **2019**, *133*, 106277. [CrossRef]
19. Shafer, M.W.; Garcia, E. The Power and Efficiency Limits of Piezoelectric Energy Harvesting. *J. Vib. Acoust.* **2014**, *136*. [CrossRef]
20. Liao, Y.; Sodano, H.A. Structural Effects and Energy Conversion Efficiency of Power Harvesting. *J. Intell. Mater. Syst. Struct.* **2009**, *20*, 505–514. [CrossRef]
21. Hadas, Z.; Vetiska, V.; Vetiska, J.; Krejsa, J. Analysis and efficiency measurement of electromagnetic vibration energy harvesting system. *Microsyst. Technol.* **2016**, *22*, 1767–1779. [CrossRef]
22. Beranek, L. *Acoustics*; Electrical and Electronic Engineering, American Institute of Physics: College Park, MD, USA, 1986.
23. Kunz, J.; Fialka, J.; Benes, P.; Havranek, Z. An Automated measurement system for measuring an overall power efficiency and a characterisation of piezo harvesters. *J. Phys. Conf. Ser.* **2018**, *1065*, 202008. [CrossRef]
24. Gade, S.; Herlufsen, H. *Use of Weighting Functions in DFT/FFT Analysis (Part I)*; Brüel & Kjær Technical Review; Brüel & Kjær: Nærum, Denmark, 1987; pp. 1–28.
25. Quattrocchi, A.; Freni, F.; Montanini, R. Power Conversion Efficiency of Cantilever-Type Vibration Energy Harvesters Based on Piezoceramic Films. *IEEE Trans. Instrum. Meas.* **2021**, *70*, 1–9. [CrossRef]
26. Roundy, S.; Leland, E.S.; Baker, J.; Carleton, E.; Reilly, E.; Lai, E.; Otis, B.; Rabaey, J.M.; Wright, P.K.; Sundararajan, V. Improving power output for vibration-based energy scavengers. *IEEE Pervasive Comput.* **2005**, *4*, 28–36. [CrossRef]
27. Alameh, A.; Gratuze, M.; Elsayed, M.; Nabki, F. Effects of Proof Mass Geometry on Piezoelectric Vibration Energy Harvesters. *Sensors* **2018**, *18*, 1584. [CrossRef]
28. Peralta, P.; Ruiz, R.; Natarajan, S.; Atroshchenko, E. Parametric study and shape optimization of Piezoelectric Energy Harvesters by isogeometric analysis and kriging metamodeling. *J. Sound Vib.* **2020**, *484*, 115521. [CrossRef]

sensors

Article

Load Resistance Optimization of Bi-Stable Electromagnetic Energy Harvester Based on Harmonic Balance

Sungryong Bae [1] and Pilkee Kim [2,3,*]

1   Department of Fire Protection and Disaster Management, Chosun University, 309 Pilmun-daero, Dong-gu, Gwangju 61452, Korea; sbae@chosun.ac.kr
2   School of Mechanical Design Engineering, College of Engineering, Jeonbuk National University, 567 Baekje-daero, Deokjin-gu, Jeonju-si, Jeollabuk-do 54896, Korea
3   Eco-Friendly Machine Parts Design Research Center, Jeonbuk National University, 567 Baekje-daero, Deokjin-gu, Jeonju-si, Jeollabuk-do 54896, Korea
*   Correspondence: pkim@jbnu.ac.kr

Abstract: In this study, a semi-analytic approach to optimizing the external load resistance of a bi-stable electromagnetic energy harvester is presented based on the harmonic balance method. The harmonic balance analyses for the primary harmonic (period-$1T$) and two subharmonic (period-$3T$ and $5T$) interwell motions of the energy harvester are performed with the Fourier series solutions of the individual motions determined by spectral analyses. For each motion, an optimization problem for maximizing the output power of the energy harvester is formulated based on the harmonic balance solutions and then solved to estimate the optimal external load resistance. The results of a parametric study show that the optimal load resistance significantly depends on the inductive reactance and internal resistance of a solenoid coil—the higher the oscillation frequency of an interwell motion (or the larger the inductance of the coil) is, the larger the optimal load resistance. In particular, when the frequency of the ambient vibration source is relatively high, the non-linear dynamic characteristics of an interwell motion should be considered in the optimization process of the electromagnetic energy harvester. Compared with conventional resistance-matching techniques, the proposed semi-analytic approach could provide a more accurate estimation of the external load resistance.

Keywords: electromagnetic energy harvester; bi-stable oscillator; load resistance optimization; frequency response analysis; harmonic balance method

Citation: Bae, S.; Kim, P. Load Resistance Optimization of Bi-Stable Electromagnetic Energy Harvester Based on Harmonic Balance. *Sensors* **2021**, *21*, 1505. https://doi.org/10.3390/s21041505

Academic Editors: Zdenek Hadas, Saša Zelenika and Vikram Pakrashi

Received: 10 December 2020
Accepted: 16 February 2021
Published: 22 February 2021

Publisher's Note: MDPI stays neutral with regard to jurisdictional claims in published maps and institutional affiliations.

1. Introduction

Energy harvesting is a series of processes that convert ambient energy (vibration, sunlight, wind, raindrop, earth heat, human motion, etc.), usually unused and wasted, into electric energy that is stored in a capacitor or used to supply power to micro-devices. The energy harvesting technology is considered a promising alternative to chemical batteries for autonomous wireless microdevices. In particular, vibration-based energy harvesters have been intensively studied and developed because vibration energy sources exist almost everywhere in our daily life and industrial environment, and its conversion mechanism is relatively simple.

The early design of vibration energy harvesters was based on a linear resonator coupled with electromagnetic, electrostatic, or piezoelectric transducers [1,2]. Such linear-resonant energy harvesters produce high electric power only in a narrow resonant frequency band, which is not suitable for harvesting energy from real-world vibration sources with time-varying and/or multiple frequency components [3,4]. Accordingly, energy harvesting systems of various structures (e.g., frequency-tunable oscillators [5–7], an array of multimodal oscillators [8–10], and non-linear mono-stable [11–14] or multi-stable oscillators [15–26]) have been investigated to deal with the narrow bandwidth problem of the linear system.

A bi-stable oscillator that possesses a double-well potential energy function is known to be one of the most advantageous candidates to overcome the limitation of the linear system. The dynamic behavior of the bi-stable oscillator is generally more complicated than that of mono-stable oscillators owing to the existence of a saddle barrier between the potential energy wells, but its large-amplitude oscillation (called interwell motion) tends to cover a much broader frequency range. During the past decade, various types of bi-stable energy harvester have been proposed based mainly on magnetically (repulsive/attractive) and/or mechanically buckled structures, and their performance has been evaluated under ambient vibration sources of the swept sine, impulse, and Gaussian white noise types [1,2,18]. The performance of a bi-stable energy harvester depends on the amount of ambient vibration energy transmitted to the energy transducer, that is, transmissibility, which is related to the features of a forced interwell motion, such as the required intensity of excitation, the associated frequency bandwidth, and its oscillation frequency and amplitude. Various optimization strategies have been proposed to enhance the performance of vibration energy harvesters (e.g., artificial intelligence-based optimizations, structural optimizations, and nonlinear optimizations) [27,28]. Additionally, an external load resistance connected to an energy harvester for harvesting electrical energy is also an important design factor that should be optimized for maximizing performance. The resistance-matching technique has been commonly used for estimating the optimal load resistance [29], but it applies in principle to systems where both the source and load devices are linear. For non-linear systems (including a bi-stable system considered in this study), the resistance-matching technique may give an inaccurate estimation of the optimal load resistance with significant error, depending on the system and the excitation conditions.

In this study, a semi-analytic approach based on the harmonic balance method is proposed to optimize the external load resistance of a bi-stable electromagnetic energy harvester accurately, when compared with conventional resistance-matching techniques. To this end, the frequency responses of the primary harmonic and subharmonic interwell motions are numerically obtained first, and their solution forms are assumed in accordance with the results of spectral analyses. For each interwell motion, an optimization problem was formulated and solved, based on the harmonic balance solution of the assumed Fourier series, to calculate the optimal external load resistance of the energy harvester. In addition, a parametric study was conducted to investigate the effects of the system parameters and excitation conditions on the optimal load resistance of the electromagnetic energy harvester, which is not considered in conventional resistance-matching techniques; the results are discussed.

2. Bi-Stable Oscillator Model of an Electromagnetic Energy Harvester

Figure 1 shows the schematic diagram of an electromagnetic energy harvester considered in this study, which is composed of two parts: a clamped-clamped beam oscillator and an external solenoid coil structure. The inertial mass, along with two permanent magnets, is attached to the middle of the stainless-steel beam to adjust the natural frequency of the beam oscillator. The solenoid coil structure is fixed to the external space and aligned with the inertial mass, so that the permanent magnet can oscillate through the inner hole of the solenoid coil when the base of the energy harvesting system is excited by a certain ambient vibration. The coil is connected to an external load resistance for harvesting electrical energy. With this structure for electromagnetic induction, the energy harvesting system can convert ambient vibration energy to electrical energy that would be used (or stored in a capacitor) for powering micro-devices.

The performance of the energy harvesting system depends significantly on the transmissibility of the system, that is, the amount of ambient vibration transmitted through the beam oscillator to the permanent magnet traveling in the inner hole of the coil. Thus, the electromagnetic energy harvester is designed based on a statically bi-stable oscillator to better respond to various types of ambient vibration with wide-band frequency components. To this end, an axial load is imposed on the beam in the longitudinal direction by moving the right clamped end slightly to the left, which causes the buckling phenomenon of the beam. It is well known that at the moment when buckling occurs, new symmetric non-trivial equilibria ($x = \pm x_0$) start bifurcating from the trivial one ($x = 0$), which is called the pitch-fork bifurcation of equilibrium. After the buckling, the beam structure becomes unstable at the trivial equilibrium, and is stable at the two non-trivial upper and lower equilibria depicted in Figure 1.

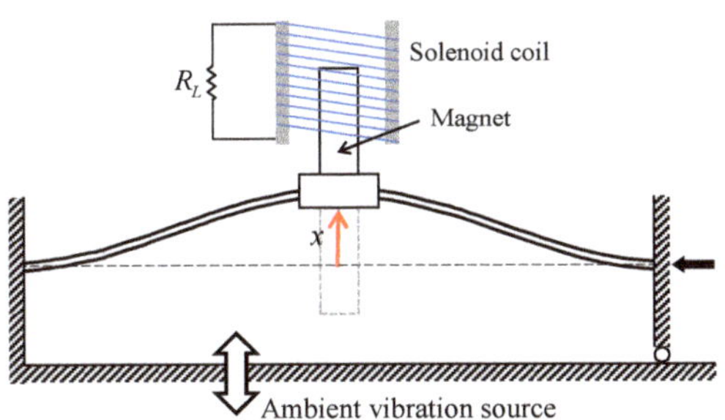

Figure 1. Schematic diagram of an electromagnetic bi-stable energy harvester composed of a clamped-clamped beam and an external solenoid coil. The right end of the beam is movable in the horizontal direction but can also be fastened at a certain position. The gravitational field is assumed to be in the perpendicular direction to the given plane.

The mathematical model for the aforementioned energy harvesting system is given by the following electro-magneto-mechanical oscillator model [30–32]:

$$\ddot{x} + 2\zeta\omega_0\dot{x} - \gamma x_0^2 x + \gamma x^3 + \kappa_0 I = -A_b\cos\Omega t, \tag{1a}$$

$$L_0\dot{I} + (R_e + r_L)I = \beta\dot{x}, \tag{1b}$$

where x is the displacement of the inertial mass attached to the middle of the beam; x_0 indicates the distance of the non-trivial equilibria measured from the trivial one; γ is the coefficient given by $(\omega_0/4x_0)^2$; I is the electrical current flowing across the load resistance R_e; ω_0 and ζ are the natural angular frequency and damping ratio, respectively; κ_0 and β are the electro-magneto-mechanical coupling constants; r_L and L_0 are the internal resistance and inductance of the coil, respectively; and A_b and Ω are the magnitude and frequency of the base acceleration, respectively. The baseline parameter values of the oscillator model are listed in Table 1. In addition, for the purpose of generality, the following dimensionless variables and parameters are introduced to Equations (1a) and (1b):

$$y = \frac{x}{x_0},\ J = \frac{I}{I_0},\ \tau = \omega_0 t,\ A = \frac{A_b}{x_0\omega_0^2},\ \omega_r = \frac{\omega}{\omega_0},\ \kappa = \frac{\kappa_0\beta L_0}{\alpha^2 r_L^2},\ \alpha = \frac{\omega_0 L_0}{r_L},\ R = \frac{R_e}{r_L}, \tag{2}$$

and the resulting dimensionless forms of governing equations can be obtained in the form:

$$\ddot{y} + 2\zeta\dot{y} - \frac{1}{2}y + \frac{1}{2}y^3 + \kappa J = -A\cos\omega\tau, \tag{3a}$$

$$\alpha\dot{J} + (R+1)J = \dot{y}, \tag{3b}$$

where ω are and the dimensionless forcing frequency (or the frequency ratio); α is the dimensionless resonant inductive reactance (or the ratio of the resonant inductive reactance of the coil, $\omega_0 L_0$, to the internal resistance of the coil, r_L); and R is the dimensionless external load resistance (or the ratio of the external load resistance to the internal resistance of the coil).

Table 1. Parameter values of the electro-magneto-mechanical oscillator [30,32].

Parameter	Value
x_0	0.29 mm
R	18 Ohm
ω_0	365 Hz
ζ	0.004
κ_0	16.7
β	0.5
r_L	18 Ohm
L_0	5 mH

3. Numerical Frequency Response Analysis of an Electro-Magneto-Mechanical Oscillator Model

3.1. Steady-State Response and Its Bifurcation

In this work, a series of numerical simulations of the electro-magneto-mechanical oscillator model expressed in Equations (3a) and (3b) are performed to investigate the dynamic behaviors of the system and the associated energy harvesting performance. By introducing the state vector $[y_1, y_2, y_3, y_4] = [\dot{y}, y, J, \theta]$, where $\theta = \omega\tau$, into Equations (3a) and (3b), the dimensionless governing differential equations can be rewritten into an autonomous system of first-order differential equations in the following vector form:

$$\dot{\mathbf{y}} = f(\mathbf{y}, t) = \begin{bmatrix} -2\zeta y_1 + \frac{1}{2}y_2 - \frac{1}{2}y_2^3 - \kappa y_3 - A\cos y_4 \\ y_1 \\ -\frac{R+1}{\alpha}y_3 + \frac{1}{\alpha}y_1 \\ \omega \end{bmatrix}. \tag{4}$$

Frequency response analyses are carried out by the direct numerical integration of Equation (4) based on the Runge–Kutta methods. For all the numerical simulations, the baseline parameter values given in Table 1 are used, unless otherwise stated. The steady-state forced responses of the electromagnetic energy harvester are evaluated first, while the frequency of the base excitation varies by means of frequency marching or frequency sweeping, and then their Floquet multipliers are obtained by solving the eigenvalue problem of the following monodromy matrix:

$$\mathbf{M} = \prod_{i=N}^{1} \exp\left[\int_{\tau_{i-1}}^{\tau_i} \mathbf{J}(\tau)d\tau\right], \tag{5}$$

where $\mathbf{J}$ is the Jacobian matrix, and N is the number of equally spaced time intervals included within one period. The bifurcation type of the stationary oscillation can be determined from the investigation of its Floquet multipliers. When the stability of the steady-state solution changes, the accompanying bifurcations can be classified into three different types: saddle-node bifurcation (if the largest Floquet multiplier μ is positive 1, i.e., $\mu = +1$), period-doubling bifurcation ($\mu = -1$), and Neimark–Sacker bifurcation ($|\mu| = 1$ and $\text{Im}[\mu] \neq 0$).

3.2. Non-Linear Oscillations Owing to Double-Well Potential: Intrawell and Interwell Motions

The electromagnetic bi-stable energy harvester possesses a double-well potential energy function, as shown in Figure 2a. There are one trivial saddle point and two non-trivial center points in the potential function. The two potential wells are in symmetric configuration with respect to the potential saddle barrier formed at the trivial point of $y = 0$. This potential saddle barrier leads to two different types of forced response: intrawell and interwell motions, depending on the intensity of the base excitation. When the base excitation is weak, the system only oscillates with small amplitude inside one of the two potential wells (as demonstrated by the thin line in Figure 2b), which is called intrawell motion, owing to the existence of the potential barrier. When the base excitation becomes relatively strong, the system gains enough kinetic energy to cross the potential barrier (the thick line in Figure 2b). Accordingly, it begins oscillating with a much larger amplitude, producing higher electrical power, and it is called interwell motion. The interwell motion tends to occur over a much broader frequency band than the resonant frequency band of linear oscillators. Using this interwell motion, the electromagnetic bi-stable energy harvester can harvest energy from ambient vibration energy sources, including wide-band frequency components. The aforementioned high-energy interwell motion of the electromagnetic energy harvester appears in various types of oscillations: primary harmonic motion, its subharmonic and superharmonic motions, and even non-periodic chaotic motion. Among these interwell motions, the primary harmonic motion of period $T = 2\pi/\omega$ has been used the most for broadband energy harvesting applications. However, it was recently reported that subharmonic motions are also useful and suitable [29].

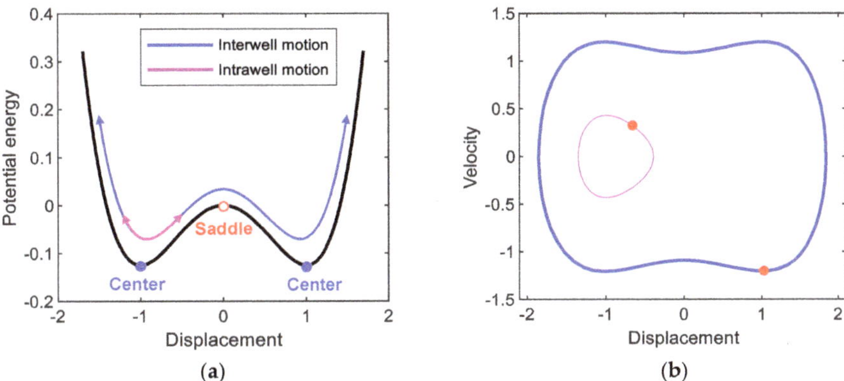

Figure 2. (a) Double-well potential energy function of the electromagnetic bi-stable energy harvester. The open and solid circles indicate the saddle and center points, respectively. (b) Phase plane representation of the intrawell and interwell oscillations (of period $T = 2\pi/\omega$), which are denoted by thick (blue) and thin (magenta) lines, respectively, obtained when $A = 0.02$ and $\omega = 0.82$. In (b), the red points indicate the stroboscopic projections of the oscillations, which are synchronized with the period T of the base excitation.

Figure 3 shows the frequency responses for (a) displacement and (b) output power of the electromagnetic energy harvester obtained when the base acceleration is 0.02. In this figure, the subharmonic interwell motions of periods $3T$ and $5T$ and the primary harmonic interwell motion of period T are presented along with the intrawell motion for comparison purposes. The subharmonic interwell motions are much larger in amplitude than the intrawell motion (as shown in Figure 3a), although they are smaller than the primary harmonic interwell motion, and accordingly can produce relatively higher output powers (as shown in Figure 3b). This is more clearly demonstrated in Figure 4 by directly comparing the period-1T, 3T, and 5T interwell motions on the state plane. The period-3T interwell motion has a notably wide frequency band, similar to that of the period-T interwell motion. It is interesting that the frequency bands of the period-1T, 3T, and 5T interwell motions are continuously connected to each other and tend to form a whole wide frequency band. The above observations indicate that the subharmonic interwell motions in addition to the primary harmonic interwell motion can play an important role in enhancing the operating frequency band of electromagnetic energy harvesters.

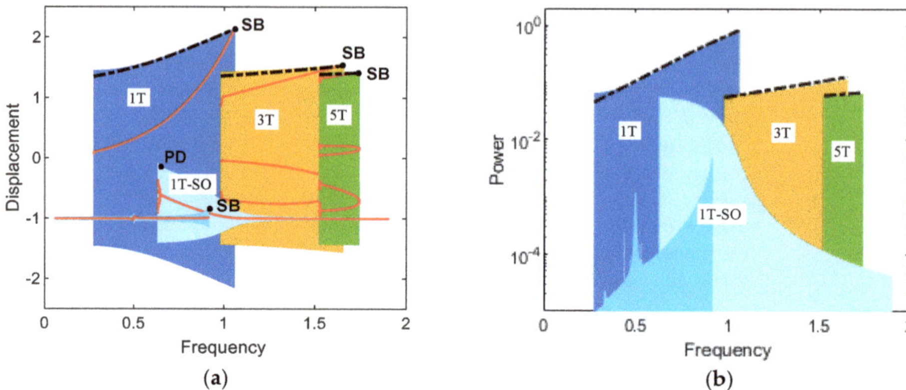

Figure 3. Frequency responses for (**a**) displacement and (**b**) output power of the electromagnetic bi-stable energy harvester obtained when $A = 0.02$: ($1T$) primary harmonic period-T interwell motion, ($1T$-SO) period-T intrawell motion, ($3T$) subharmonic period-$3T$ interwell motion, and ($5T$) period-$5T$ interwell motion. The saddle-node bifurcations and period-doubling bifurcation are designated by points SB and PD, respectively. In (**a**), the red dots indicate the stroboscopic projections of the oscillations, which are synchronized with the period T of the base excitation. The dashed-dotted lines are the harmonic balance solutions for the primary harmonic and subharmonic interwell motions, which will be discussed in Section 4.

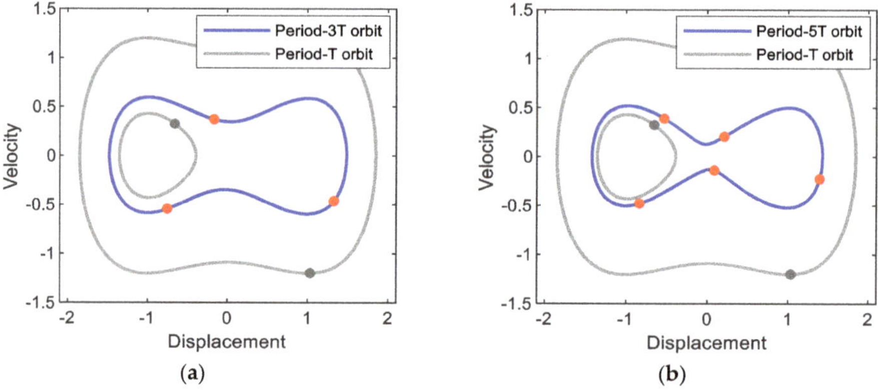

Figure 4. Phase plane representations of (**a**) period-$3T$ and (**b**) period-$5T$ motions obtained with the base acceleration of $A = 0.02$ and forcing frequencies of (**a**) $\omega = 1.37$ and (**b**) $\omega = 1.65$, respectively. The period-T interwell and intrawell motions of Figure 2b are also presented for comparison purpose. The solid circle dots indicate the stroboscopic projections of the oscillations, which are synchronized with the period T of the base excitation.

4. Semi-Analytic Approach to Optimizing the External Resistance

The external load resistance of an electromagnetic energy harvester, together with the transmissibility, is an important design factor, which must be optimized to maximize its output electrical power. An impedance-matching technique has been commonly used to solve this optimization problem, with the assumption that the inductance of the coil is negligible. In this case, the external load resistance is actually matched with the internal resistance of the solenoid coil; thus, it will be called 'resistance matching' in this study. However, it should be noted that the impedance- (or resistance-) matching technique only applies when both the source and load devices are linear, although they are frequently used for non-linear systems at the cost of accuracy. In this work, a semi-analytical approach based on the method of harmonic balance is implemented to evaluate the optimal value of the external load resistance accurately, with which the interwell motion of the system can most efficiently produce electrical power.

First, Fourier analyses are performed to find the appropriate solution forms of the steady-state primary harmonic and subharmonic interwell motions, which depend on the type and intensity of non-linearity [33], and the results of the time responses and the corresponding amplitude spectra are presented in Figure 5. The forcing period of the base excitation is given by $T = 2\pi/\omega$, and accordingly the oscillation frequencies of the subharmonic period-3T and 5T motions tend to be lower than the forcing frequency ω, as shown in Figure 5c,e. Such subharmonic behavior is known to be a typical feature of a bi-stable system. Therefore, the fundamental oscillation frequencies of the interwell motions can be defined by $\omega_k = \omega/k$, where k = 1, 3, and 5 for the period-1T, 3T, and 5T motions, respectively. From Figure 5b,d,f, it can be observed that the interwell motions only have two or three important frequency components: ω_1- and $3\omega_1$-components for the period-T motion, ω_3- and $3\omega_3$-components for the period-3T motion, and ω_5-, $3\omega_5$- and $5\omega_5$-components for the period-5T motion, as shown in Figure 5b,d,e, respectively. Thus, the steady-state periodic solutions can be assumed in the following forms of the Nth-order Fourier series:

$$(\text{Period} - T \text{ motion}) \; y = \sum_{n=1,3} a_n \cos n\omega\tau + b_n \sin n\omega\tau, \; J = \sum_{n=1,3} c_n \cos n\omega\tau + d_n \sin n\omega\tau \tag{6a}$$

$$(\text{Period} - 3T \text{ motion}) \; y = \sum_{n=1,3} a_n \cos \frac{n\omega}{3}\tau + b_n \sin \frac{n\omega}{3}\tau, \; J = \sum_{n=1,3} c_n \cos \frac{n\omega}{3} + d_n \sin \frac{n\omega}{3} \tag{6b}$$

$$(\text{Period} - 5T \text{ motion}) \; y = \sum_{n=1,3,5} a_n \cos \frac{n\omega}{5}\tau + b_n \sin \frac{n\omega}{5}\tau, \; J = \sum_{n=1,3,5} c_n \cos \frac{n\omega}{5} + d_n \sin \frac{n\omega}{5} \tag{6c}$$

where the highest order of harmonics is given as N = 3 for the period-1T and 3T motions and N = 5 for the period-5T motion, and all the constant terms are set to zero as the main interest is on the interwell motion, of which the center of oscillation is the trivial point (y = 0).

Substituting Equations (6a)–(6c) into Equations (3a) and (3b), followed by equating the coefficients of like harmonic components, leads to a system of non-linear algebraic equations for unknown variables a_n, b_n, c_n, and d_n, which are numerically solved by the Newton–Raphson method. As shown in Figure 3, the harmonic balance solutions (dashed-dotted lines) for primary harmonic and subharmonic interwell motions are well matched with the numerical solutions. Additionally, time-domain comparisons are made in Figure 6, where numerical and analytical results are also in a good agreement with each other.

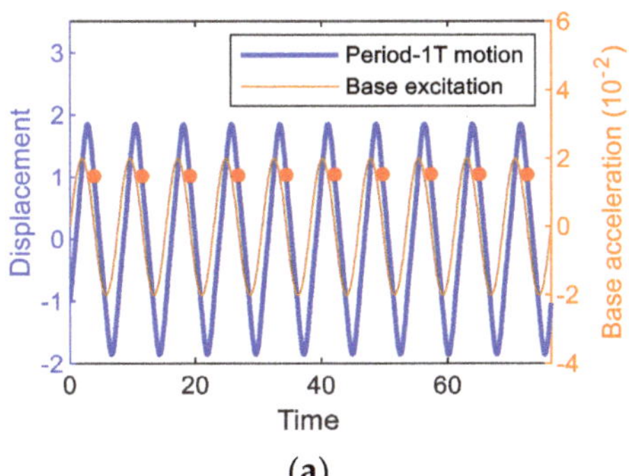

(a)

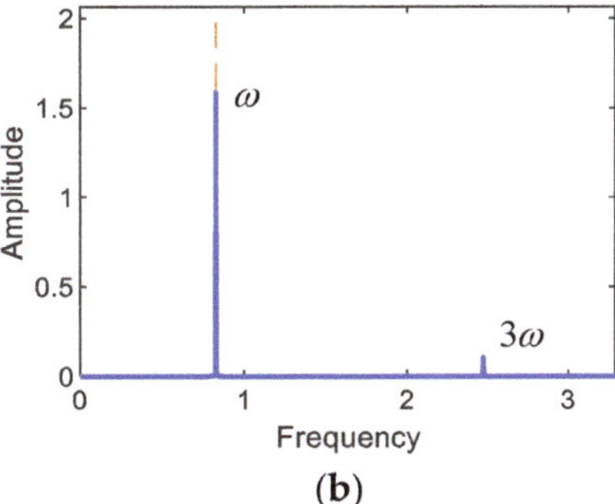

(b)

Figure 5. *Cont.*

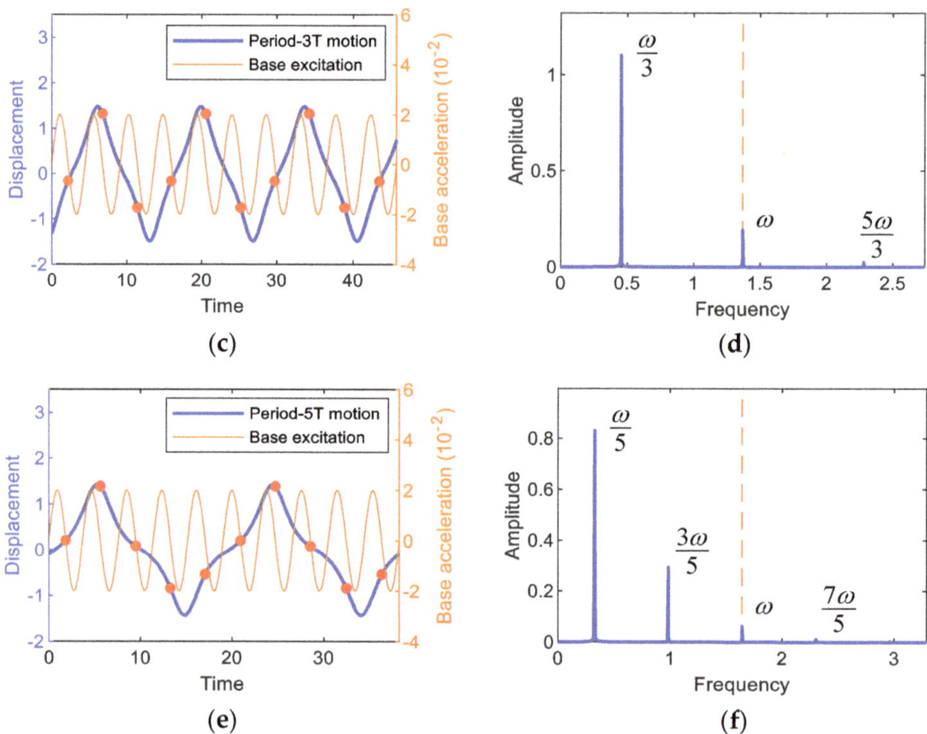

Figure 5. (**first column**) Time responses and (**second column**) the associated amplitude spectra, obtained by the fast Fourier transform. In (**a**,**b**) the period-1*T* motion when $A = 0.02$ and $\omega = 0.82$, in (**c**,**d**) the period-3*T* motion when $A = 0.02$ and $\omega = 1.37$, and in (**e**,**f**) the period-5*T* motion when $A = 0.02$ and $\omega = 1.65$. The base excitation used for this simulation is also presented for comparison. In (**a**,**c**,**e**), the red dots indicate the stroboscopic projections of the oscillations, which are synchronized with the period T of the base excitation.

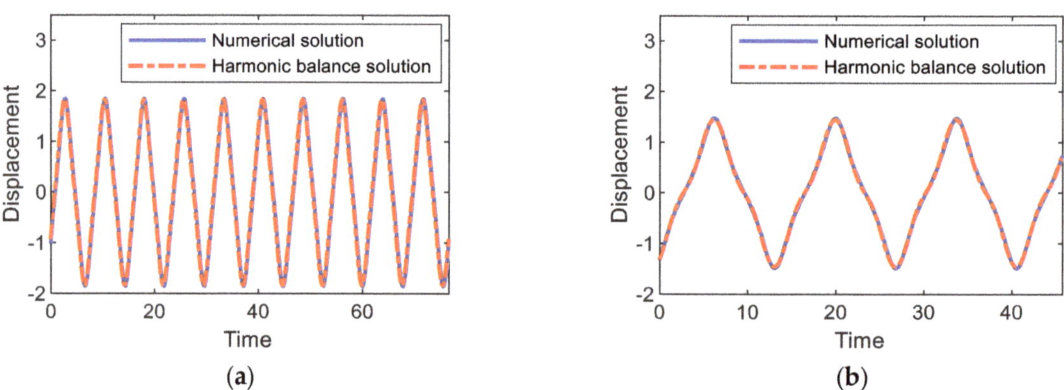

Figure 6. *Cont.*

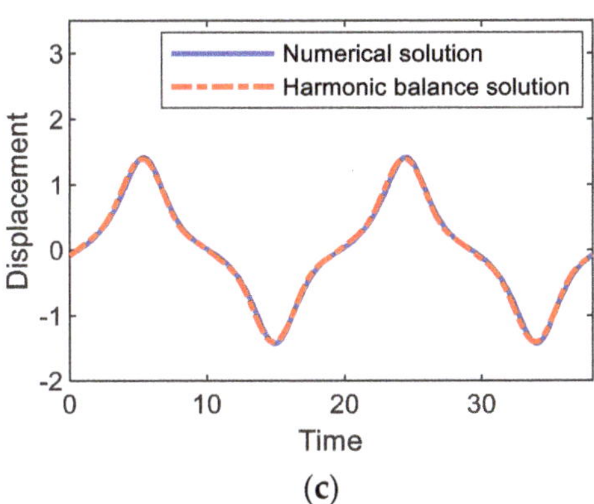

(c)

Figure 6. Comparisons between numerical and analytical solutions for (**a**) period-1*T*, (**b**) 3*T*, and (**c**) 5*T* interwell motions, which are obtained when the amplitude and frequency of the base acceleration are given as *A* = 0.02 and ω = (**a**) 0.82, (**b**) 1.37, and (**c**) 1.65.

Using the obtained harmonic balance solutions, the average output power P_{av} of the electromagnetic energy harvester can be evaluated as $P_{av}^2 = \sum (c_n^2 + d_n^2)/2$, and the optimal value of the external load resistance is estimated by solving the optimization problem of $\partial P_{av}/\partial R = 0$. The optimal external load resistance reduces to 1 for the same problem when the resistance matching technique is used.

Figure 7 shows the average output powers, harvested from (**a**) period-1*T*, (**b**) period-3*T*, and (**c**) period-5*T* motions, with respect to the external load resistance. The numerical and semi-analytical results are compared to each other in Figure 7. As shown in this figure, the analytic and numerical results for the average output powers are well matched with each other. Furthermore, the optimal values of the external load resistance were evaluated, and it was observed that the analytic results are in an excellent agreement with numerical results with a maximum relative error of less than (**a**) 0.3%, (**b**) 2.0%, and (**c**) 1.5%. This observation supports the validity of the semi-analytic approach to optimize the external load resistance implemented in this study.

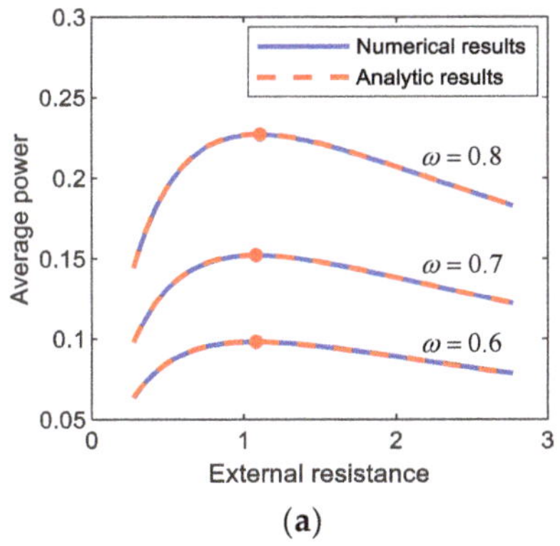

(a)

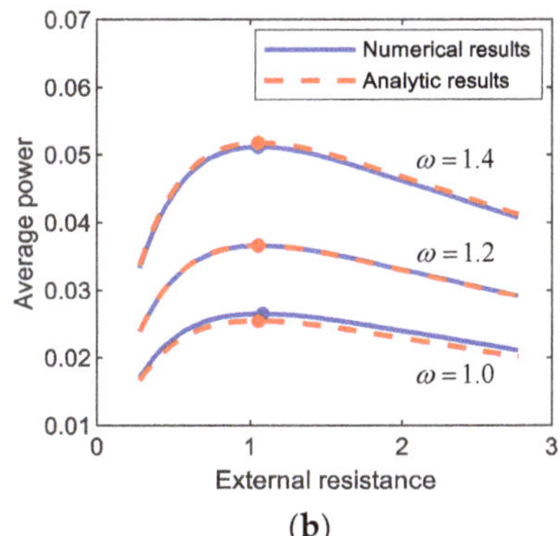

(b)

Figure 7. *Cont.*

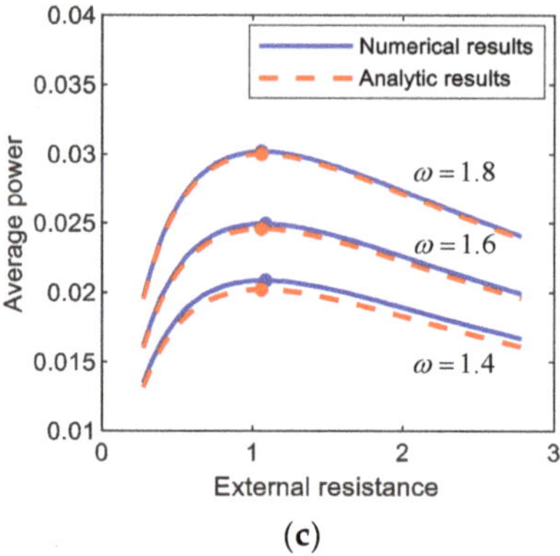

(**c**)

Figure 7. Plots of the external load resistance versus the average output power harvested from (**a**) period-1T, (**b**) period-3T, and (**c**) period-5T motions of the electromagnetic energy harvester. The results are obtained by numerical integration (blue solid lines) and the semi-analytic approach (red dashed lines) and compared to each other. For each motion, the comparisons in between are made with three different forcing conditions: (**a**) (A, ω) = (0.012, 0.6), (0.02, 0.7), (0.03, 0.8), (**b**) (A, ω) = (0.012, 1.0), (0.02, 1.2), (0.03, 1.4), (**c**) (A, ω) = (0.012, 1.4), (0.02, 1.6), (0.03, 1.8). The solid circle points indicate the optimal conditions for the external load resistance with which the output power is maximized.

In general, the subharmonic period-3T and 5T interwell motions tend to produce relatively lower electric power than the primary harmonic period-1T interwell motion (but relatively higher than intrawell motions). However, the frequency bands of the period-1T, 3T, and 5T interwell motions are possibly different from each other (as shown in Figure 3), and each of the interwell motions could be used to harvest energy in the associated frequency band, depending on excitation conditions.

5. Effects of Inductive Reactance and Excitation Conditions on Optimal External Resistance

Using the semi-analytic method mentioned in the previous section, a parametric study is conducted to investigate the effects of the dimensionless resonant inductive reactance α (defined by the ratio of $\omega_0 L_0$ to r_L in Equation (3b)) and the excitation conditions (the frequency and intensity of the base excitation) on the optimal load resistance of the electromagnetic energy harvester, which is not considered in the resistance-matching technique.

Figure 8 shows the optimal values of the external load resistance for (a) period-1T, (b) period-3T, and (c) period-5T motions of the electromagnetic energy harvester, obtained by using the semi-analytic method, with respect to the resonant inductive reactance. In this figure, the results clearly indicate that the optimal load resistance of the electromagnetic energy harvester can be significantly affected by the resonant inductive reactance and excitation frequency. For all the period-1T, 3T, and 5T motions, the optimal load resistance increases as the dimensionless resonant inductive reactance increases——in other words, as the inductance (L_0) of the solenoid coil becomes relatively larger——and accordingly, it deviates from the optimal value (i.e., 1) estimated by the resistance-matching technique, in which the inductance of the coil is assumed to be zero. Such a deviation tends to be larger when the frequency (ω) of the base excitation is higher, which is more remarkable for the interwell motion of the shorter oscillation period (in the order of periods (a) 1T, (b) 3T, and (c) 5T in Figure 8) or higher oscillation frequency (in the order of (a) ω_1, (b) ω_3, and (c) ω_5). From these observations, it is deduced that the optimal load resistance depends on the inductive reactance of the coil, which is defined by $\omega_k L_0$, in addition to the internal resistance (r_L) of the coil.

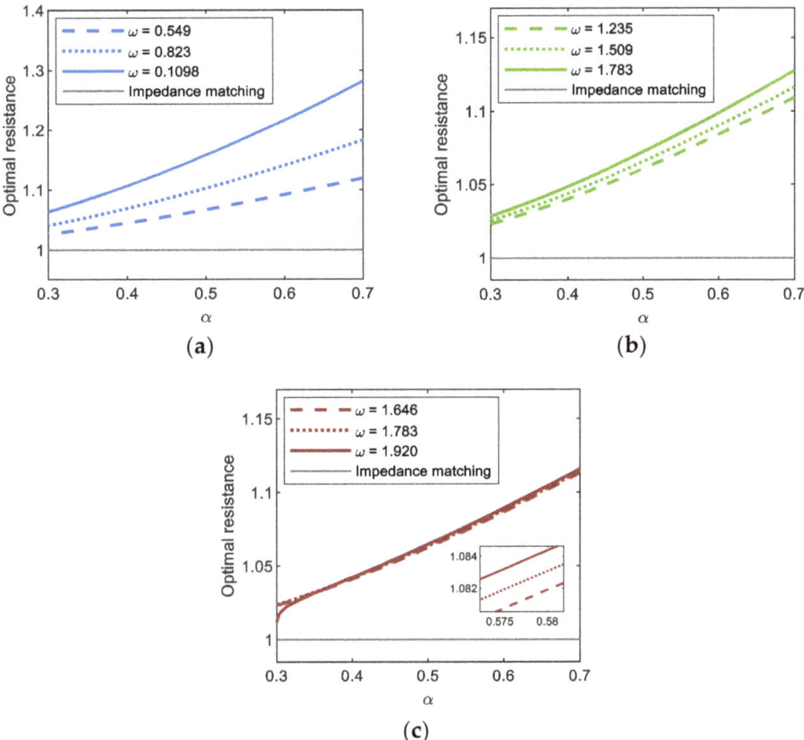

Figure 8. Plots of the resonant inductive reactance versus the optimal load resistance for (**a**) period-1*T*, (**b**) period-3*T*, and (**c**) period-5*T* motions of the electromagnetic energy harvester. For each motion (colored thick lines), the results are calculated by using the semi-analytic method, with three different values of excitation frequency and a constant base acceleration of 0.02, and compared with that obtained by the resistance-matching technique (grey thin line).

Figure 9 shows the optimal values of the external load resistance for the period-1*T*, 3*T*, and 5*T* motions of the electromagnetic energy harvester, obtained with three different values of base acceleration: (a) $A = 0.012$, (b) $A = 0.02$, and (c) $A = 0.03$. The calculations are made only in the range of the excitation frequency in which the interwell (period-1*T*, 3*T*, and 5*T*) motions exist. As shown in this figure, the period-1*T* and period-3*T* motions occur in a broad range of excitation frequencies, whereas the period-5*T* motion occurs in a relatively narrow range, which means that the former are more important design factors in broadband energy harvesting applications. The frequency range for each interwell motion tends to be broader as the intensity of the base excitation increases, slightly shifting to a higher frequency. For the period-1*T* motion, the optimal load resistance obviously increases with excitation frequency, whereas for other interwell motions, only small variations in the optimal load resistance are observed in the overall range of excitation frequency (particularly for period-5*T* motion, the resistance is almost constant). This is because the lower the oscillation frequencies of the subharmonic period-3*T* and 5*T* motions ($\omega_3 = \omega/3$ and $\omega_5 = \omega/5$, respectively) are, the weaker the effects of the inductive reactance of the solenoid coil on the optimal resistance. Furthermore, the optimal load resistance is not sensitive to changes in the intensity of the base excitation. As shown in Figure 10, the optimal load resistance for the period-1*T* and 3*T* motions tends to increase slightly with the base acceleration and, contrarily, the one for the period-5*T* motion decreases slightly. However, these changes in the optimal resistance are not significant (less than 0.01%).

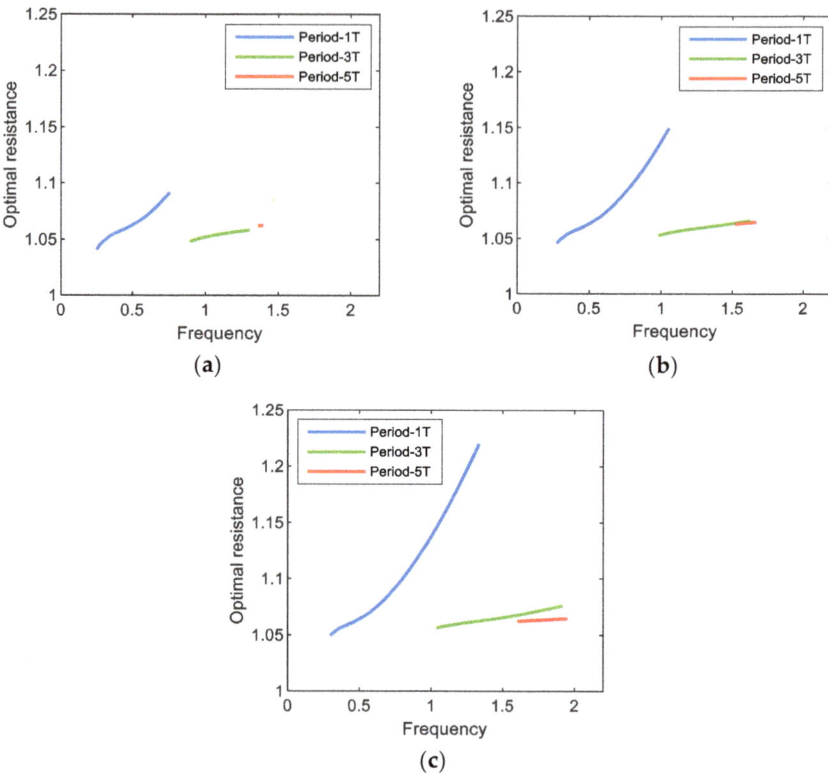

Figure 9. Plots of the excitation frequency versus the optimal load resistance for the interwell motions (with periods $1T$, $3T$, and $5T$) of the electromagnetic energy harvester. In this simulation, the acceleration of base excitation is set to (**a**) 0.012, (**b**) 0.02, and (**c**) 0.03.

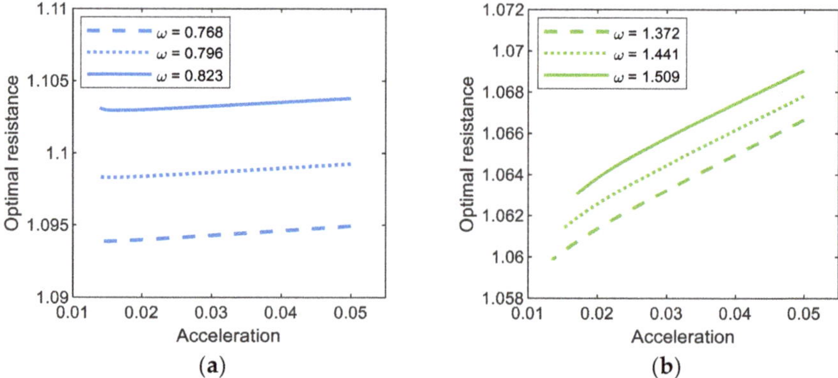

Figure 10. *Cont.*

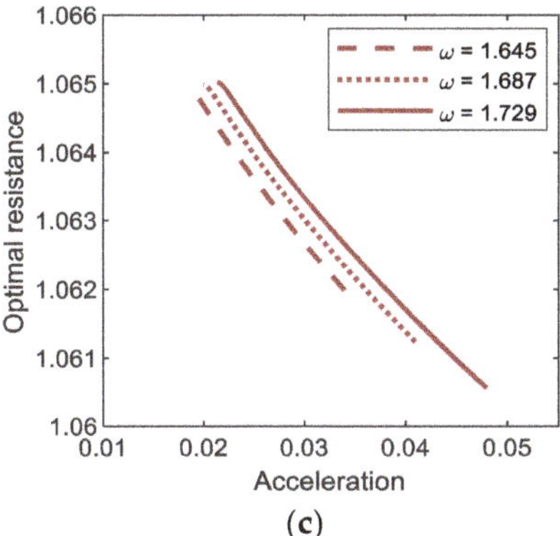

(c)

Figure 10. Base acceleration versus optimal load resistance for (**a**) period-1*T*, (**b**) period-3*T*, and (**c**) period-5*T* motions of the bi-stable electromagnetic energy harvester.

6. Broadband Energy Harvesting Applications

The output power of the bi-stable electromagnetic energy harvester is evaluated, while the frequency of the harmonic base excitation varies by means of frequency marching, and compared with that of its linear counterpart in Figure 11 to demonstrate its broadband energy harvesting performance. As shown in this figure, for the linear system, a sharp resonant peak of the output power response appears in a very narrow frequency band. In this case, the natural frequency of the system must be designed to be synchronized with the forcing frequency of the harmonic excitation, otherwise it is likely to fail to produce high output power. On the contrary, the bi-stable system possesses the multiple branches of the interwell motions (including the period-1*T*, *3T*, and *5T* motions) in a broad frequency band and thereby can keep normally operating with high output power, regardless of the synchronization of the frequencies. Therefore, the bi-stable energy harvester is definitely more advantageous in real applications, in which the ambient harmonic vibration possibly varies, than the linear system. Actually, there are various types of ambient vibration source such as harmonic, periodic, impulsive, and random excitations. The external load resistance of the bi-stable electromagnetic energy harvester should be optimized in an appropriate manner that depends on the type of the ambient vibration source.

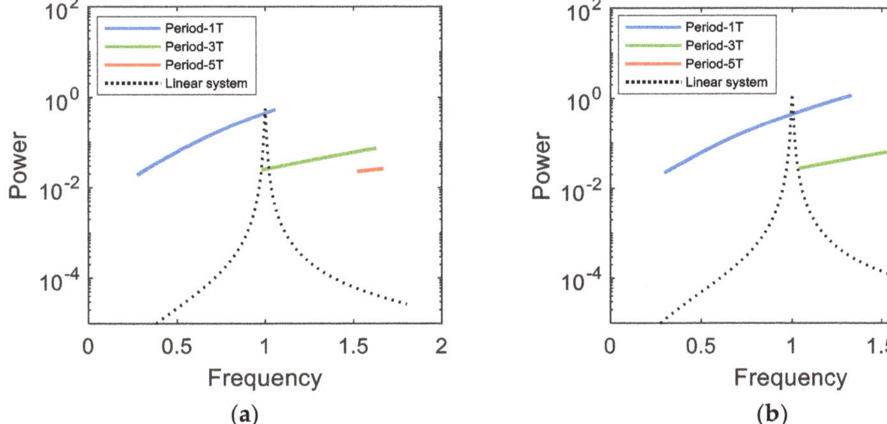

(a)   (b)

Figure 11. Frequency marching responses for the output powers of the electromagnetic bi-stable energy harvester and its linear counterpart, which are obtained when (**a**) *A* = 0.02 and (**b**) 0.03.

Basically, the semi-analytical optimization approach proposed in this study is suitable for harmonic vibration sources, since it is based on the harmonic balance solution. For all the above simulations, the optimal load resistance was evaluated and investigated, when the frequency of the harmonic base excitation was constant, to demonstrate the dynamic characteristics of the optimal resistance. In this section, the proposed optimization approach is further extended for piecewise harmonic excitations with constant amplitude but stepwise changes in frequency.

The general form of the average output power is given in the form:

$$P_{av} = J_{RMS}^2 R = \frac{1}{\Delta T} \int_{T_0}^{T_N} J^2 R d\tau, \tag{7}$$

where J_{RMS} is the root mean square (RMS) of the output current flowing across the external load resistance over the whole-time interval $\Delta T (= T_n - T_0)$. For most base excitations, the numerical integration method can be commonly used to evaluate the average power of Equation (7). Particularly for the piecewise harmonic excitation, the frequency of which is defined on a sequence of sub-time intervals $\Delta T_k = T_k - T_{k-1}$ $(k = 1, 2, \cdots, N)$, the resulting average output power is rewritten as the weighted sum of N RMS values:

$$P_{av} = \frac{\Delta T_1}{\Delta T} \left(\frac{1}{\Delta T_1} \int_{T_0}^{T_1} J^2 R d\tau \right) + \frac{\Delta T_2}{\Delta T} \left(\frac{1}{\Delta T_2} \int_{T_1}^{T_2} J^2 R d\tau \right) + \cdots + \frac{\Delta T_N}{\Delta T} \left(\frac{1}{\Delta T_N} \int_{T_{N-1}}^{T_N} J^2 R d\tau \right), \tag{8a}$$

or

$$P_{av} = \sum_{k=1}^{n} r_k J_k^2 R \text{ with } r_k = \frac{\Delta T_k}{\Delta T} \text{ and } J_k = \sqrt{\frac{1}{\Delta T_k} \int_{T_{k-1}}^{T_k} J^2 d\tau}, \tag{8b}$$

where J_k is the RMS output current on the subinterval ΔT_k, and r_k is the weighting factor. Assuming that the base excitation is harmonic on each subinterval ΔT_k, the RMS output current (J_k) and the associated average power ($J_k^2 R$) can be evaluated by using the harmonic balance solutions of Equations (6a)–(6c). The optimization problem for maximizing the total output power given by Equation (8) is readily formulated and solved to estimate a single optimal value of the external load resistance.

Figure 12a shows the amplitude of the output current response to (red line) piecewise harmonic excitation, which is obtained numerically for the period-1T interwell motion of the bi-stable electromagnetic energy harvester. In this simulation, the forcing frequency of the piecewise base excitation decreased from 1.0 by 0.05 to 0.5 for every 1000 cycles, but the amplitude of the base acceleration is set to be 0.02 g. The resulting output current response tends to intermittently vary with stepwise change in excitation frequency. Additionally, the average output power is evaluated with the variation of the external load resistance, as illustrated in Figure 12b. The optimal load resistance, 1.111, is estimated and compared with the analytical one, 1.104, evaluated by using Equations (6) and (8). The relative error of the analytical optimal resistance is very small (less than 0.7%), which supports the assertion that the proposed optimization approach is also valid for the piecewise harmonic excitation. In Figure 12a,b, the results for (blue line) swept-sine excitation are presented for comparison purposes. The swept-sine excitation continuously decreased with the slow sweep rate of 5×10^{-6} in the same frequency range. The output current response to the swept-sine excitation gradually decreases in the similar tendency to the piecewise harmonic case but with the difference in amplitude, which results in the difference in average output power. However, the optimal resistance, 1.139, for the swept-sine excitation (particularly with a slow sweep rate) is very similar to that for the piecewise harmonic case, with a small difference (less than 2.6%). This means that the semi-analytical optimal resistance is possibly used as an approximate for slowly swept excitation.

As another example, Figure 13 shows the output current responses to piecewise harmonic and swept-sine excitations, which is obtained numerically for the period-3T interwell motion. The forcing frequency of the piecewise base excitation increased from 1.0 by 0.05 to 1.5 for every 1000 cycles. The frequency of the swept-sine excitation increased with the sweep rate of 8.9×10^{-6}. The numerical optimal resistances for the piecewise harmonic and swept-sine excitations are nearly close to each other, approximately 1.056.

The semi-analytical optimal resistance, 1.059, is in good agreement with the numerical results for both the piecewise harmonic and swept-sine excitations.

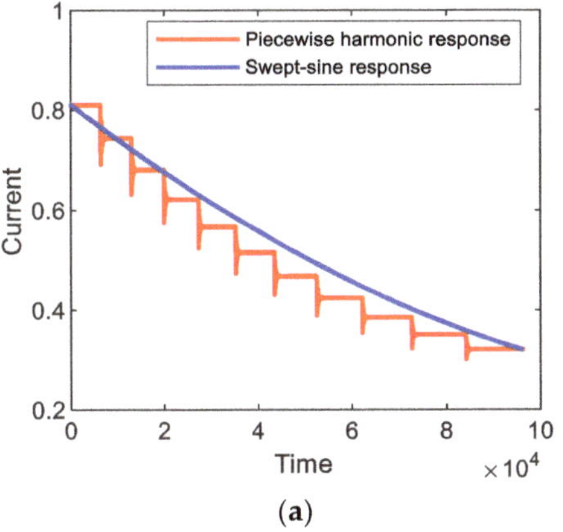
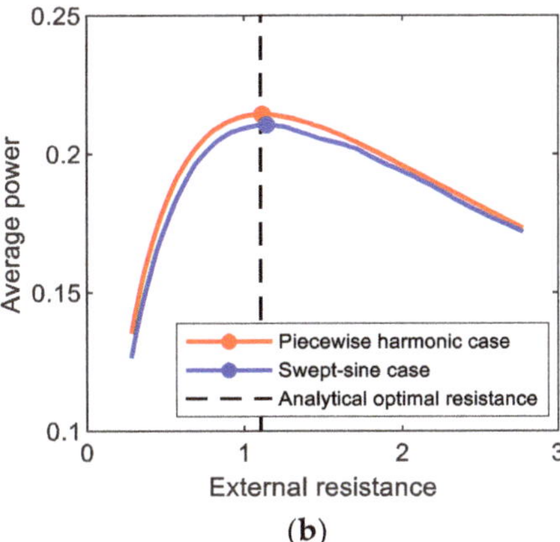

Figure 12. (**a**) Amplitudes of the output current responses to (red line) piecewise harmonic and (blue line) swept-sine excitations, which are obtained for the period-$1T$ interwell motions of the bi-stable electromagnetic energy harvester, and (**b**) average output powers evaluated with the variation of the external load resistance.

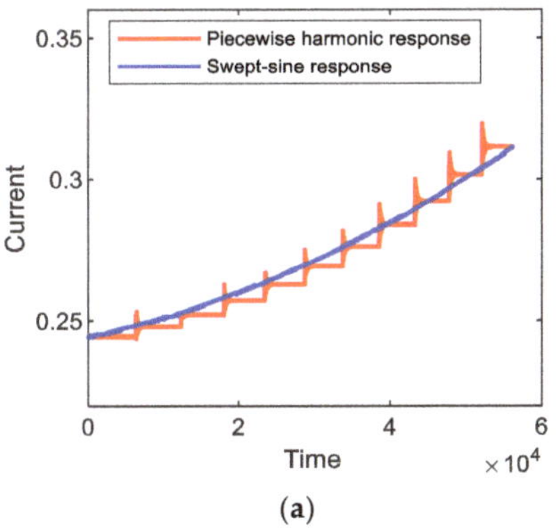
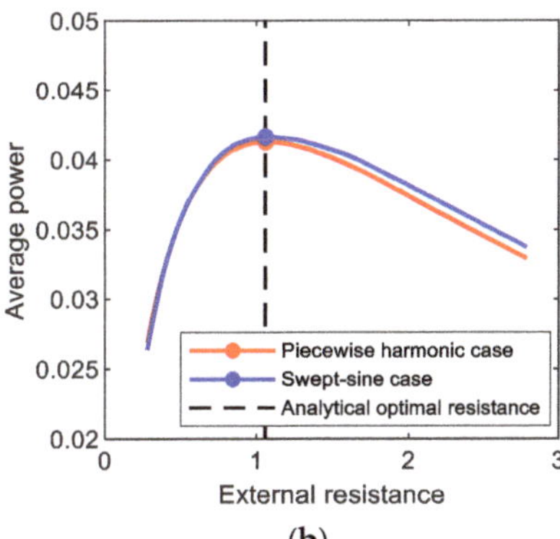

Figure 13. (**a**) Amplitudes of the output current responses to (red line) piecewise harmonic and (blue line) swept-sine excitations, which are obtained for the period-$3T$ interwell motions of the bi-stable electromagnetic energy harvester, and (**b**) average output powers evaluated with the variation of the external load resistance.

7. Conclusions

In this study, a semi-analytic approach to optimizing the external load resistance of an electromagnetic energy harvester was proposed for primary harmonic (period-$1T$) and subharmonic (period-$3T$ and $5T$) interwell motions. The harmonic balance solutions of three interwell motions were obtained first and used to formulate an optimization problem for the external load resistance. The optimal load resistance was obtained, by solving the formulated problem, and investigated. A parametric study was performed to investigate the effects of the system parameters and excitation conditions on the optimal load resistance of the electromagnetic energy harvester. The results showed that for all the period-$1T$, $3T$, and $5T$ motions, the optimal load resistance tended to increase as the inductance of a solenoid coil was larger and as the frequency of the base excitation was higher, but it was not sensitive to the intensity of the base excitation. Additionally, the optimal load resistance significantly depended on the oscillation frequency (or period) of the interwell motion—the higher the oscillation frequency (or the shorter the period) is, the larger the optimal resistance; thus, for the period-1T, 3T, and 5T motions in order. All of the above observations consistently indicate that the effect of the inductive reactance of the solenoid coil on the optimal load resistance becomes significant, particularly when the frequency of ambient vibration is relatively high. In this case, the non-linear dynamic behaviors of the interwell motions and the associated inductive reactance of the solenoid coil should be considered as important design factors in the optimization process of the electromagnetic energy harvester. Therefore, when compared with conventional resistance-matching techniques, the semi-analytic approach presented in this study could provide a more accurate estimation of the optimal load resistance.

Author Contributions: Conceptualization, S.B. and P.K.; methodology, P.K.; software, S.B. and P.K.; validation, S.B.; formal analysis, P.K.; writing—original draft preparation, S.B. and P.K.; supervision, P.K.; project administration, P.K. All authors have read and agreed to the published version of the manuscript.

Funding: This work was supported by the National Research Foundation of Korea (NRF) grant funded by the Korean Government (MSIT) (No. NRF-2019R1C1C1009732). This research was also supported by "Research Base Construction Fund Support Program" funded by Jeonbuk National University in 2017.

Institutional Review Board Statement: Not applicable.

Informed Consent Statement: Not applicable.

Data Availability Statement: Data sharing not applicable.

Conflicts of Interest: The authors declare no conflict of interest.

References

1. Harne, R.L.; Wang, K.W. A review of the recent research on vibration energy harvesting via bistable systems. *Smart Mat. Struct.* **2013**, *22*, 023001. [CrossRef]
2. Nabavi, S.; Zhang, L. Portable Wind Energy Harvesters for Low-Power Applications: A Survey. *Sensors* **2016**, *16*, 1101. [CrossRef] [PubMed]
3. Roundy, S.; Wright, P.K.; Rabaey, J. A study of low level vibrations as a power source for wireless sensor nodes. *Comput. Commun.* **2003**, *26*, 1131–1144. [CrossRef]
4. Mann, B.P.; Sims, N.D. Energy harvesting from the nonlinear oscillations of magnetic levitation. *J. Sound Vib.* **2009**, *319*, 515–530. [CrossRef]
5. Ibrahim, S.W.; Ali, W.G. A review on frequency tuning methods for piezoelectric energy harvesting systems. *J. Renew. Sustain. Energy* **2012**, *4*, 062703. [CrossRef]
6. Zhu, D.; Roberts, S.; Tudor, M.J.; Beeby, S.P. Design and experimental characterization of a tunable vibration-based electromagnetic micro-generator. *Sens. Actuator A Phys.* **2010**, *158*, 284–293. [CrossRef]
7. Sun, W.; Jung, J.; Wang, X.Y.; Kim, P.; Seok, J.; Jang, D.-Y. Design, simulation, and optimization of a frequency-tunable vibration energy harvester that uses a magnetorheological elastomer. *Adv. Mech. Eng.* **2015**, *7*, 147421. [CrossRef]
8. Shahruz, S.M. Limits of performance of mechanical band-pass filters used in energy scavenging. *J. Sound Vib.* **2006**, *293*, 449–461. [CrossRef]
9. Abdelkefi, A.; Najar, F.; Nayfeh, A.H.; Ben Ayed, S. An energy harvester using piezoelectric cantilever beams undergoing coupled bending-torsion vibrations. *Smart Mat. Struct.* **2011**, *20*, 115007. [CrossRef]
10. Wu, H.; Tang, L.; Yang, Y.; Soh, C.K. A novel two-degrees-of-freedom piezoelectric energy harvester. *J. Intell. Mat. Syst. Struct.* **2012**, *24*, 357–368. [CrossRef]
11. Stanton, S.C.; Erturk, A.; Mann, B.P.; Inman, D.J. Nonlinear piezoelectricity in electroelastic energy harvesters: Modeling and experimental identification. *J. Appl. Phys.* **2010**, *108*, 074903. [CrossRef]
12. Kim, P.; Bae, S.; Seok, J. Resonant behaviors of a nonlinear cantilever beam with tip mass subject to an axial force and electrostatic excitation. *Int. J. Mech. Sci.* **2012**, *64*, 232–257. [CrossRef]
13. Abdelkefi, A.; Nayfeh, A.H.; Hajj, M.R. Global nonlinear distributed-parameter model of parametrically excited piezoelectric energy harvesters. *Nonlinear Dyn.* **2012**, *67*, 1147–1160. [CrossRef]
14. Kim, P.; Yoon, Y.-J.; Seok, J. Nonlinear dynamic analyses on a magnetopiezoelastic energy harvester with reversible hysteresis. *Nonlinear Dyn.* **2016**, *83*, 1823–1854. [CrossRef]
15. Cottone, F.; Vocca, H.; Gammaitoni, L. Nonlinear energy harvesting. *Phys. Rev. Lett.* **2009**, *102*, 080601. [CrossRef] [PubMed]

16. Stanton, S.C.; McGehee, C.C.; Mann, B.P. Nonlinear Dynamics for broadband energy harvesting: Investigation of a bistable piezoelectric inertial generator. *Phys. D* **2010**, *239*, 640–653. [CrossRef]
17. Erturk, A.; Inman, D.J. Broadband piezoelectric power generation on high-energy orbits of the bistable Duffing oscillator with electromechanical coupling. *J. Sound Vib.* **2011**, *330*, 2339–2353. [CrossRef]
18. Daqaq, M.F.; Masana, R.; Erturk, A.; Quinn, D.D. On the role of nonlinearities in vibratory energy harvesting: A critical review and discussion. *Appl. Mech. Rev.* **2014**, *66*, 040801. [CrossRef]
19. Nguyen, M.S.; Yoon, Y.-J.; Kwon, O.; Kim, P. Lowering the potential barrier of a bistable energy harvester with mechanically rectified motion of an auxiliary magnet oscillator. *Appl. Phys. Lett.* **2017**, *111*, 253905. [CrossRef]
20. Nguyen, M.S.; Yoon, Y.-J.; Kim, P. Enhanced Broadband Performance of Magnetically Coupled 2-DOF Bistable Energy Harvester with Secondary Intrawell Resonances. *Int. J. Precis Eng. Manuf-Green Technol.* **2019**, *6*, 521–530. [CrossRef]
21. Kim, P.; Seok, J. A multi-stable energy harvester: Dynamic modeling and bifurcation analysis. *J. Sound Vib.* **2014**, *333*, 5525–5547. [CrossRef]
22. Cao, J.; Inman, D.J.; Lin, J.; Liu, S.; Wang, Z. Broadband tristable energy harvester: Modeling and experiment verification. *Appl. Energy* **2014**, *133*, 33–39.
23. Kim, P.; Seok, J. Dynamic and energetic characteristics of a tri-stable magnetopiezoelastic energy harvester. *Mech. Mach. Theory* **2015**, *94*, 41–63. [CrossRef]
24. Jung, J.; Kim, P.; Lee, J.-I.; Seok, J. Nonlinear dynamic and energetic characteristics of piezoelectric energy harvester with two rotatable external magnets. *Int. J. Mech. Sci.* **2015**, *92*, 206–222. [CrossRef]
25. Cao, J.; Zhou, S.; Wang, W.; Lin, J. Influence of potential well depth on nonlinear tristable energy harvesting. *Appl. Phys. Lett.* **2015**, *106*, 173903. [CrossRef]
26. Kim, P.; Son, D.; Seok, J. Triple-well potential with a uniform depth: Advantageous aspects in designing a multistable energy harvester. *Appl. Phys. Lett.* **2016**, *108*, 243902. [CrossRef]
27. Nabavi, S.; Zhang, L. Design and optimization of a low-resonant-frequency piezoelectric MEMS energy harvester based on artificial intelligence. *Proceedings* **2018**, *2*, 930. [CrossRef]
28. Zhang, B.; Zhang, Q.; Wang, W.; Han, J.; Tang, X.; Gu, F.; Ball, A.D. Dynamic modeling and structural optimization of a bistable electromagnetic vibration energy harvester. *Energies* **2019**, *12*, 2410. [CrossRef]
29. Safaei, M.; Sodano, H.A.; Anton, S.R. A review of energy harvesting using piezoelectric materials: State-of-the-art a decade later (2008–2018). *Smart Mat. Struct.* **2019**, *28*, 113001. [CrossRef]
30. Huguet, T.; Badel, A.; Lallart, M. Exploiting bistable oscillator subharmonics for magnified broadband vibration energy harvesting. *Appl. Phys. Lett.* **2017**, *111*, 173905. [CrossRef]
31. Arroyo, E.; Badel, A.; Formosa, F.; Wu, Y.; Qiu, J. Comparison of electromagnetic and piezoelectric vibration energy harvesters: Model and experiments. *Sens. Actuators A Phys.* **2012**, *183*, 148–156. [CrossRef]
32. Huguet, T.; Badel, A.; Druet, O.; Lallart, M. Drastic bandwidth enhancement of bistable energy harvesters: Study of subharmonic behaviors and their stability robustness. *Appl. Energy* **2018**, *226*, 607–617. [CrossRef]
33. Fu, C.; Xu, Y.; Yang, Y.; Lu, K.; Gu, F.; Ball, A. Dynamics analysis of a hollow-shaft rotor system with an open crack under model uncertainties. *Commun. Nonlinear Sci. Numer. Simulat.* **2020**, *83*, 105102. [CrossRef]

Article

A Numerical Model for Experimental Designs of Vibration-Based Leak Detection and Monitoring of Water Pipes Using Piezoelectric Patches

Favour Okosun [1,2,3]**, Mert Celikin** [4,5] **and Vikram Pakrashi** [1,2,3,]*

[1] Dynamical Systems and Risk Laboratory, School of Mechanical and Materials Engineering, University College Dublin, D04 Dublin, Ireland; favour.okosun@ucdconnect.ie

[2] Science Foundation Ireland (SFI) MaREI Centre, University College Dublin, D04 Dublin, Ireland

[3] The Energy Institute, University College Dublin, D04 Dublin, Ireland

[4] Materials Design and Processing Laboratory, School of Mechanical and Materials Engineering, University College Dublin, D04 Dublin, Ireland; mert.celikin@ucd.ie

[5] I- Form, the SFI Research Centre for Advanced Manufacturing, D04 Dublin, Ireland

* Correspondence: vikram.pakrashi@ucd.ie

Received: 3 November 2020; Accepted: 21 November 2020; Published: 24 November 2020

Abstract: While the potential use of energy harvesters as structural health monitors show promise, numerical models related to the design, deployment and performance of such monitors often present significant challenges. One such challenge lies in the problem of leak detection in fluid-carrying pipes. Recent advances in experimental studies on energy harvesters for such monitoring has been promising but there is a paucity in existing literature in linking relevant fluid–structure interaction models around such applications. This paper addresses the abovementioned issue by developing a numerical model with Computational Fluid Dynamics (CFD) and Finite Element (FE) tools and carries out extensive analyses to compare it with existing experiments under controlled laboratory conditions. Conventional Polyvinylidene Fluoride (PVDF) films for leak detection and monitoring of water pipes were considered in this regard. The work provides guidelines on parameter selection and modeling for experimental design and repeatability of results for these types of experiments in future, around the demands of leak monitoring. The usefulness of such models is also demonstrated through the ability to estimate the optimum distribution frequency of these sensors that will enable the detection of the smallest leak of consequence under a known or established flow condition.

Keywords: PVDF patches; structural health monitoring; sensing; energy harvesting; pipe leak detection; computational fluid dynamics; optimum sensor distribution

---

1. Introduction

Vibration leak detection methods have been identified as effective for early leak detection in pipes. They are a popular choice for any leak detection set-up because they are non-invasive and more suited for monitoring than inspection [1–3]. The principle of vibration-based leak detection is anchored to the Fluid–Structure Interaction (FSI) and Negative Pressure Wave Propagation Attenuation Mechanisms (NPWPAM) phenomena [4–6]. Research has been carried out using commercially available accelerometers as sensors for vibration pipe leak detection [7,8]. However, there are some established downsides to their use for such applications, ranging from them being costly [9,10], requiring an external power source to operate, to being generally rigid, making it difficult to achieve excellent conformance with the cylindrical pipes they are bonded to [11]. The need for cheap, output-only and flexible vibration sensors for pipe leak detection is what motivated the development of patches made from piezoelectric materials as alternative sensors to commercial accelerometers. PVDF patches are relatively cheaper (when compared to commercial accelerometers), flexible and responsive to leak-induced changes to pipe surface vibration levels. Okosun et al. 2018 [12], presented the fabrication experimental validation for metal pipe leak detection and monitoring of these PVDF patches. Despite the opportunities presented by various experiments, there is a gap in the literature around numerical modeling for such systems. Development of a reasonable fluid–structure interaction model connected to the energy harvesting based monitoring framework can guide future experiments and also help in sensor placement strategies. This paper addresses this gap and puts current experimental results and future experimental designs into context.

Obtaining the pipe surface vibration levels numerically for the healthy pipe state and a number of leak states are essential for experimental design and numerical validation of the PVDF patches for water pipe leak detection, and this task is a two-pronged complex problem. The first phase of the problem deals with the turbulent flow dynamics, Fluid–Structure Interaction (FSI), and leak induced Negative Pressure Wave (NPW; for leak pipe states) with the output from this step being the internal pipe wall pressure fluctuations. The second addresses the propagation of resulting internal pipe wall pressure fluctuation to the external pipe surface exciting vibration response with the output being either pipe external surface strain or acceleration (in this case strain, since piezoelectric patches are strain based vibration sensors). After obtaining the pipe surface strain level for the all the simulation cases, and knowing the properties of the PVDF patches, the theoretical voltage output from the PVDF patches can be calculated for positions of interest along the pipe length, using the already established strain–voltage relationship for PVDF films [13].

In addition to the numerical validation of these piezoelectric patch sensors, this paper also presents a numerical methodology for determining the optimum frequency distribution of these sensors (i.e., their maximum distance apart) that will enable the detection of the smallest conceivable leak of consequence under any known/established flow condition. To be able to detect a small pipe leak under low flow rate conditions using vibration patch sensors, the distance between two sensors must be equal to or less than the length of the portion of the pipe that will be influenced by the induced NPW due to the onset of the leak at that flow condition. Hence, the lowest flow rate, the smallest expected leak size, and the length of pipe influenced by the leak-induced NPW before its complete decay are crucial information for determining the distribution frequency of sensors adopted for a pipe leak vibration monitoring application. Such guidance cannot be obtained from the experimental validation exercise presented in [12] because of the limited size of a typical fluid test rig. The test pipe section of the rig consisted of 100 cm long steel test pipes and the influence of the smallest test leak size (a 2 mm hole) travelled through the entire length of the pipe irrespective of the position the leak was introduced along the pipe length (details of this test rig can be found in [12]).

The results and findings from this paper guide the procedure of creating a numerical framework for interpretation of existing or earlier experiments within the context of the Fluid–Structure Interaction for pipeline leaks in various sectors of application. The work also helps in designing future experiments and provides some quantitative estimates on the choice of parameters for modeling, measuring and comparing along with their quantitative. Finally, the work can be used to obtain the minimum number of sensors required to detect a certain level of leakage, for a given flow-rate. The findings can be easily adapted to a range of sensors and can thus be useful for development of novel sensors and measurement chains around this topic.

2. Modeling Turbulent Fluid Flow

2.1. Detection Context

Fluid flow regimes are mainly laminar or turbulent [14]. Laminar flow is characterized by fluid particles flowing in orderly streamlines, with each layer moving smoothly past the adjacent layer with little or no mixing, whereas turbulent flow is chaotic where the fluid particles have random motions in all three dimensions. Turbulence leads to irregular and unsteady flow dynamics characterized by the fluctuations of transported quantities (mass, momentum and scalar species), in time and space. In turbulent flow, eddies or vortices are generated by the relative motion of fluids near the boundary layer. These eddies are characterized by identifiable swirling patterns and the energy dissipated by these eddies is converted to heat and wall pressure [15].

The ratio of inertial forces to viscous forces within a flowing fluid, can tell if the flow is laminar or turbulent. This ratio is given by a dimensionless quantity called the Reynolds number (Re) [16], and the relationship is given below.

$$Re = \frac{\rho U d}{\mu} \tag{1}$$

where ρ is the fluid density, U is the mean velocity of the flow, d is the diameter of the pipe inner cross-section and μ is the fluid dynamic viscosity.

Laminar flows can be described completely mathematically by the continuity, Navier–Stokes, energy conservation equations and the equation of state. However, in the case of turbulent flows, in addition to the aforementioned equations, the turbulence transport properties must also be accounted for [17]. Commercial Computational Fluid Dynamics (CFD) codes provide models that utilizes additional terms (other than those provided by the governing CFD equations) to account for these transport properties but great care has to be taken in the modeling of the problem, which can be a rigorous undertaking. Flows in pipes in real life applications are normally turbulent, which is why they are very complicated [18–20]. Fluid Dynamics problems involving flow-induced pressure fluctuations or wall pressure fluctuations caused by turbulence are very complex, often proving difficult to model and solve. It exists over a range of frequencies; hence, it can be termed a broadband phenomenon [21]. This wall pressure fluctuation is the desired output from the first phase of the simulation process in the validation of PVDF patch sensors for pipe leak monitoring and the subsequent section of this paper will provide the details of the solution methodology adopted.

2.2. Overview of Modeling Methods

In practice, there are three main methods for the analysis of turbulent flows in commercial CFD codes, namely: Direct Numerical Simulation (DNS), Large Eddy Simulation (LES) and Reynolds-Averaged Navier-Stokes (RANS) models. With DNS being the most accurate and RANS the least accurate.

Direct numerical simulation (DNS) solves these equations numerically in a rigorous way to a desired accuracy without any additional model or correlation. However, its application is still limited because existent problems require a large amount of computational resources, exceeding the capacity of conventional computers, hence it is not a very practical model. For this reason, DNS was not employed in solving the FSI problem of this research [16,18].

Reynolds Averaged Navier–Stokes (RANS) employs equations in modeling the turbulent flow. These models do not provide instantaneous values for the flow and are based on time averages, e.g., they do not compute the pressure fluctuations at the Fluid-Structure Interface [22,23]. The k–ε model and the k–ω model are the commonly used RANS-based two equations turbulence models. The two extra transport equations accounts for the turbulent properties of the flow. Depending on the chosen model, the transported variables are most often the turbulent kinetic energy k and turbulent dissipation rate ε or specific turbulence dissipation rate ω. The scale and energy of turbulence are determined by solving the two transport equations [24]. The k–ω model has an advantage of having an improved performance for near wall boundary layer regions of the flow under adverse pressure gradients when compared to the k–ε model. The k–ε model on the other hand, is more robust in the free shear flows and mainstream regions [25]. An integrated model that takes advantages of both models is known as the shear stress transport k–ω (SST k–ω) model [20]. The SST k–ω turbulence model operates by employing the k–ω model in the near-wall region and by employing a blending function, switches to the k–ε model in the free shear flow turbulent region [26].

In this study, internal pipe wall pressure fluctuations or variations is the output quantity of interest from the CFD simulations and it is required as input for the second phase of FE simulations to obtain the pipe surface vibrations, hence, RANS based models cannot accomplish the central purpose of this study. However, the SST k–ω RANS model was employed in the process of selection of a mesh for the pipe models before simulation, as one of the selection criteria requires time-averaged solutions.

As described above, the reasons that DNS and the RANS models cannot be employed for obtaining the internal pipe wall pressure fluctuations from the turbulent fluid flow simulation are clear. Here, the LES model, which models the actual physics of the flow better when compared to the RANS models was employed. The LES approach is a hybrid model derived from a combination of DNS and the RANS models. In contrast to a time-averaged approach, LES provides a model that computes the instantaneous velocity and pressure field, in contrast to the time-averaged approach adopted by RANS and it is not as computationally expensive as DNS [16]. In LES, the flow is resolved to a characteristic scale, usually taken to be the size of the grid, and then modeled on the smaller scales. The idea for the LES model stems from the fact that large eddies possess an anisotropic behavior and at the smallest scales, the turbulence is isotropic. Hence, while the large eddies need to be resolved the smallest scales can be solved adequately statistically. Grid scales (GSs) are length scales the size of the grid or larger and scales smaller than that are referred to as subgrid scales (SGSs). The model of a turbulent flow problem should be such that the grid spacing results in most of the total turbulent kinetic energy contained in the large eddies being directly computed, and the remaining fraction of the kinetic energy that is not resolved to the GS modeled [27]. A variety of SGS eddy viscosity models for LES have been detailed in literature including the Wall Adapting Local Eddy-Viscosity (WALE) model, the Smagorinsky and Smagorinsky–Lilly models [16,20]. The LES WALE model was employed for the LES simulation runs in this study since it is known that it performs significantly better in the near wall and boundary layer region when compared to Smagorinsky models [28,29].

In this study, the commercial codes employed were ANSYS FLUENT for the CFD (first phase) simulations and ANSYS Transient Structural for the (FE second phase) simulations. The governing equations and basic formulations of the LES model, which is the primary turbulence model employed in this study can be found in [28,30]. That of the SST k–ω RANS model employed in the mesh selection process can be found in [20,26].

2.3. Important Parameters in Turbulent Flow Modeling

To get accurate results from the Fluent turbulent simulation, the pipe flow model and simulation set up must at least come close to satisfying certain conditions. There are some parameters that can guide the preparation and validation of the model before it is employed for simulation runs.

2.3.1. Length of the Pipe Domain and Near-Wall Treatment

One of such parameters is the length of the streamwise pipe domain. For a good FLUENT solution, it is advised that the flow should be fully developed. For this to occur, the length of pipe should be at least 5 times the internal or hydraulic pipe diameter [31]. Other important considerations are the inner wall coordinate (y^{+}) of the first mesh cell from the pipe wall and fineness of the mesh in the boundary layer regions (near the wall) [32].

$$y^{+} = \frac{u^{*}y}{v} \tag{2}$$

where u^{*} is the friction velocity at the wall, y is the normal distance from the wall and v is the kinematic viscosity.

$$u^{*} = \sqrt{\frac{\tau_w}{\rho}} \tag{3}$$

where, τ_w is the wall shear stress and ρ is the fluid density.

$$\tau_w = 0.5 C_f \rho U^2 \tag{4}$$

where C_f is the skin friction coefficient and U is the average fluid velocity. For pipe flow, C_f is given as:

$$C_f = 0.027 Re^{-\frac{1}{7}} \tag{5}$$

To deal with near wall turbulence, the way the near wall flow is treated is important, and this is done using wall functions.

The dimensionless velocity u^+ is related to the inner wall coordinate, y^+.

$$u^+ = F\!\left(y^+\right) \tag{6}$$

$$u^+ = \frac{u}{u^*} \tag{7}$$

where, u is the local velocity.

The laminar sublayer is characterized by small values of y^+, i.e., $y^+ < 5$, and in this region the velocity reduces to:

$$u^+ = y^+ \tag{8}$$

For larger $y^+ > 30$, the velocity is given as:

$$u^+ = \frac{1}{\kappa} \ln\!\left(y^+\right) + B \tag{9}$$

where $\kappa = 0.419$ (Von Karman constant) and $B = 5.1$.

The local y^+ values determines the layer the first local mesh cell is located and its distance away from the pipe wall, hence the pipe wall treatment applied. For $y^+ < 5$, the first cell away from the pipe wall is in the laminar sublayer known as the linear region, for $5 < y^+ < 30$, it is in the buffer region and for $y^+ > 30$, it is in the mainstream layer of flow that is predominantly turbulent. This region ($y^+ > 30$) is known as the log-law layer due to the logarithmic relationship between u^+ and y^+. The buffer region is influenced both by the linear and logarithmic regions.

There are two common choices for the wall function: standard wall function and the enhanced wall function. In the standard wall function, the first grid is located within the range $30 < y^+ < 150$ and it is in the predominantly turbulent layer. This wall function is employed in simulations where the flow model is large with a very high Reynolds number, making it difficult for the turbulent boundary layer to be resolved due to lack of computer resources. For the enhanced wall function, at least 10 cells should be in the viscosity affected region (laminar sublayer) to be able to resolve it and the first cell should be in the order of y^+ almost equal to 1 [33].

2.3.2. Mesh Grid Size

To resolve the high energy containing eddies using the LES turbulent model, the mesh of the fluid flow model must be sufficiently fine. Therefore, the meshing of the turbulent flow problem in LES is crucial as it has significant influence on the results. The size of the largest eddies is described by the turbulent length scale, L_t, in pipe flow problems, hence, L_t, must be considered in deciding the grid size. L_t is approximately 7% of the diameter of the pipe inner cross-section. Eddies of roughly half the size of the turbulent length scale must be resolved to resolve 80% of the turbulent kinetic energy [26]. Hence, the turbulent length scale should serve as a guide in determining the mesh grid size.

2.3.3. LES Time-Step and Courant Number

In selecting transient simulation conditions like the time-step, it is important to consider the characteristic time of transit of a fluid element across a volume. To thoroughly resolve a turbulent flow, the ratio of the time step to the time of transit of the fluid element known as the courant number should be less than 1. The courant number, which provides insights of the fluid movement through the computational cells, is a dimensionless quantity, and can be calculated from:

$$Courant = \frac{\Delta t}{\Delta x_{cell}} U \tag{10}$$

where, Δt is the time step size, U is the average fluid velocity and Δx_{cell} is the length of the mesh cell.

A courant number ≤ 1 means that within one-time step (at most), the fluid particles move from one cell to another. This is the ideal case scenario; hence the time step should be chosen such that courant < 1. The two options for reducing the time step if it is greater than one is reducing the time-step and/or coarsening the mesh, if possible.

3. Methods

This section first presents an overview of the scope of simulations in the numerical model, identifying the range of flow rates, defining the healthy benchmark and outlining the types of analyses that are carried out subsequently. The validation of the selected mesh is presented next and the wall function of the models is established. Matching of pressure drop through modeling is carried out next, before the final choice of mesh. Considerations of turbulence and mass flow rates due to the leak are considered next. Finally, the vibrations from fluid flow are estimated from the Fluid-Structure Interaction model and subsequently converted to pipe strains, and linked to energy harvesting.

3.1. Overview of Numerical Modeling

For the numerical validation of the PVDF patches, the pipe model and simulation conditions were designed to replicate the experimental conditions detailed in [12]. Hence, pipe flow simulations were carried out on 5 states of the pipe i.e., the healthy pipe, 2 mm, 5 mm, 7 mm leak and 10 mm leak states. These simulations were done at 5 different flow rates (ranging from high to low) of 90.85 L/min (24 gpm), 71.92 L/min (19 gpm), 56.78 L/min (15 gpm), 45.42 L/min (12 gpm) and 26.50 L/min (7 gpm), for each pipe state. This resulted in a total of 25 scenarios and independent simulations. Similar to the experimental campaign, the material and properties specified for the pipe model was that of galvanized steel with a length of 1 m. Before the simulation runs, three different pipe mesh models were prepared for the healthy pipe state and a model selection exercise was performed to determine the model, if any, that is best suited for the exercise.

Three healthy pipe models with different mesh models were prepared and an analysis was carried out to validate these meshes and determine the best one to be adopted for the simulation runs. Simulation runs were then performed on the chosen pipe flow model using the LES WALE model of ANSYS FLUENT for the healthy state and leak pipe states of the pipe at various flow conditions. On obtaining the solution of the flow field, the pressure field on the pipe wall was exported to a Finite Element (FE) package (ANSYS Transient Structural) to calculate the pipe response in the form of the pipe surface strain. This pipe surface strain at the positions along the pipe length where the PVDF patch sensors were bonded in the experimental campaign is then used to calculate the theoretical voltage output of the PVDF patch sensors.

3.2. Mesh Validation and Selection

Three different pipe mesh models were prepared for the healthy pipe state and a model selection exercise was performed to determine the model, if any, that is best suited for the exercise. The validation exercise was carried out for the healthy pipe state only, at all 5 test flow rates, therefore, it involved 5 simulation scenarios out of the 25 total scenarios. In addition to the important modeling parameters mentioned in the previous section, a comparison of the static pressure gradients (ΔP) from each validation simulation and all three models with the theoretical pressure gradient obtained using the Darcy-Weisbach equation was employed in the mesh selection process. The Darcy-Weisbach equation, given below, relates the pressure loss, due to friction along the length of the pipe to the average velocity of the fluid flow for an incompressible fluid.

$$\frac{\Delta P}{L} = f\frac{\rho U^2}{2D} \tag{11}$$

where L is the length of the pipe, D is the internal pipe diameter, U is the mean flow velocity and f is the Darcy friction factor, obtained from the Colebrook equation given below:

$$\frac{1}{\sqrt{f}} = -0.869 \ln\left(\frac{e_{roughness}}{3.7D} + \frac{2.523}{Re\sqrt{f}}\right) \tag{12}$$

where $e_{roughness}$ is the pipe wall roughness, which is equal to 1.5×10^{-4} m for the galvanized steel pipe material.

The closer the FLUENT pressure gradient is to the theoretical pressure gradient for the different test scenarios, the more suitable the mesh model is. The Darcy–Weisbach equation is only valid for the steady state fully developed pipe flow, hence the SST k–ω RANS model was employed for the mesh selection exercise as LES models are employed for transient simulations. This section of the paper discusses the pipe model and mesh preparation, the conditions of the simulations, the setup for the SST k–ω RANS model employed for this exercise, and comparison of the simulation results of the three mesh models.

The operational pipe model and meshing was done in ANSYS ICEM CFD. ICEM CFD provides advanced geometry/mesh generation and mesh diagnosis and repair functions necessary for the in-depth analysis [34]. The pipe material was galvanized steel with properties presented in Table 1 below. Bearing the important turbulent model parameters in mind, three mesh models of the healthy pipe were prepared such that each model had cell elements starting at different distances from the pipe wall. The models also had varying sizes of the mesh grid and different mesh growth rate as you move from the pipe wall to the main turbulent flow stream. Simulations were performed on these models using the SST k–ω RANS model at the 5 different flow rates.

Table 1. Properties of the pipe material.

Parameter	Value	Unit
Internal pipe diameter	37.3	mm
Pipe wall thickness	2.5	mm
Bulk Modulus	160	GPa
Modulus of Elasticity	200	GPa
Poisson ratio	0.29	NA
Density	7850	kg/m^3

For the setup of each model, the fluid domain was set to be water (density of 1000 kg/m^3 and dynamic viscosity of 1.0×10^{-3} Ns/m^2). The velocity-inlet Boundary Condition (BC) was selected for the pipe inlet, the value of velocity of flow is calculated from the flow rate for each of the simulation scenario. The pressure outflow BC was selected for the pipe outlet. This boundary condition was selected over the pressure outlet BC because it allows Fluent to calculate the pressure gradient (ΔP) along the pipe length without imposing a pressure value at the outlet, which the pressure outlet BC requires. This provided a better representation of the experimental setup that we are hoping our model replicates. This is because, in the experimental validation setup [12], the test pipe outlet does not discharge to the atmosphere, as the water returns back to the reservoir and is circulated in a cyclic manner, hence, we cannot specify a gauge pressure of zero as the outlet pressure. Additionally, although there is a pressure gauge at the outlet of the test pipe section of the test rig, the resolution of the gauge and errors associated with such measuring devices means that whatever pressure read-out by the gauge is only an approximation of the actual outlet pressure. So, since the output of interest was not in the point-by-point static pressure along the pipe length but rather the pressure gradient along the length of the pipe, it was decided that it was best to use the pressure outflow BC and allow Fluent calculate this pressure gradient. Stationary wall with "no slip" was employed for the pipe wall.

Table 2 below provides details of the mesh of the three candidate mesh models being evaluated. From the table, HEX cells and QUAD faces refers to meshing elements hexahedral cells and quadrilateral faces, respectively. The table shows that mesh model 3 had the finest meshing with the most nodes and boundary faces while mesh model 1 had the coarsest meshing.

Table 2. Mesh description of candidate models.

Mesh	Fluid Domain	Interior Faces	Inlet Faces	Outlet Faces	Wall Faces
	(HEX Cells)			(QUAD Faces)	
Model 1	496,545	1,187,992	1754	1754	21,368
Model 2	671,553	1,897,520	2697	2697	7652
Model 3	787,089	2,343,664	3161	3161	38,884

After running the simulations for the 3 models at all 5 flow rates, the wall function of the models was established from the simulation results and this was the first consideration in the mesh selection process. The wall y^+ values are directly proportional to the average velocity of flow (see Equation (2)), hence results from the simulation at the highest flow rate of 90.85 L/min will be employed for ascertaining the maximum possible y^+ values and the wall-function applied by the candidate models. The contours of wall y^+ for the pipe models is shown in Figure 1 below.

From Figure 1 above, Mesh 1 had y^+ values in the range of about 31–51, meaning that the closest cells to the pipe wall for this mesh is in the turbulence dominant region of the flow, hence, near wall treatment was what was applied for this model. Mesh 2 had y^+ values in the range of about 6.5–11, meaning that the closest cells to the pipe wall were in the buffer region, and Mesh 3 had values in the range of about 0.16–1, meaning that the closest cells to the pipe wall was in the laminar sublayer, and consequently satisfying the conditions for enhanced wall treatment, even at the highest simulation flow rate. Based on these findings, Mesh 3 best satisfied this selection criteria as the objective was to implement an enhanced wall treatment wall function in the model to better resolve the turbulence close to the pipe wall and obtain more accurate wall pressure fluctuations.

A straight line through the centre of the internal pipe wall along the entire pipe length is shown in Figure 2. Figure 3 subsequently presents curves of wall y^+ for model 3 and all 5 test flow rates. These curves show the wall y^+ values recorded along below.

Figure 3 shows that the range of y^+ values recorded along the pipe wall for model 3 decreased with decreasing mean flow velocity, justifying the earlier assertion that the maximum possible y^+ values will be obtained from the simulation at the highest flow rate. Additionally, it can also be seen from the figure that the y^+ values was highest at the pipe inlet and it fell steeply just after the pipe inlet as you move along the pipe length until it became almost stable. This is because y^+ values tend to be highest in the developing regions of the flow, hence it is highest at the uniform velocity pipe inlet and decreases as the velocity profile develops [35].

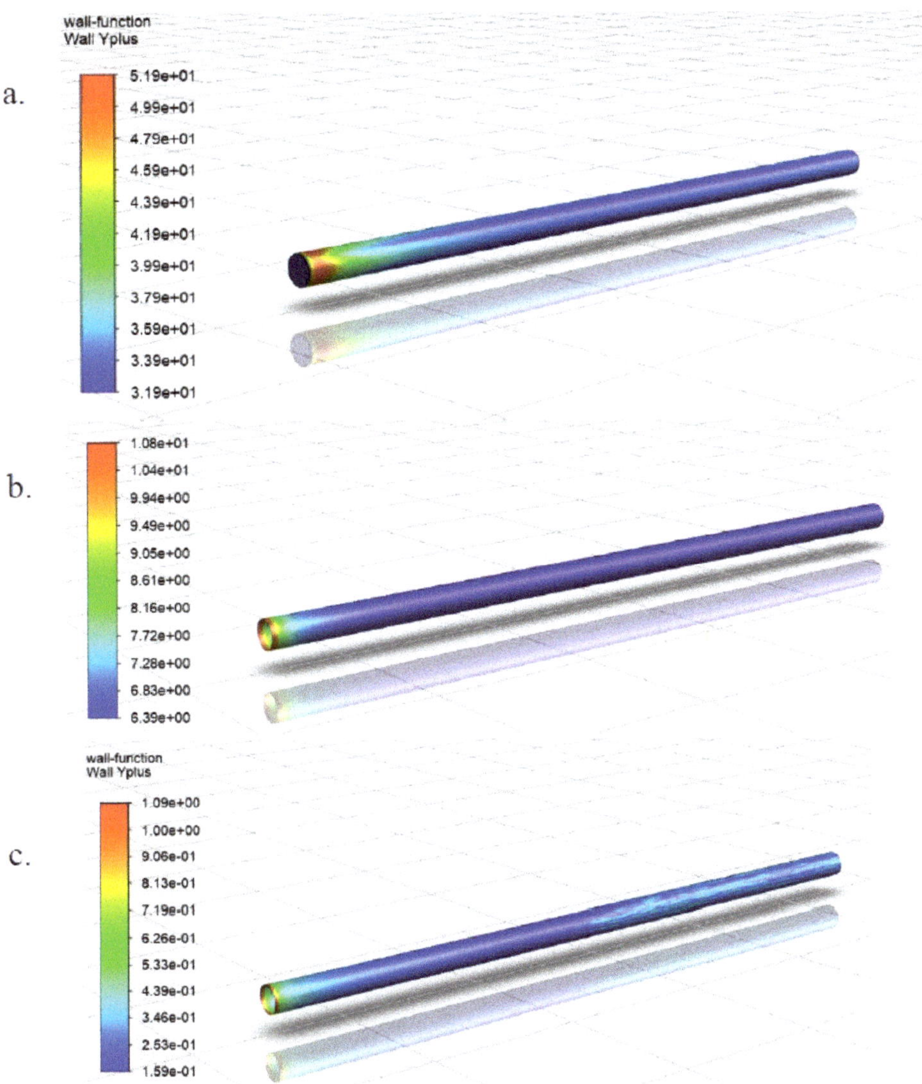

Figure 1. Wall y^+ contours at highest test flow rate of 90.85 L/min for: (**a**) Mesh Model 1; (**b**) Mesh Model 2 and (**c**) Mesh Model 3.

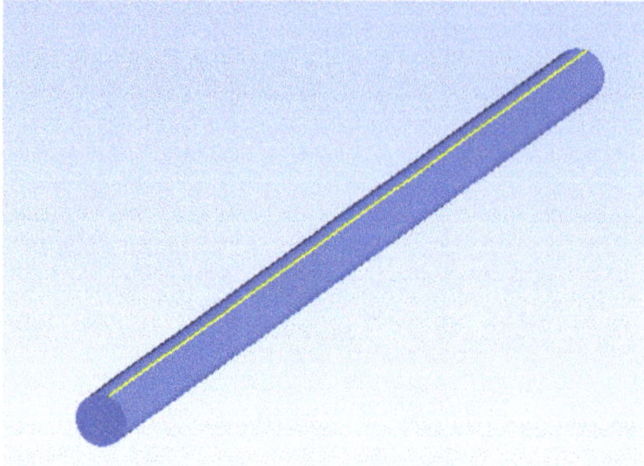

Figure 2. Straight line through the centre of the internal pipe wall.

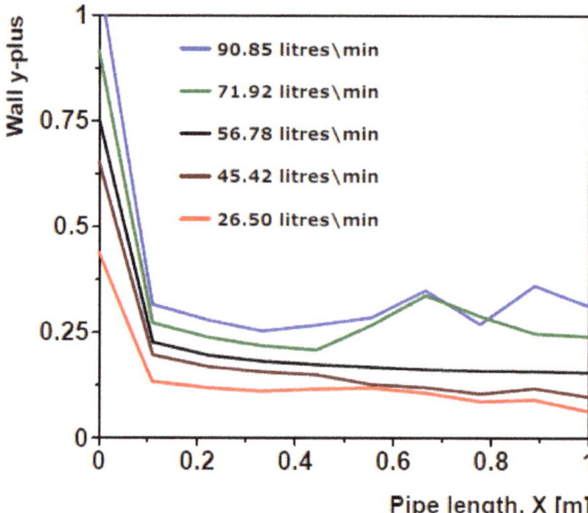

Figure 3. Wall y^+ curves for mesh model 3 at all simulation flow rates.

3.3. Consideration of Pressure Drop

The second consideration in this selection exercise involved calculating the theoretical pressure drop (ΔP) due to friction along the pipe length for the 5 flow rates from Equations (11) and (12), and comparing the values to the pressure gradient obtained for the three models for the 5 simulation scenarios. In the calculation of the theoretical pressure gradient, the Darcy friction factor, f was first solved for implicitly using Equation (12). Figure 4 below shows representative contour diagrams from the lowest simulation flow rate for all three candidate models.

From these contours, the fluent pressure drop was calculated for all test flow rates (including the higher flow rate simulation cases whose contour diagrams were not presented). The table below shows a summary of the theoretical and Fluent pressure gradients for all simulation cases and candidate models.

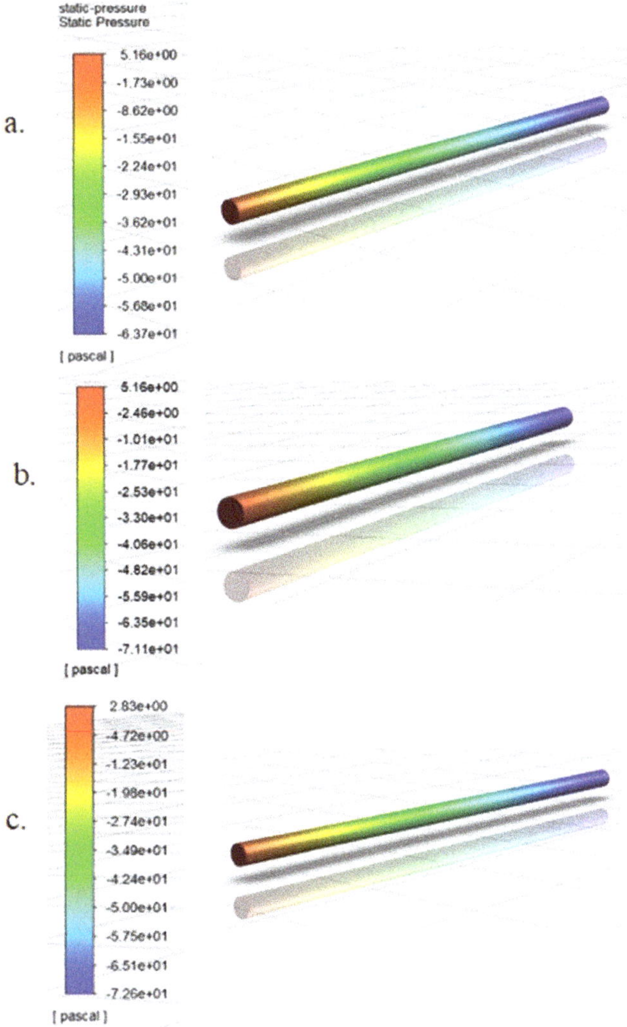

Figure 4. Static pressure contours at the lowest simulation flow rate of 26.50 L/min for: (**a**) Mesh Model 1; (**b**) Mesh Model 2 and (**c**) Mesh Model 3.

Table 3 presents the theoretical pressure gradients (Theo. ΔP) for all the flow rates, the simulation pressure gradients (FLUENT ΔP) for the three candidate mesh models, andquantifies the percentage deviation (in parenthesis) of the FLUENT ΔP from the theoretical ΔP for all simulation cases and mesh models. From the table, the simulation pressure gradients obtained from the results of mesh model 3 had the least deviation from the theoretical pressure gradient for all simulation cases while Mesh 1 consistently registered the highest deviation. The values from mesh 3 closely matched the theoretical values that the maximum deviation recorded was under 1%. The deviation for mesh 2 was consistently higher than that of mesh 3, but less than that of mesh 1. Another noteworthy feature of the table is that the Darcy friction factor (f) slightly increased with decreasing flow velocity. This is because f is inversely proportional to the Reynolds number. However, this does not translate to an increase in the pressure drop (ΔP) with decreasing flow rates or Reynolds number. The reality is quite the contrary, as (ΔP) will continue to decrease with decreasing flow rates, because the flow velocity and turbulence that is directly proportional to ΔP has a higher influence on frictional losses than the friction factor, f. ΔP varies directly with the second power of the fluid velocity and the first power of f (see Equation (11)). Additionally, the friction factor is analogous kinetic or sliding friction, which is much less than static friction [36,37]. Table 3 also shows the Reynolds number and friction factor values at each simulation flow rate, the Reynolds numbers were calculated from Equation (1) and it shows that the flow is turbulent ($Re > 4200$) for all simulation cases.

Table 3. Comparison of theoretical and simulation pressure gradients.

Flow Rate	Re	f	Theo. ΔP	FLUENT ΔP (Pa) Outlet Faces		
(liters/min)			(Pa)	Mesh 1	Mesh 2	Mesh 3
90.85	51,686.80	0.0305	785.05	751.26 (4.30%)	777.62 (0.94%)	783.88 (0.15%)
71.92	40,918.72	0.0309	498.46	462.52 (7.21%)	483.00 (3.10%)	499.61 (0.23%)
56.78	32,304.35	0.0315	316.74	287.88 (9.11%)	303.27 (4.26%)	317.94 (0.38%)
45.42	25,843.47	0.0322	207.23	193.91 (6.43%)	213.80 (3.17%)	209.13 (0.92%)
26.50	15,075.35	0.0342	74.90	68.85 (8.07%)	76.26 (1.81%)	75.43 (0.71%)

3.4. Choice of Mesh

From the result of the mesh validation exercise, it is clear that mesh 3 (the finest mesh) satisfied both selection criteria best and will afford the best opportunity of obtaining accurate results from the simulations. The performance of mesh 2 was fair, but it did not satisfy the enhanced wall treatment condition and the difference in the number of cells and nodes when compared to mesh 3 was not large enough to present a substantial gain in computation time. Hence, mesh 3 was selected for this exercise. Figure 5 below presents simple wire-frame display of the selected mesh model.

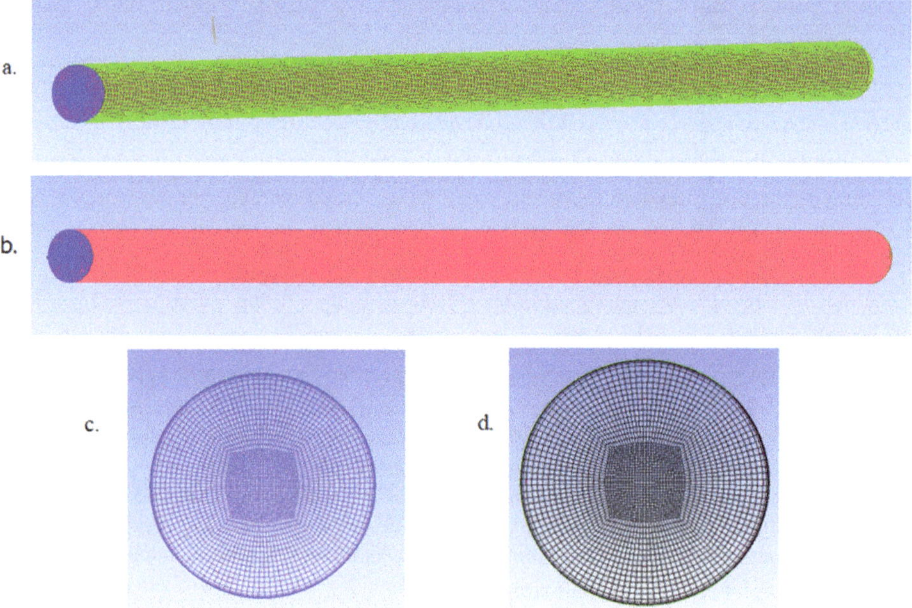

Figure 5. Selected pipe mesh model (mesh 3) showing: (**a**) the internal pipe wall and inlet; (**b**) fluid domain (pipe wall suppressed); (**c**) pipe inlet and (**d**) pipe outlet.

A close look at Figure 5c,d shows a concentration of mesh cells close to the pipe wall. It is also worthy to note that the largest cell in the selected mesh model, i.e., model 3, had a maximum volume of 1.165×10^{-9} m^3, which gives a maximum length dimension of 1.05×10^{-3} m. The recommendation for the maximum length of a node as stated earlier is that it should not be more than half of the turbulent length scale, $L_t = 0.07D$. For this pipe flow model, half of L_t was 1.31×10^{-3} m, which was more than the maximum length of any cell in the mesh model. This further validated the selected model as ideal for this exercise.

3.5. LES Setup to Obtain Internal Pipe Wall Pressure Fluctuations

After selection of the mesh model, the first phase for the validation of the PVDF sensors was executed using the LES WALE turbulence model. As earlier stated, the LES turbulent model is the best practical model for solving transient turbulence problems. This phase involved running simulations for the healthy state pipe and damaged state pipes for all 5 flow rates to obtain the internal wall pressure fluctuations due to FSI for the healthy state pipe and FSI and NPW for the leak state pipe. The idea here is that the influence of the leak will cause an increase in the pipe pressure fluctuation due to the leak induced NPW, and this increase in NPW will in turn reflect as an increase in the pipe surface strain for the leak state pipe when compared to the healthy state pipe at the same flow rate.

For the simulations, the healthy pipe model was that of the selected mesh model and the BCs also remained the same as those employed in the mesh selection exercise. For the leak states of the pipe, leak being represented as 2 mm, 5 mm, 7 mm and 10 mm diameter circular holes were introduced at the 60 cm mark along the pipe length of the selected model without really changing the other model mesh properties, the only modification from the healthy state pipe model is that the mesh nodes closer to the leak was made finer. The leaks have a small leak wall of 2.5 mm, which is the same as the thickness of the test pipe, this is because they were modeled such that they represent the physical pipe leak state as best as possible. Consequently, in addition to the inlet, outlet and pipe wall boundaries, there were two additional boundaries, namely the leak wall, and the leak outlet. The BC for the leak wall was the same as the pipe wall, i.e., stationary wall with no slip, and mass flow outlet BC was employed for the leak outlet. Since the leak discharges to the atmosphere, atmospheric condition was selected as the operating condition at the leak outlet. The velocity-inlet and pressure outflow BCs remained the pipe inlet and pipe outlet BCs.

The theoretical leak mass flow rates (Q_l) employed for the simulations were derived from the leak orifice equation presented in Equation (13) below [38].

$$Q_l = \frac{C_K A_K \sqrt{\rho\left[8A^2(\overline{P}_0 - \left(\frac{l}{L_p}\right)(\overline{P}_0 - \overline{P}_{L_p}) + C_K^2 A_K^2 \rho a^2\right] - C_K^2 A_K^2 \rho a}}{2A} - 2\frac{A}{a}P_g \tag{13}$$

where C_K and A_K are the discharge coefficient and area of the leakage orifice respectively, A is the area of the pipe inner cross-section, ρ is the density of fluid in the pipe and a is the speed of propagation of the NPW in the pipe medium. $\overline{P}_0$ and $\overline{P}_{L_p}$ are the steady pressures at inlet and outlet of the pipeline before leakage respectively, L_P is the length of the pipeline, l denotes the distance of the leakage site from the inlet and P_g denotes the pressure relative to barometric pressure around the outside pipe wall (which is zero in this case as the leak discharges to the atmosphere).

The relationship for calculating a in the pipe medium can be found in [38], and was determined as 1.383 km/s. For pipe flow, $C_K = 0.6$ [39]. Table 4 below shows the leak mass flow rate values for all the simulation pipe flow rates and pipe leak states.

Table 4. Leak mass flow rate for all leak pipe state simulation cases.

Flow Rate	Leak Mass Flow Rate, Q_l (Kg/s) Outlet Faces			
(liters/min)	2 mm Leak	5 mm Leak	7 mm Leak	10 mm Leak
90.85	0.0054	0.026	0.053	0.11
71.92	0.0045	0.021	0.042	0.086
56.78	0.0028	0.016	0.033	0.068
45.42	0.0021	0.013	0.027	0.054
26.50	0.0013	0.0079	0.016	0.032

Figure 6 below shows a representative mesh for the leak pipe states with a focus on the leak area.

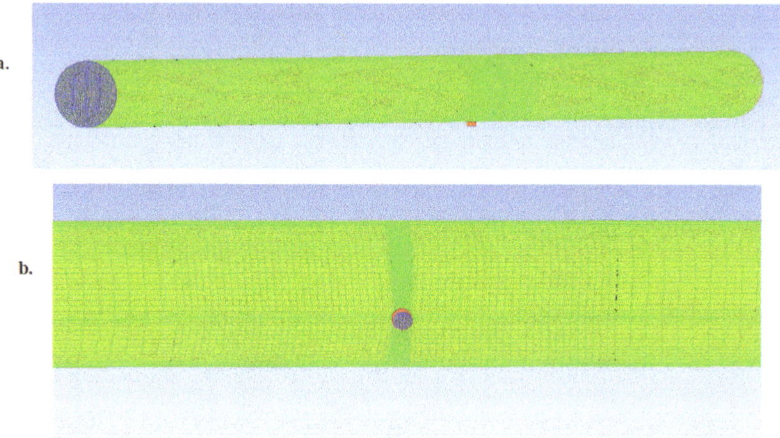

Figure 6. A 5 mm leak pipe mesh model showing: (**a**) the full pipe length and (**b**) the leak area.

For the LES simulations, a time-step size of 0.001 s was selected, and the simulation was run for 2000 time-steps for each simulation case. This amounted to a total flow time of 2 s. The courant number criterion was used to determine if the selected time-step size and mesh model was adequate for this transient simulation exercise. As mentioned earlier, for a good solution the Courant number is less than one. Using the above transient conditions, the healthy pipe simulations were first conducted for all 5 flow rates to determine the courant number.

Similar to the wall y^+, the courant number is also directly proportional to the average velocity of flow (see Equation (10)), hence results from the simulation at the highest flow rate of 90.85 L/min will give the highest courant number range and thus will be employed in evaluating the chosen time-step size and pipe mesh model.

Figure 7 above shows that the range of values of courant number from the simulation is between 0.00192 and 0.564 at 90.85 L/min and 0.000281 and 0.234 L/min at 26.50 L/min. These representative flow rates were the highest and lowest simulation flow rates. From these results, we could tell that the maximum obtainable courant number from all the simulation cases when employing the selected pipe mesh model and time-step size was 0.564, which was less than 1. This validated the chosen time-step size of 0.001 s and the selected pipe mesh model.

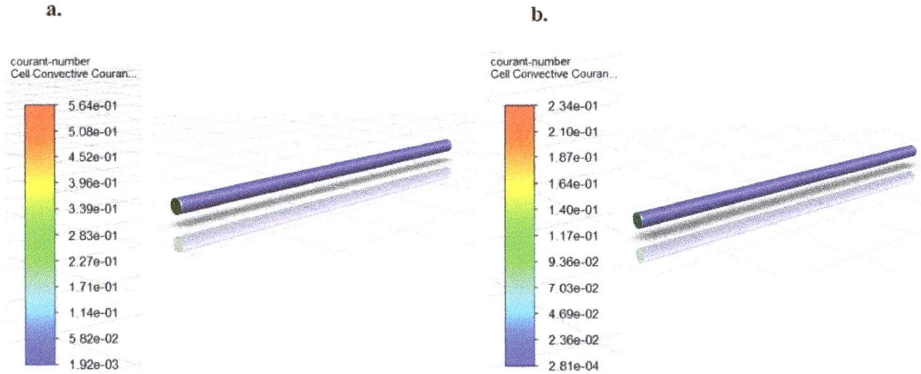

Figure 7. Courant number contours for healthy pipe transient simulations at (**a**) the highest simulation. Flow rate of 90.85 L/min and (**b**) the lowest simulation flow rate of 26.50 L/min.

Additionally, another noteworthy point is that the range of values (maximum and minimum) obtained from the courant number contours for the simulation at the highest flow rate was higher than that recorded for the lowest flow rate, confirming that courant number was directly proportional to the average flow velocity. This justifies the decision to use the highest flow rate courant number contour for validating the time-step and pipe mesh model.

After the successful validation of the transient simulation conditions, the simulations were conducted for all the pipe states (healthy and leak states) at all 5 flow rates using a time-step size of 0.001 s, and 2000 time-steps. This amounted to a total of 25 simulation cases. The average time to complete one simulation ranged from 26 to 30 h. During the simulation, the data sampling for time statistics was turned on, and set to every time step, meaning that all 2000 time-steps were employed in calculating the transient results from the FLUENT simulation, After the simulations have been completed, the fluctuating pressure of the internal pipe wall was then extracted and exported into a transient finite element model of the pipe to obtain the external pipe surface strain conditions for each pipe state and flow condition.

3.6. Determination of Pipe Surface Strain Conditions and Theoretical PVDF Patch Voltage Output

Although all 2000 time-steps were employed in calculating the FLUENT transient results, the pressure fluctuations of the internal pipe wall were extracted for every 10 time-steps for each simulation case and imported into the pipe FE model in ANSYS Transient Structural to determine the structural response (pipe surface strain fluctuation) of the pipe using ANSYS CFD-Post. This meant that a total of 200 time-step pressure fluctuation results were imported into ANSYS Transient Structural per simulation case. This is because extracting these results is very laborious and time consuming as it must be done one time-step at a time for each case. There is also the problem of file size restrictions when saving extracted data. The extraction of the file was done with the aid of the record session option in ANSYS CFD-post.

Before importing the pressure fluctuations, the FE pipe model was set up and validated. For the model set up, after meshing, the pipe support type was specified to be fixed supports as both inlet and outlet boundary conditions. The distance between these supports being the length of the test pipe, i.e., 1 m. This best represents the physical condition of the test pipe during the experimental validation exercise (as it was clamped to the test rig at both inlet and outlet) [12]. The pipe material was specified to be galvanized steel.

After validating the FE model, it was adopted for the transient structural simulations. As earlier stated, the pressure fluctuations were imported into the transient structural model for all 25 simulation cases from FLUENT. The pipe surface strain response was recorded and the theoretical performance of the PVDF sensor was evaluated for positions of interest along the pipe length, employing the PVDF film sensor–voltage relationship established by [13] and shown below. Table 5 below presents a summary of the fabricated sensor properties. The relationship between the strain acting on the piezoelectric patch and the resulting voltage is presented as

$$\varepsilon_1 = \frac{V_p C_P}{S_q} \tag{14}$$

where ε_1 is the strain acting on the sensor, V_p represents the voltage generated by the sensor, C_P is capacitance of the sensor and S_q is a sensitivity parameter $= d_{ij} Y A_p$. Additionally, d_{ij} is the piezoelectric constant, Y = Young's modulus of the Piezoelectric material and A_p = Area of the sensor.

Table 5. PVDF patch sensor properties.

Parameter	Value	Unit
Area, A_p	0.00175	m^2
Capacitance, C_p	Approx. 3.30	nF
Thickness, t_p	52	μm
Resistance, R	2.66	MΩ
Modulus of Elasticity, Y	8.3	GPa
Piezoelectric strain constant, d_{31}	30	PC/N

3.7. Determination of the Optimal Distribution of PVDF Patches to Detect the Smallest Pipe Leak

Ensuring that there is optimum distribution of vibration sensors along the pipe length for any pipe leak vibration monitoring application is very important. If the distance between the sensors is much less than what is optimal to detect the smallest expected leak size, although leak detection might be achieved, it will lead to deploying more sensors than what is required thereby driving costs up. This will have a significant financial impact in extensive applications involving many sensors. On the other hand, the distance between the sensors being more than what is optimal will affect the performance of the monitoring system negatively, as small leaks might not be detected. This paper attempts to establish a simple method for determining this optimum sensor distribution. Here, we relied on CFD modeling using ANSYS FLUENT to achieve this by modeling the smallest leak in the pipe model adopted for the numerical validation of the PVDF patches and running simulations for the lowest operating flow rate. Since leak detection is reliant on transient leak induced NPW altering the pipe surface strain conditions, the length of the area of pipe affected by this NPW obtained from simulations based on the above leak size and flow rate conditions, can be adopted as the maximum allowable distance between two sensors.

To demonstrate this idea, we took the pipe model and simulation conditions employed in the preceding subsection of this paper for validating PVDF patch sensors for leak detection into consideration. The lowest flow rate in this case was 26.50 L/min and the smallest leak size 2 mm. The pipe model being a 1 m long galvanized steel pipe. From the results obtained, it was clear that that the influence of the small leak at that flow rate was prominent throughout the entire pipe length, because although the leak-induced NPW started decaying away from the leak, it still had positive values both at the pipe inlet and pipe outlet (more details on this is provided in the next section), this shows that the pipe length was too short to determine the maximum distance between two sensors to detect the smallest leak size of 2 mm at the lowest pipe flow rate.

Here, the solution to this problem was attempted by creating a healthy pipe mesh model with the same internal diameter but a longer length. The mesh of this model was validated following the steps employed in the Section 3.1. By observing the trend of decay of the leak induced NPW at the simulation case of interest (i.e., 26.50 L/min flow rate and 2 mm leak size) from the validation exercise results, a pipe length of 4 m was deemed sufficient for this investigation. The BCs and model setup adopted was the same as those employed in Section 3.1. A bid to satisfy the conditions of $y^+ \approx 1$ and courant < 1 that was met in the preceding subsection resulted in too many mesh nodes due to the 4 m length of the pipe, and an attempt at simulation kept crashing FLUENT. Care was taken to mesh the model with the above turbulent parameter conditions relaxed a bit. This resulted in an effective mesh with wall y^+ range of 0.0597–10.5, indicating the first cells from the pipe wall were partly in the buffer layer and the laminar layer. The courant number was found to range from 0.829 to 10.50. These can be seen from the Figure 8 below.

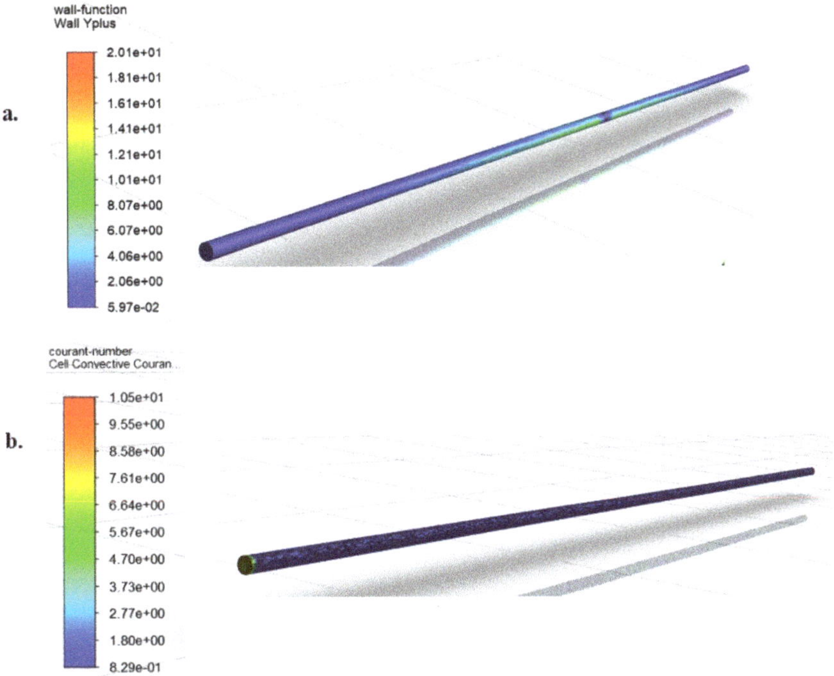

Figure 8. Contours of turbulent parameters for a 4 m length pipe model showing: (**a**) wall y^+ and (**b**) courant number.

Similarly, the 2 mm leak pipe model was also created, and it had the same properties as the healthy pipe model, except for the additional leak outlet and leak wall BCs. This time, the leak was introduced at the half-way along the pipe length, i.e., 2 m from the inlet and outlet. The BCs employed here was also the same as that employed in the preceding Section 3.2. FLUENT simulations were run for both the healthy and leak pipe models, and the results were analyzed to determine the length of pipe influenced by the leak induced NPW before it completely decayed, and consequently, the optimum distance between two PVDF patch sensors for the subject pipe model and flow conditions.

4. Results and Discussion

This section demonstrated the impact of the model developed in this paper. The simulations first established the root-mean-square estimates of negative pressure waves in pipes to be a valid indicator of the leak. The effect of presence, location and the extent of leakage and its interaction with the distance from the sensor were investigated next extensively for various flow rates and levels of leakage. The effective calibrations of markers of such detection are presented, which is particularly relevant for any future experimental design. The section subsequently demonstrated how the spacing of the sensors could be determined through the developed numerical method as a function of the smallest size of leak that is intended to be detection.

4.1. Numerical Validation of PVDF Sensor Patches

4.1.1. Pipe Flow Simulations (FLUENT)

Post Fluent simulation, the pipe wall fluctuating pressure for the simulation time, at any point along the pipe length of the pipe, could be obtained. The frequency of sampling being the inverse of the time-step size, hence 1000 Hz. The instantaneous pipe wall fluctuating pressure (P_f), at any point along the pipe length and at any time of sampling, t is given by the relationship below (Bai et al., 2019).

$$P_f = P - P_m \tag{15}$$

where P is the instantaneous static pressure and P_m is the mean static pressure calculated over the sampling time.

When P_f plotted against time, the result is a random curve with alternating negative and positive values. Plots of P_f against time for a point at the 0.6 cm mark along the pipe length just above the leak for a two of the simulation cases is presented in Figure 9 below. The figure shows that for the same flow condition and position on the pipe, the 5 mm leak pipe state recorded higher amplitude of P_f over the simulation flow time than the healthy pipe. This is because the P_f is solely a consequence of FSI alone for the healthy pipe, while it is due to FSI plus NPW for the leak pipe states.

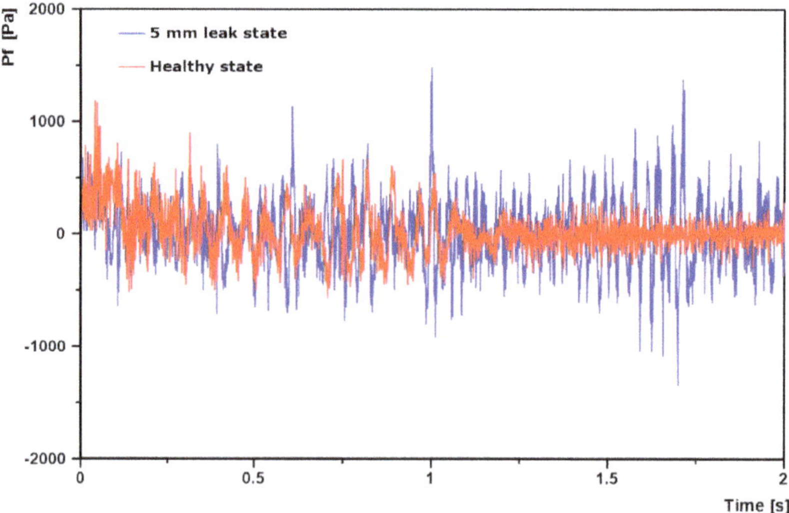

Figure 9. Plots of P_f against time, for the healthy pipe state and 5 mm leak pipe state at 90.85 L/min.

From the FLUENT simulation results, time-averaged statistics was calculated for all simulation cases. One of those statistics important to this analysis is the Root Mean Square Error (RMSE) of the static pressure. The RMSE of static pressure is the same as the root mean square of the fluctuating pressure (P_frms). The calculated rms fluctuating pressure can be obtained for any point along the pipe length, and it provides a single representative value for the magnitude of the time varying pressure fluctuation recorded at that point over the entire simulation time. That way P_frms along any line running through the pipe wall can be plotted against the pipe length. The value of P_frms can be obtained for as many points as possible. This is not possible for experimental data, which can only be recorded in positions where the sensors are bonded. Here, P_frms is recorded along a straight line through the centre of the internal pipe wall along the entire pipe length (see Figure 2), for all simulation cases. For each flow condition, the difference in the $P_f$rms obtained for each pipe state and that obtained for the healthy pipe state at that same flow rate was calculated. This represents the contribution of the *NPW* to the fluctuating pressure recorded for that simulation case. Here we refer to this difference as the *NPWrms*.

Figure 10 below shows plots of *NPWrms* against the pipe length (X). In the figure, each subfigure shows curves of *NPWrms* for the different pipe states at a common flow rate. From the plots, it can be seen that in all cases of the healthy pipe, there was no *NPW*, this is expected as the healthy pipe states are the baseline conditions. For all the leak pipe states, the influence of the *NPW* is prominent throughout the pipe length, even at the furthest sections from the leak (inlet and outlet), the *NPW* have positive values for all simulation cases. This is the case for the smallest leak size 2 mm at the lowest flow rate that unsurprisingly records the least *NPWrms* curve in terms of magnitude. An important trend to notice is that the *NPW* was consistently highest at the leak position (i.e., 60 cm mark), and in all simulation cases it gradually decayed in both directions away from the leak in an almost identical manner. Furthermore, observing individual subfigures, it was observed that the value of *NPW* with decreasing flow rates, i.e., the curves for 90.85 L/min had the highest magnitudes and 26.50 L/min the least magnitudes. This is because as the flow rate increased, the flow turbulence increased and consequently the FSI. Additionally, flow also had a directly proportional relationship with the leak induced *NPW*.

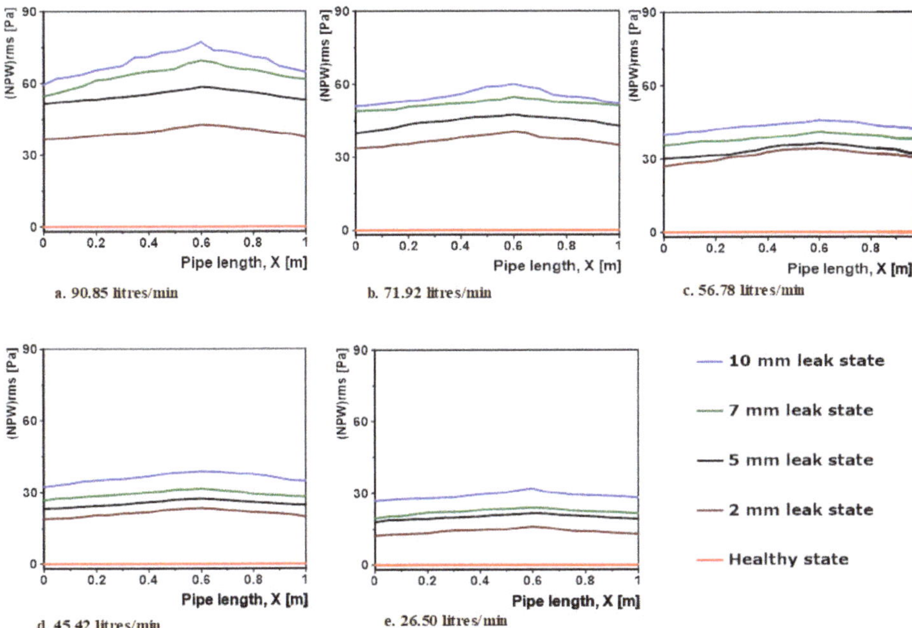

Figure 10. Plots of *NPWrms* recorded along the pipe length for all simulation cases for flow rates (**a**) 90.85 litres/per minute (**b**) 71.92.85 litres/per minute; (**c**) 56.78 litres/per minute; (**d**) 45.42 litres/per minute; (**e**) 26.50 litres/per minute.

The findings from this analysis served to confirm that an *NPW* was induced on the onset of a leak, this *NPW* increased with increasing leak size and it contributed to the internal pipe wall pressure fluctuation of any leaking pipe. Therefore, monitoring any parameter that was influenced by this *NPW* could prove an effective method for leak detection and monitoring.

4.1.2. Transient Structural Simulations and Determination of Theoretical PVDF Patch Output

Post transient structural simulations, the pipe vibration response in terms of surface strain had to be obtained at specific positions along the pipeline to be able to calculate the theoretical voltage output from PVDF patch sensors. The positions selected were 20 cm, 60 cm and 80 cm from the pipe inlet, named sensor position 1 (SP1), sensor position 2 (SP2) and sensor position 3 (SP3) for the purpose of analysis (see Figure 11). Sensor position 2 (SP2) was directly above the leak.

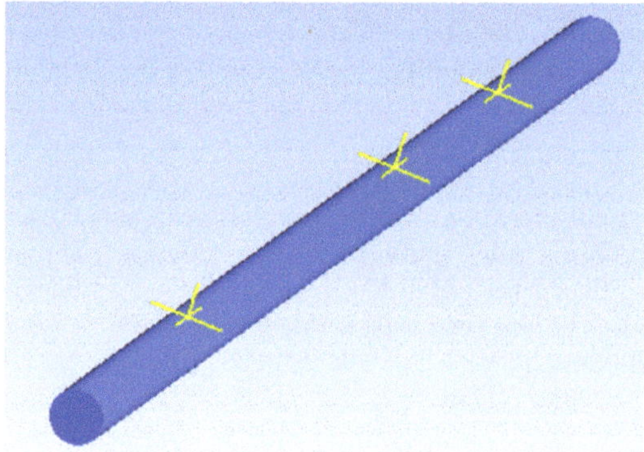

Figure 11. The three PVDF patch sensor positions from the pipe inlet.

These are the same positions where the patches were bonded for the experimental campaign detailed in [12]. The pipe surface strain conditions at these three points were extracted for all simulation cases and the data adopted for analysis. Figure 12 below shows representatives of this data.

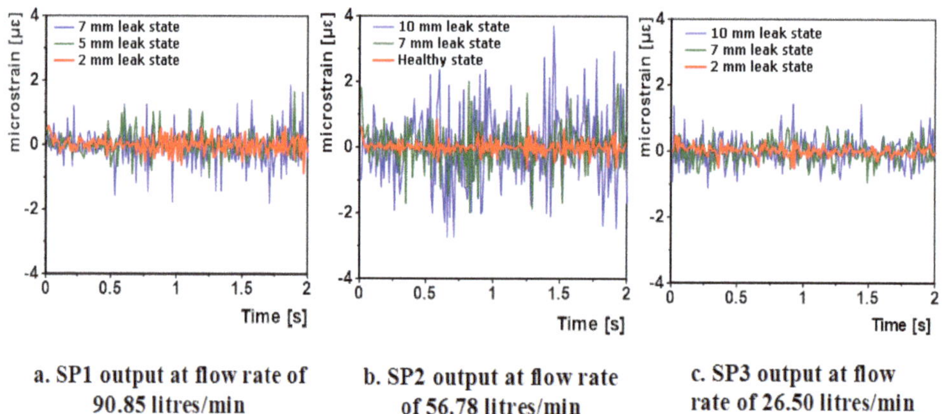

a. SP1 output at flow rate of 90.85 litres/min   **b. SP2 output at flow rate of 56.78 litres/min**   **c. SP3 output at flow rate of 26.50 litres/min**

Figure 12. Samples of strain data extracted at SP1, SP2 and SP3 for a number of simulation cases, at different flow rates (**a**) 90.85 litres/per minute; (**b**) 56.78 litres/per minute; (**c**) 26.50 litres/per minute.

The above figure shows that at a given flow rate, the amplitude of the recorded strain increased with the introduction of a 2 mm small leak and this amplitude increased even further with increasing leak severity.

To determine the theoretical performance of the PVDF patch sensors, the theoretical root mean square voltage, *Vrms* from the sensor for each of the three sensor positions and all simulation cases must be calculated. This is the same analysis adopted for the experimental validation. Additionally, it was the only way to see if the trend of the theoretical and experimental sensor outputs was in tandem. To do this, the root mean square of the strain data was first calculated for all simulation cases and sensor positions. Then using the established voltage–strain relationship for PVDF patches (Equation (14)), the theoretical *Vrms* for each sensor was calculated and it is presented in the Figure 13 below. The sensors are named PS1, PS2 and PS3 in line with sensor positions 1, 2 and 3. They were also named this way for the experimental validation exercise [12].

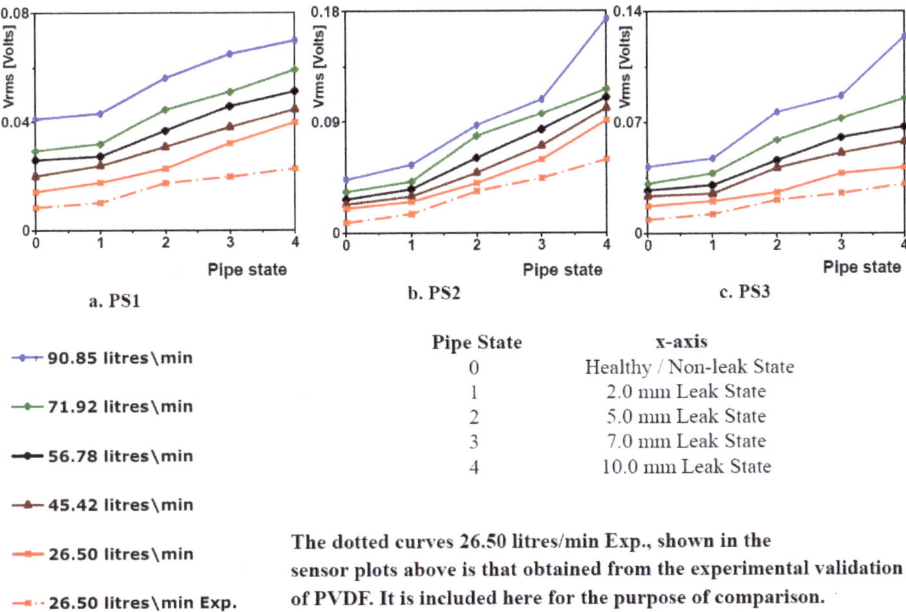

Pipe State	x-axis
0	Healthy / Non-leak State
1	2.0 mm Leak State
2	5.0 mm Leak State
3	7.0 mm Leak State
4	10.0 mm Leak State

The dotted curves 26.50 litres/min Exp., shown in the sensor plots above is that obtained from the experimental validation of PVDF. It is included here for the purpose of comparison.

Figure 13. Theoretical Vrms output of PVDF patches test scenarios (**a**) PS1; (**b**) PS2; (**c**) PS3 with the locations of sensors indicated as SP1, SP2 and SP3 respectively in Figure 12.

The horizontal (x) axis from Figure 13 below represented the health of the pipe, where 0 is the healthy pipe state, 1 is the 2 mm leak pipe state, 2 is the 5 mm leak pipe state, 3 is the 7 mm leak pipe state and 4 is the 10 mm leak pipe state.

It is important to note that, the Vrms values of the pipe states simulated for are represented on the plots by markers (see figure legend). The Vrms values of the intermediate pipe states inferred by joining the markers to form curves do not represent data from this exercise as simulations were not conducted at these intermediate states. Joining the markers to form curves is for visual aid, to help appreciate the trend of the theoretical sensors' output with the worsening pipe leak state.

From Figure 13, the introduction of 2 mm leak resulted in an increase in the voltage output (*Vrms*) of the PVDF sensors. This can be observed by the upward tilt of the curves moving from points 0–1 on the horizontal axis representing the state of the pipe, and with increasing severity of the leak, there is a corresponding increase in the sensor voltage output. This is depicted by the rise in the curves moving from left to right of the plots. This shows that at a given flow rate, the pipe experiences more vibration with the introduction of leaks due to the induced NPW and furthermore vibration as the leak size increased. This increasing vibration was reflected consistently from the simulation results and consequently in the theoretical output of all the PVDF patches.

The trend of the curves obtained from the results of this numerical validation exercise explained above mirrors that obtained from the experimental validation exercise detailed in [12], and both numerical and experimental results validates PVDF patches as effective for pipe leak detection and monitoring. The disparity being that the theoretical *Vrms* of the PVDF sensors was consistently higher than the experimental output of the sensors. This is the case for all simulation cases and sensor positions. Included in Figure 13 plots are the curves from the experimental results of the 26.50 L/min test scenarios, for illustration and comparison. The intention was to show more curves from the replica experimental results in the above figure, but due to almost coincident points and curves crossing, the idea was shelved. The representative experimental 26.50 L/min curves were lower in magnitude when compared to their theoretical counterparts for all three sensors, and it is a representative of the relationship between the theoretical and experimental curves of the other flow rates (see [12] for plots of curves from the experimental sensor output). The maximum deviation was observed for PS1 at a simulation test flow rate of 26.50 L/min and 10 mm leak pipe state, where the theoretical *Vrms* was 1.92 times the experimental *Vrms*. There might be a few reasons for this, one being that there are likely small measuring losses through the piezoelectric measuring chain, which may lower the experimental *Vrms*. Another one is that although, there are no vibration losses in the finite bond layer thickness between PVDF patch sensors and the pipe if perfectly bonded [13], any imperfection, however small, while adhering the patches to the pipe surface will result in little losses, and this only affects the experimental results as no correction factor was applied to the strain–voltage relationship when calculating the theoretical *Vrms*. A final important consideration is that, although a lot of consideration and checks were implemented in executing this numerical validation exercise to ensure reliable simulation results, the fluid flow turbulence problem is such a complex one that a numerical solution that approximates and gives verifiable information about the practical flow solution can be deemed reliable.

Additionally, since the leak induced NPW that aids leak detection was highest at the leak position and decayed away from the leak in both directions, the area of the pipe closest to the leak could be identified, by monitoring the leak induced vibration at various sections of the pipe. The fact that the sensors were at different distances from the leak position was utilized in evaluating the theoretical performance of the PVDF sensors for leak localization. The approach employed here is same as that employed in the experimental validation [12], where the leak induced vibration is quantified by a theoretical leak index that measures how much the theoretical *Vrms* for a PVDF sensor patch at a particular flow rate deviates from the healthy state theoretical *Vrms* at the same flow rate.

The leak index was calculated for all simulation cases and the three PVDF patch sensors. Figure 14 below show the plots of theoretical leak index against pipe state for the three sensors at two different flow rates.

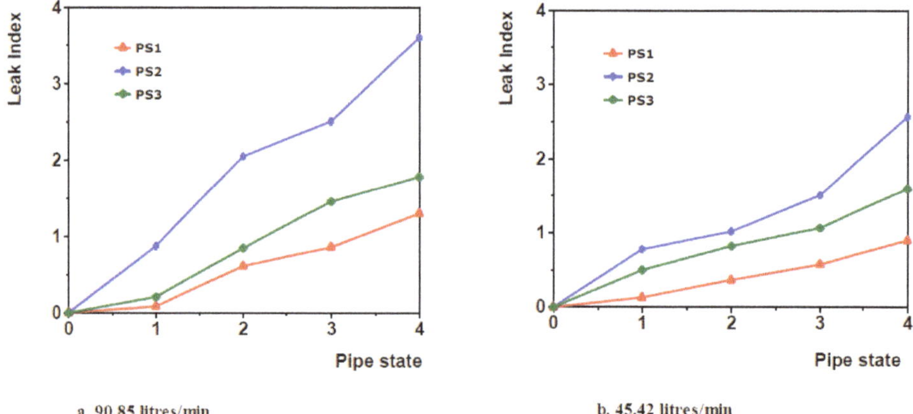

Figure 14. Representative lots of the theoretical leak index against the pipe state at two flow rates (**a**) 90.85 litres/per minute and (**b**) 45.42 litres/per minute.

The horizontal (x)-axis representation is same as that employed for Figure 14. The plots in Figure 14 show that the theoretical leak index curves for PS2, which was bonded just above the leak was the highest while that of PS1, which was 40 cm from the leak was the lowest at both flow rates. The curves for PS3 located 20 cm from the leak were in-between those of PS2 and PS1 at both flow rates shown. This is the same for all flow rates, and consistent with the trend of the experimental leak index presented in [12].

The results numerically validate the experimental results from the PVDF patch sensors obtained in [12], as the theoretical results and performance indicate that they could detect, monitor and localize a leak.

4.2. Determination of the Optimal Distribution of PVDF Patches to Detect the Smallest Pipe Leak

As mentioned earlier, the healthy pipe and 2 mm leak pipe states at a flow rate of 26.50 L/min simulation conditions from the numerical validation exercise were adopted for this study. The 2 mm leak pipe state at flow rate of 26.50 L/min simulation condition corresponds to the smallest leak pipe state at the lowest flow rate from the numerical exercise. For this study, the length of the pipe model was 4 m as opposed to the 1 m length from the numerical validation study. Figure 10e shows the *NPWrms* curve for 2 mm leak size at 26.50 L/min for the numerical validation study. From the curve, we can see that the influence of the small leak at that flow rate was prominent throughout the entire 1 m pipe length, because although it started decaying away from the leak, it still had positive values both at the pipe inlet and pipe outlet, this shows that the 1 m pipe length was too short to determine the maximum distance between two sensors to detect the smallest leak size of 2 mm at the lowest pipe flow rate. This justifies using a longer pipe for this study, and by observing the trend of decay of the NPW in Figure 10e, a pipe length of 4 m was deemed sufficient for this investigation.

After FLUENT simulations, the $P_f$rms curves along a straight line (similar to Figure 2), on the internal pipe wall was obtained for healthy pipe state and the 2 mm leak pipe state. The *NPWrms* curve of the 2 mm leak state pipe was derived by subtracting the $P_f$rms curve of the healthy pipe from the $P_f$rms curve of the leak pipe state. Plot of *NPWrms* against pipe length is shown in Figure 15 below.

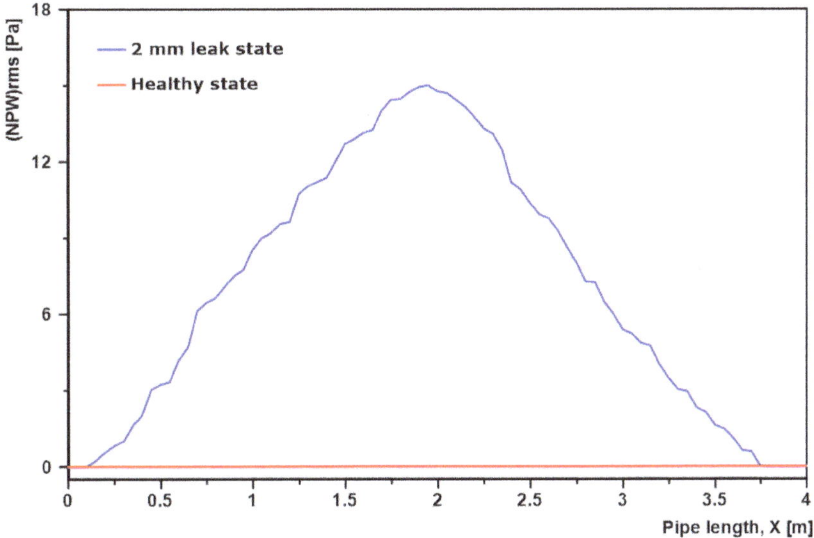

Figure 15. Plot of *NPWrms* for the 2 mm leak pipe state at 26.50 L/min.

From Figure 15 above, the value of NPW was zero along the pipe length for the healthy pipe state, because the healthy pipe is the baseline condition and there was no leak inducing NPW. For the 2 mm leak pipe state the NPW was at its highest at the position of the leak, i.e., the 2-m mark along the pipe length. This NPW gradually decayed in both directions away from the leak finally falling to zero at the 0.15 m mark and the 3.75 m mark. The area under the curve represents the area of the pipe affected by the NPW induced due to the 2 mm smallest leak size and the least simulation flow rate of 26.50 L/min. The length of pipe in this area was 3.60 m and it represents the maximum distance between two piezoelectric strain-based sensors if the 2 mm leak was to be detected at a flow rate of 26.50 L.

Therefore, for the pipe model in this exercise and our simulation condition, the maximum distance between two sensors for the smallest leak size to be detected at the least operating pressure was approximately 3.60 m.

The presented numerical framework and related experiments could thus adapt itself to not just detection hardware [40] but also for other output-only detection or monitoring frameworks [41–43], degradation [44] or understanding of computing demands [45].

5. Conclusions

This paper presented a numerical model, combining the Fluid–Structure Interaction to estimate the vibrations and dynamic strains on a fluid-carrying pipe and estimated the impact of leaks. The work then linked this model to energy harvesting based monitoring of such leaks through a PVDF patch, in the context of existing experimental results. This method relies on monitoring the leak induced NPW and its attenuation away from the leak. The results from the numerical validation exercise corroborated the experimental results presented in [12], providing a first comprehensive bench marked evidence base and implementation protocol on this topic. The model was then able to estimate the maximum spacing of sensors that still can detect the minimum leak of consequence. The results provided guidelines for future experimental designs through the type model presented here and established sensor placement strategies. The model can be adapted easily to new sensors and detection algorithms and can thus be used for assessing the performance of a sensor or a method in future. The results from the exercise reinforced the confidence in piezoelectric patch sensors as being effective for pipe leak detection and monitoring.

Despite these advantages, the study naturally has some limitations. The energy-harvesting model is relatively simplistic and variations of harvesting circuits have not been explored, limiting the discussions to the open circuit voltages. The electromechanical coupling is kept constant, but in low powers, this coefficient may vary. The challenge of potential false alarms caused by environmental perturbations unrelated to the leak has not been investigated here either. Additionally, effects of temperature, chemical exposure, humidity and material degradation and durability have not been considered in this model.

Author Contributions: Conceptualization, F.O. and V.P.; Methodology, F.O., M.C., V.P.; Software, F.O., V.P.; Validation, F.O., M.C. and V.P.; Formal Analysis, F.O.; Investigation, F.O.; Resources, F.O., M.C. and V.P.; Data Curation, F.O.; Writing-Original Draft Preparation, F.O., M.C. and V.P.; Writing-Review & Editing, F.O., M.C. and V.P.; Visualization, F.O., M.C. and V.P.; Supervision, M.C. and V.P.; Project Administration, V.P.; Funding Acquisition, F.O., M.C. and V.P. All authors have read and agreed to the published version of the manuscript.

Funding: This research was funded by the Nigerian government through the Niger Delta Development Commission (NDDC), the Irish Research Council and Environmental Protection Agency Award IRC-GOIPG/2019/2184 and EU Cooperation in Science and Technology (COST) Action CA18203 Optimizing Design for Inspection (ODIN).

Acknowledgments: The authors acknowledge the support of ANSYS UK, that provided commercial CFD and FE codes needed to carry out this study and acknowledge the strategic support of SFI centre Marine and Renewable Energy Ireland (MaREI).

Conflicts of Interest: The authors declare no conflict of interest.

References

1. Choi, J.; Shin, J.; Song, C.; Han, S.; Park, D., II. Leak detection and location of water pipes using vibration sensors and modified ML prefilter. *Sensors* **2017**, *17*, 2104. [CrossRef] [PubMed]
2. Gao, Y.; Muggleton, J.M.; Liu, Y.; Rustighi, E. An analytical model of ground surface vibration due to axisymmetric wave motion in buried fluid-filled pipes. *J. Sound Vib.* **2017**, *395*, 142–159. [CrossRef]
3. Adegboye, M.A.; Fung, W.K.; Karnik, A. Recent advances in pipeline monitoring and oil leakage detection technologies: Principles and approaches. *Sensors* **2019**, *19*, 2548. [CrossRef] [PubMed]
4. Keramat, A.; Tijsseling, A.S.; Hou, Q.; Ahmadi, A. Fluid-structure interaction with pipe-wall viscoelasticity during water hammer. *J. Fluids Struct.* **2012**, *28*, 434–455. [CrossRef]
5. Zhang, T.; Tan, Y.; Zhang, X.; Zhao, J. A novel hybrid technique for leak detection and location in straight pipelines. *J. Loss. Prev. Process. Ind.* **2015**, *35*, 157–168. [CrossRef]
6. Li, S.; Karney, B.W.; Liu, G. FSI research in pipeline systems—A review of the literature. *J. Fluids Struct.* **2015**, *57*, 277–297. [CrossRef]
7. Ismail, M.; Dziyauddin, R.A.; Salleh, N.A.A. Performance evaluation of wireless accelerometer sensor for water pipeline leakage. In Proceedings of the International Symposium on Robotics and Intelligent Sensors (IRIS), Langkawi, Malaysia, 18–20 October 2015; pp. 120–125.
8. El-Zahab, S.; Mohammed Abdelkader, E.; Zayed, T. An accelerometer-based leak detection system. *Mech. Syst. Signal Process.* **2018**, *108*, 58–72. [CrossRef]
9. Wang, Y.H.; Song, P.; Li, X.; Ru, C.; Ferrari, G.; Balasubramanian, P.; Amabili, M.; Sun, Y.; Liu, X. A paper-based piezoelectric accelerometer. *Micromachines* **2018**, *9*, 19. [CrossRef]
10. Varanis, M.; Silva, A.; Mereles, A.; Pederiva, R. MEMS accelerometers for mechanical vibrations analysis: A comprehensive review with applications. *J. Braz. Soc. Mech. Sci. Eng.* **2018**, *40*, 1–18. [CrossRef]
11. Mahmood, M.S.; Celik-Butler, Z.; Butler, D.P. Design, fabrication and characterization of flexible MEMS accelerometer using multi-Level UV-LIGA. *Sens. Actuators A Phys.* **2017**, *263*, 530–541. [CrossRef]
12. Okosun, F.; Cahill, P.; Hazra, B.; Pakrashi, V. Vibration-based leak detection and monitoring of water pipes using output-only piezoelectric sensors. *Eur. Phys. J. Spec. Top.* **2019**, *228*, 1659–1675. [CrossRef]
13. Sirohi, J.; Chopra, I. Fundamental understanding of piezoelectric strain sensors. *J. Intell. Mater. Syst. Struct.* **2000**, *11*, 246–257. [CrossRef]
14. Hellum, A.M.; Mukherjee, R.; Hull, A.J. Dynamics of pipes conveying fluid with non-uniform turbulent and laminar velocity profiles. *J. Fluids Struct.* **2010**, *26*, 804–813. [CrossRef]
15. Escue, A.; Cui, J. Comparison of turbulence models in simulating swirling pipe flows. *Appl. Math. Model.* **2010**, *34*, 2840–2849. [CrossRef]
16. Argyropoulos, C.D.; Markatos, N.C. Recent advances on the numerical modelling of turbulent flows. *Appl. Math. Model.* **2015**, *39*, 693–732. [CrossRef]
17. Wang, X.; Wang, L.; Zheng, G.; Sun, S.; Yang, R. Simulation of leak-age model of long-distance oil pipeline based on FLUENT. *Appl. Mech. Mater.* **2013**, *310*, 280–286. [CrossRef]
18. Pittard, M.T.; Evans, R.P.; Maynes, R.D.; Blotter, J.D. Experimental and numerical investigation of turbulent flow induced pipe vibration in fully developed flow. *Rev. Sci. Instrum.* **2004**, *75*, 2393–2401. [CrossRef]
19. Alfredsson, P.H.; Örlü, R.; Segalini, A. A new formulation for the streamwise turbulence intensity distribution in wall-bounded turbulent flows. *Eur. J. Mech. B Fluids* **2012**, *36*, 167–175. [CrossRef]
20. Khalili, F.; Gamage, P.P.T.; Mansy, H.A. Verification of Turbulence Models for Flow in a Constricted Pipe at Low Reynolds Number. In Proceedings of the 3rd Thermal and Fluids Engineering Conference (TFEC) Florida, Fort Lauderdale, FL, USA, 4–7 March 2018; pp. 1865–1874.
21. Kamruzzaman, M.; Bekiropoulos, D.; Wolf, A.; Lutz, T.; Würz, W.; Krämer, E. Study of turbulent boundary layer wall pressure fluctuations spectrum models for trailing-edge noise prediction. In Proceedings of the 15th International Conference on the Methods of Aerophysical Research (ICMAR), Novosibirsk, Russia, 1–6 November 2010; pp. 1–11.
22. Catalano, P.; Wang, M.; Iaccarino, G.; Moin, P. Numerical simulation of the flow around a circular cylinder at high Reynolds numbers. *Int. J. Heat. Fluid. Flow.* **2003**, *24*, 463–469. [CrossRef]
23. Nishino, T.; Roberts, G.T.; Zhang, X. Unsteady RANS and detached-eddy simulations of flow around a circular cylinder in ground effect. *J. Fluids Struct.* **2008**, *24*, 18–33. [CrossRef]

24. Kaur, K.; Annus, I.; Vassiljev, A.; Kändler, N. Determination of Pressure Drop and Flow Velocity in Old Rough Pipes. *Proceedings* **2018**, *2*, 590. [CrossRef]
25. Wilcox, D.C. *Turbulence Modeling for CFD*, 3rd ed.; DCW Industries: La Cañada, CA, USA, 2006.
26. Wei, Z.; Yang, W.; Xiao, R. Pressure fluctuation and flow characteristics in a two-stage double-suction centrifugal pump. *Symmetry* **2019**, *11*, 65. [CrossRef]
27. Pope, S.B. *Turbulent Flows*; Cambridge University Press: Cambridge, UK, 2000.
28. Weickert, M.; Teike, G.; Schmidt, O.; Sommerfeld, M. Investigation of the LES WALE turbulence model within the lattice Boltzmann framework. *Comput. Math. Appl.* **2010**, *59*, 2200–2214. [CrossRef]
29. Koukouvinis, P.; Naseri, H.; Gavaises, M. Performance of turbulence and cavitation models in prediction of incipient and developed cavitation. *Int. J. Engine Res.* **2017**, *18*, 333–350. [CrossRef]
30. Wang, Y.; Wang, S.-X.; Liu, Y.-H.; Chen, C.-Y. Influence of cavity shape on hydrodynamic noise by a hybrid LES-FW-H method. *China Ocean Eng.* **2011**, *25*, 381–394. [CrossRef]
31. Zeng, Y.; Luo, R. Numerical analysis on pipeline leakage characteristics for incompressible flow. *J. Appl. Fluid Mech.* **2019**, *12*, 485–494. [CrossRef]
32. Liu, S. Implementation of a Complete Wall Function for the Standard k—ε Turbulence Model in OpenFOAM 4.0. In *Proceedings of CFD with OpenSource Software*. Nilson, H., Ed.; 2016. Available online: http://www.tfd.chalmers.se/~{}hani/kurser/OS_CFD_2016 (accessed on 6 October 2016).
33. Olivares, P.A.V. Acoustic Wave Propagation and Modeling Turbulent Water Flow with Acoustics for District Heating Pipes Water Flows with Acoustics for District Heating Pipes. Ph.D. Thesis, Uppsala University, Uppsala, Sweden, 2009.
34. Loh, S.K.; Faris, W.F.; Hamdi, M. Fluid-structure interaction simulation of transient turbulent flow in a curved tube with fixed supports using LES. *Prog. Comput. Fluid Dyn.* **2013**, *13*, 11–19. [CrossRef]
35. Ahsan, M. Numerical analysis of friction factor for a fully developed tur- bulent flow using k–ε turbulence model with enhanced wall treatment. *Beni-Suef Univ. J. Basic Appl. Sci.* **2014**, *3*, 269–277. [CrossRef]
36. Mirmanto, M. Developing Flow Pressure Drop and Friction Factor of Water in Copper Microchannels. *J. Mech. Eng. Autom.* **2013**, *3*, 641–649.
37. Valizadeh, K.; Farahbakhsh, S.; Bateni, A.; Zargarian, A.; Davarpanah, A.; Al- Izadeh, A.; Zarei, M. A parametric study to simulate the non-Newtonian turbulent flow in spiral tubes. *Energy Sci. Eng.* **2020**, *8*, 134–149. [CrossRef]
38. Ge, C.; Wang, G.; Ye, H. Analysis of the smallest detectable leakage flow rate of negative pressure wave-based leak detection systems for liquid pipelines. *Comput. Chem. Eng.* **2008**, *32*, 1669–1680. [CrossRef]
39. Van Zyl, J.E.; Clayton, C.R.I. The effect of pressure on leakage in water distribution systems. *Water Manag.* **2007**, *160*, 109–114. [CrossRef]
40. Okosun, F.; Pakrashi, V. Experimental validation of a piezoelectric measuring chain for monitoring structural dynamics. In Proceedings of the 2020 IEEE International Instrumentation and Measurement Technology Conference (I2MTC), Dubrovnik, Croatia, 25–28 May 2020.
41. Krishnan, M.; Bhowmik, B.; Hazra, B.; Pakrashi, V. Real time damage detection using recursive principal components and time varying autoregressive modeling. *Mech. Syst. Signal. Process.* **2020**, *101*, 549–574.
42. Bhowmik, B.; Tripura, T.; Hazra, B.; Pakrashi, V. First order eigen perturbation techniques for real time damage detection of vibrating systems: Theory and applications. *Appl. Mech. Rev.* **2020**, *71*, 060801. [CrossRef]
43. Cahill, P.; Pakrashi, V.; Sun, P.; Mathewson, A.; Nagarajaiah, S. Energy Harvesting Techniques for Health Monitoring and Indicators for Control of a Damaged Pipe Structure. *Smart Struct. Syst.* **2018**, *21*, 287–303.
44. Srbinovski, B.; Magno, M.; Edwards Murphy, F.; Pakrashi, V.; Popovici, E. An Energy Aware Adaptive Sampling Algorithm for Energy Harvesting WSN with Energy Hungry Sensors. *Sensors* **2016**, *16*, 448. [CrossRef]
45. Tripura, T.; Bhowmik, B.; Hazra, B.; Pakrashi, V. Real time damage detection of Degrading Systems. *Struct. Health Monit.* **2020**, *19*, 810–837. [CrossRef]

Publisher's Note: MDPI stays neutral with regard to jurisdictional claims in published maps and institutional affiliations.

Article

Dynamic Modeling and Experimental Validation of an Impact-Driven Piezoelectric Energy Harvester in Magnetic Field

Chung-De Chen *, Yu-Hsuan Wu and Po-Wen Su

Department of Mechanical Engineering, National Cheng Kung University, Tainan City 701, Taiwan; n16064593@gs.ncku.edu.tw (Y.-H.W.); n16084462@gs.ncku.edu.tw (P.-W.S.)
* Correspondence: cdchen@mail.ncku.edu.tw

Received: 29 September 2020; Accepted: 26 October 2020; Published: 29 October 2020

Abstract: In this study, an impact-driven piezoelectric energy harvester (PEH) in magnetic field is presented. The PEH consists of a piezoelectric cantilever beam and plural magnets. At its initial status, the beam tip magnet is attracted by a second magnet. The second magnet is moved away by hand and then the beam tip magnet moves to a third magnet by the guidance of the magnetic fields. The impact occurs when the beam motion is stopped by the third magnet. The impact between magnets produces an impact energy and causes a transient beam vibration. The electric energy is generated by the piezoelectric effect. Based on the energy principle, a multi-DOF (multi-degree of freedom) mathematical model was developed to calculate the displacements, velocities, and voltage outputs of the PEH. A prototype of the PEH was fabricated. The voltages outputs of the beam were monitored by an oscilloscope. The maximum generated energy was about 0.4045 mJ for a single impact. A comparison between numerical and experimental results was presented in detail. It showed that the predictions based on the model agree with the experimental measurements. The PEH was connected to a diode bridge rectifier and a storage capacitor. The charges generated by the piezoelectric beam were stored in the capacitor by ten impacts. The experiments showed that the energy stored in the capacitor can light up the LED.

Keywords: piezoelectric energy harvesting; vibration; frequency-up conversion

---

1. Introduction

The piezoelectric energy harvesters (PEHs) have received attention in the past two decays due to the high cost for battery replacements in the wireless sensor networks [1]. In the early development of PEHs, the resonators, such as cantilever beams, are used to harvest vibration energy from the environments [2–4]. The narrow bandwidth limits the application of the resonance-based energy harvesters. Many researchers have developed broad-band strategies to broaden the bandwidth of the resonance-based energy harvester, such as nonlinear oscillators, arrayed oscillators and multi-mode coupled oscillators [5].

In recent years, harvesting energy from human motions has become a hot topic because of the increasing demands of wearable devices [6]. The frequency of the human motion is extreme low, typically less than 5 Hz. For such low frequency, the energy conversion efficiency is also low for resonance-based energy harvesters. Although some broad-band strategies are adopted to enhance the efficiency, it is difficult to widen the bandwidth covering a frequency less than 5 Hz [7]. As an alternative, the concept of frequency-up conversion has been proven to be able to harvest energy at high efficiency from energy sources with extreme low frequencies. Umeda et al. presented a concept for energy harvesting based on the frequency up-conversion [8]. They investigate a ball free falling onto a piezoelectric beam to produce a transient vibration. Gu and Livemore presented a two-beam assembly [9]. The long beam vibrates in responding to the environment with a low frequency and hits the short piezoelectric beam with a high natural frequency. The electrical energy can be extracted from the transient vibration of the short beam with high efficiency. In recent years, some similar studies in energy harvesting based on frequency-up conversion with impact type can be also seen in [10–13].

In the above mentioned frequency up-conversion energy harvesters, the conversions from low frequency to high frequency through mechanical contact. Wickenheiser and Garcia presented a cantilever beam with magnet plucked by magnetic forces [14]. Although it avoids some problems such as surface wear and noises due to contact, the energy generated by the magnetic plucking is usually smaller than those based on mechanical contact. The frequency up-conversion driven by magnetic plucking forces was also adopted in some papers [15,16].

The frequency up-conversion concept has been proven having potential to harvest energy with extreme low frequency, and hence can be adopted to harvest energy from human motion. Pozzi and Zhu presented a device mounted at the knee to generate electrical energy from knee motion [17]. Their device is composed of a stator with plectra and a rotor with cantilever piezoelectric beams. As the knee moves, the stator and rotor have a relative rotational motion and the plectra pluck the beams. A refined device to harvest energy from knee motion was proposed by Kuang et al. [18]. The device is similar to that in [17] except that the beams were plucked by magnetic force. Their experiments showed that a maximum power of 4.5 mW can be harvested from walking. Wei et al. proposed an impact type energy harvester [19]. The presented this device mounted on a human leg to harvest energy from human walk. Their experiments showed that a power of 51 μW was generated for a walking speed of 5 km/h.

The magnetic forces are commonly used in the PEH designs. In the early development, the magnetic forces were assumed as simple forms such as inverse-square model [20–22]. These simple models are easy to implement. However, these forms valid only for some specific displacement range. For the case that the distance between two magnets are short or long, these simple models give inaccurate predictions. Some researchers assumed that the magnet as dipole without occupying a volume [23–26]. For the case of the large distance between two magnets, the model gives an accurate magnetic force. However, when one magnet approaches another, the assumption of the point-dipole become doubtful. Some 3D models have been proposed to calculate the interaction force between two magnets [27,28]. Comparing to the assumed form model and the point-dipole model, the 3D model requires more computation resources as it gives a more accurate prediction in magnetic force. In recent years, the 3D model has been introduced in developing the mathematical models of PEH [16,29–31].

In this paper, an impact-driven piezoelectric energy harvester (PEH) in the magnetic field is presented. A multi-DOF mathematical is developed to investigate the dynamic behaviors of the PEH. A 3-D magnetic force model is also introduced to calculate the magnetic forces between magnets. The voltage responses and energy harvested by the PEH can be calculated by the model. A prototype of the proposed design is fabricated to verify the numerical results of the model. Human motions, such as finger pressing, have been proven to trigger an PEH for driving a batteryless switch [26]. In this study, the PEH is driven by finger pressing and can be applicable to such applications.

This paper is organized as follows. In Section 1, the background of this study is presented. Some selected papers for piezoelectric energy harvesting are reviewed. The working principle of the PEH is mentioned in Section 2. In Section 3, the mathematical model for calculating the dynamic behaviors of the PEH are presented in detail. The fabrication and experimental setup of the PEH prototype are presented in Section 4. The discussions on the numerical and experimental results are presented in Section 5. The conclusions and findings of this study are summarized in Section 6.

2. Working Principle of the PEH

Figure 1a shows a conceptual drawing of the PEH, which includes a bimorph piezoelectric cantilever beam and three magnets. The magnetizations of the magnets are marked by N and S. The moving directions of the magnet A and the beam are also marked in Figure 1a. Figure 1b shows the side view and front view of the PEH. The magnet B is mounted at the beam's tip. At its initial position, the beam magnet B is attracted on the moving magnet A. When magnet A moves along y-direction, the magnetic force acting on magnet B become smaller. At a critical position u_0, the elastic force of the beam equals the magnetic force, and the beam magnet B is about to separate from magnet A, as shown in Figure 1c. The upward motion of the beam is driven by an upward force F_b, which is the resultant of the magnetic forces from magnets A and C, and the elastic force of the beam bending. It should be noted that at the critical position u_0, the beam magnetic force could produce a minor torque on the beam. It is assumed that the torque is small so that the induced torsion deformation can be neglected. During its upward-moving, the beam magnet B has a position u_z and a velocity $\dot{u}_z$. Meanwhile, the magnet A has a position u_y and a velocity $\dot{u}_y$, as illustrated in Figure 1d. Finally, the beam magnet B collides magnet C. After the impact, a transient vibration in the beam occurs and the electric energy is then extracted from the vibration due to the piezoelectric effect.

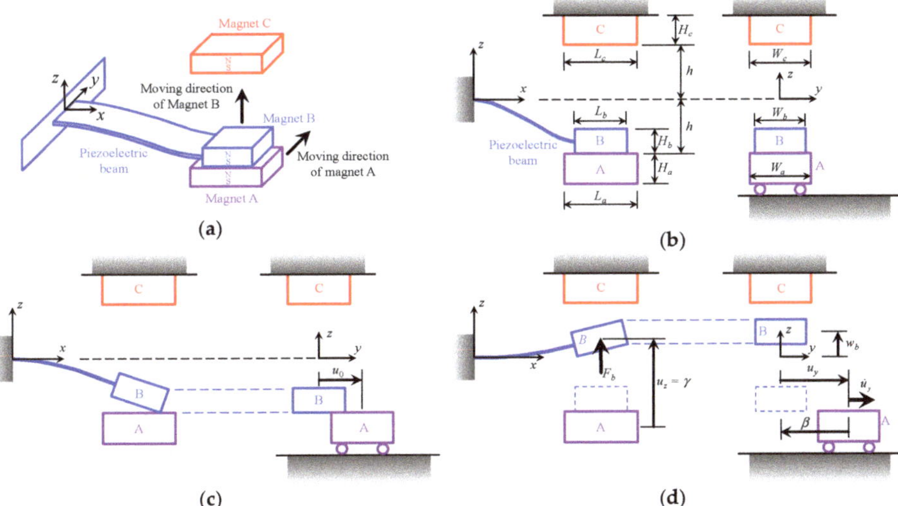

Figure 1. The conceptual drawing of the piezoelectric energy harvester (PEH). (**a**) The iso-view of the PEH at the initial status; (**b**) The side view and front view of the PEH at the initial status; (**c**) The PEH at the time that magnet B separates from magnet A; (**d**) The beam deflection before impact.

In the proposed PEH, the magnet B is moved by finger pressing. It means that the PEH is driven by human motion and could be applicable for driving a batteryless switch [26].

3. Dynamic Model of the PEH

As shown in Figure 2, the piezoelectric beam is composed of a pair of PZT (lead zirconate titanate) and a middle metal shim. The beam is divided by 7 sub-beams. According to the motion sequence of the PEH, the analysis is divided into two phases. In the first phase, we consider the dynamic responses of the beam motion before the impact. In the second phase, the transient vibration after the impact is solved. In the rest of this section, the mathematical procedures of the two phases will be derived separately.

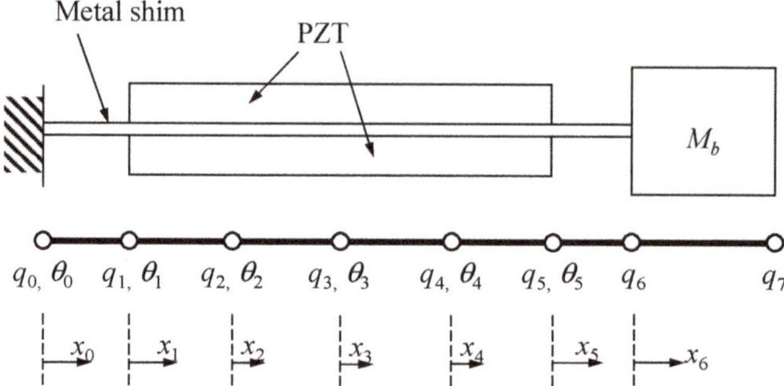

Figure 2. The beam model of the piezoelectric beam.

3.1. Analysis before Impact

The first phase of the model is to consider the beam motion before impact. In this phase, the acceleration is much smaller than that after impact so that we only consider the fundamental vibration mode of vibration, which is assumed to be the same as the static beam deflection subjected to a concentrated force at its end. Based on the Euler beam theory, the static beam deflection can be easily determined by the flexural formula. In a general form, the deflection can be written as

$$w_k(x_k, t) = w_k^{(1)}(x_k, t) = \frac{w_b(t)}{U_b} W_k^{(1)}(x_k), \ k = 0, 1, \cdots 6 \tag{1}$$

where the superscript (1) denotes the first phase, $w_k^{(1)}(x_k, t)$ denotes the deflection of the kth sub-beam before impact, $W_k^{(1)}(x_k)$ are the normalized deflection of the kth sub-beam under a unit force applied at the beam tip, and U_b is the normalized deflection at the beam tip, i.e., $U_b = W_6^{(1)}(x_6 = L_6)$. The expressions for $W_k^{(1)}(x_k)$ is given in the Appendix A. It is seen that w_b actually denotes the deflection at the beam tip, i.e., $w_b(t) = w_6^{(1)}(x_6 = L_6)$.

By neglecting the small tilt angle θ of beam magnet B and the magnetic force along x and y directions, the force F_{a-b} along z direction acting on magnet B from magnet A can be written in the form [27]

$$F_{a-b} = \frac{M_a M_b}{4\pi\mu_0}(\phi_1 + \phi_2 + \phi_3 + \phi_4) \tag{2}$$

where M_a and M_b denote the magnetizations of the magnets A and B, respectively, $\mu_0\left(= 4\pi \times 10^{-7} \ \text{N/A}^2\right)$ denotes the permeability of the air, and

$$\phi_1 = -\sum_{i=0}^{1}\sum_{j=0}^{1}\sum_{k=0}^{1}\sum_{l=0}^{1}\sum_{p=0}^{1}\sum_{q=0}^{1}\left[u_{ij}w_{pq}\ln\left(\sqrt{u_{ij}^2 + v_{kl}^2 + w_{pq}^2} - u_{ij}\right)\right] \tag{3}$$

$$\phi_2 = -\sum_{i=0}^{1}\sum_{j=0}^{1}\sum_{k=0}^{1}\sum_{l=0}^{1}\sum_{p=0}^{1}\sum_{q=0}^{1}\left[v_{kl}w_{pq}\ln\left(\sqrt{u_{ij}^2 + v_{kl}^2 + w_{pq}^2} - v_{kl}\right)\right] \tag{4}$$

$$\phi_3 = \sum_{i=0}^{1}\sum_{j=0}^{1}\sum_{k=0}^{1}\sum_{l=0}^{1}\sum_{p=0}^{1}\sum_{q=0}^{1}\left[u_{ij}v_{kl}\tan^{-1}\left(\frac{u_{ij}v_{kl}}{w_{pq}\sqrt{u_{ij}^2 + v_{kl}^2 + w_{pq}^2}}\right)\right] \tag{5}$$

$$\phi_4 = -\sum_{i=0}^{1}\sum_{j=0}^{1}\sum_{k=0}^{1}\sum_{l=0}^{1}\sum_{p=0}^{1}\sum_{q=0}^{1}\left[w_{pq}\sqrt{u_{ij}^2 + v_{kl}^2 + w_{pq}^2}\right] \tag{6}$$

In Equations (3)–(6), the parameters u_{ij}, v_{kl}, and w_{pq} are given by

$$u_{ij} = \alpha + (-1)^j L_a - (-1)^i L_b, \quad i, j = 0, 1 \tag{7}$$

$$v_{kl} = \beta + (-1)^l W_a - (-1)^k W_b, \quad k, l = 0, 1 \tag{8}$$

$$w_{pq} = \gamma + (-1)^q H_a - (-1)^p H_b, \quad p, q = 0, 1 \tag{9}$$

where (L_a, H_a, W_a) and (L_b, H_b, W_b) be the dimensions of the two magnets, (α, β, γ) is relative position vector from the center of magnet A to the center of magnet B. As shown in Figure 1d, it is seen that the components of the position vector are $\alpha = 0$, $\beta = -u_y$ and $\gamma = u_z$ in this particular case.

Following a very similar procedure, $F_{c\text{-}b}$, the z-component of magnetic force acting on the beam magnet B from magnet C, can be also calculated. The resultant force of the two magnetic forces is $F_b = F_{a\text{-}b} + F_{c\text{-}b}$. Initially, the magnetic force F_b is greater than the elastic force of the beam and the magnet B is attracted by and at rest on the magnet A. When the magnet A begins to move upward along y direction, the magnetic force F_b decreases. At a critical status $u_y = u_0$, the beam begins to move upward. During the upward motion, the kinetic energy $T^{(1)}$ elastic internal energy $V_e^{(1)}$ and electrostatic internal energy $V_s^{(1)}$ can be written in the following forms:

$$T^{(1)} = \frac{1}{2} \sum_{k=0}^{6} \int_0^{L_k} (\rho A)_k \left(\dot{w}_k^{(1)} \right)^2 dx_k \tag{10}$$

$$V_e^{(1)} = \frac{1}{2} \sum_{k=0}^{5} \int_0^{L_k} (EI)_k \left(\frac{d^2 w_k^{(1)}}{dx_k^2} \right)^2 dx_k - \frac{bd_{31}\left(t_m + t_p\right)V}{4s_{11}^E} \sum_{k=1}^{4} \int_0^{L_k} \frac{\partial^2 w_k^{(1)}}{\partial x^2} dx_k \tag{11}$$

$$V_s^{(1)} = \frac{bd_{31}\left(t_m t_p\right)V^{(1)}}{4s_{11}} \sum_{k=1}^{4} \int_0^{L_k} \frac{\partial^2 w_k^{(1)}}{\partial x_k^2} dx_k + \left(\frac{\varepsilon_{33}s_{11} - d_{31}^2}{s_{11}} \right) \frac{b(L_1 + L_2 + L_3 + L_4)\left(V^{(1)}\right)^2}{4t_p} \tag{12}$$

where $V^{(1)}$ denotes the output voltage of the PEH before impact, $(EI)_k$ is the bending rigidity of the kth sub-beam, b is the width of the beam, d_{31} is the piezoelectric constant, s_{11} is the compliance of the piezoelectric, ε_{33} is the dielectric constant of the piezoelectric, t_m is the thickness of the metal shim, and t_p is the thickness of the piezoelectric. The Lagrange equation for the 1-D motion can be written as

$$\frac{d}{dt}\left(\frac{\partial L^{(1)}}{\partial \dot{w}_b} \right) - \frac{\partial L^{(1)}}{\partial w_b} = F_b \tag{13}$$

where the Lagragian $L^{(1)}$ is defined by $L^{(1)} = T^{(1)} - V_e^{(1)} + V_s^{(1)}$. The current generated by the PEH is

$$i^{(1)} = -\frac{bd_{31}\left(t_m + 2t_p\right)}{2s_{11}} \sum_{k=1}^{4} \int_0^{L_k} \frac{\partial^2 \dot{w}_k^{(1)}}{\partial x_k^2} dx_k + \frac{b\left(d_{31}^2 - s_{11}\varepsilon_{33}\right)(L_1 + L_2 + L_3 + L_4)\dot{V}^{(1)}}{2t_p s_{11}} \tag{14}$$

If the PEH is connected to an external resistance R, the equation of circuit can be obtained by the Ohm's law, i.e.,

$$i^{(1)} = \frac{V^{(1)}}{R} \tag{15}$$

Substituting Equations (10)–(12) into Equation (13) and substituting Equation (14) into Equation (15), the equation of motion and equation of circuit before impact are

$$m_{eq}\ddot{w}_b + k_{eq}w_b + \alpha V = F(u_y, u_z) \tag{16}$$

$$\eta \alpha \dot{w}_b - C_p \dot{V} = \frac{V}{R} \tag{17}$$

where

$$u_z(t) = w_6^{(s)}(x_6 = L_6/2, t) + h + \frac{H_a}{2} \tag{18}$$

$$\alpha = -\frac{b d_{31}(t_m t_p)}{4 s_{11} U_b} \sum_{k=1}^{4} \int_0^{L_k} \frac{\partial^2 W_k^{(s)}(x_k)}{\partial x_k^2} dx_k \tag{19}$$

$$\eta = \frac{2 t_m + 4 t_p}{t_m t_p} \tag{20}$$

$$C_p = \frac{b(L_1 + L_2 + L_3 + L_4)}{2 t_p} \left(\varepsilon_{33} - \frac{d_{31}^2}{s_{11}} \right) \tag{21}$$

Assume that the magnet A is moving at a constant velocity v_a. Then the position of the magnet A is

$$u_y = u_0 + v_a(t - t_0) \tag{22}$$

where t_0 is the time that the beam magnet B is about to separate from the magnet A. The initial condition at $t = t_0$ is

$$u_y = u_0, \ w_b = -h - \frac{H_a}{2}, \ \dot{w}_b = 0 \tag{23}$$

By solving Equations (16) and (17) with the initial conditions in Equation (23), the beam tip deflection w_b and output voltage $V^{(1)}$ before impact can be solved. By substituting the solved w_b into Equation (1), the deflection curve of the piezoelectric beam before impact can be determined.

3.2. Analysis after Impact

At the instance of the impact, the deflection at the beam tip is $w_b = h - H_b/2$. By the use of Equation (1), the deflection of the beam at this instance is

$$w_k^{(1)}(x_k, t = 0) = \frac{2h - H_b}{2U_b} W_k^{(1)}(x_k), \ k = 0, 1, \cdots 6 \tag{24}$$

Figure 3a shows the illustration of the deflection curve at the impact and after the impact. By introduce the dynamic displacement $w_k^{(2)}(x_k, t)$ after the impact, the deflection of the beam $w_k^{(2)}$ after the impact can be written as

$$w_k(x_k, t) = w_k^{(2)}(x_k, t) + w_k^{(1)}(x_k, t = 0), \ k = 0, 1, \cdots 6 \tag{25}$$

Note that the second term in the right-hand-side of Equation (25) is independent of time so that the velocity is $\dot{w}_k(x_k, t) = \dot{w}_k^{(2)}(x_k, t)$. The dynamic displacement $w_k^{(2)}(x_k, t)$ can be written in linear combination of the interpolation functions $N_{k1}(x_k)$ to $N_{k4}(x_k)$, $k = 0, 1, \ldots, 6$, i.e.,

$$w_k^{(2)}(x_k, t) = q_k^{(2)}(t)N_{k1}(x_k) + \theta_k^{(2)}(t)N_{k2}(x_k) + q_{k+1}^{(2)}(t)N_{k3}(x_k) + \theta_{k+1}^{(2)}(t)N_{k4}(x_k), \ k = 0, 1, \cdots, 6 \tag{26}$$

where $q_k^{(2)}(t)$ and $\theta_k^{(2)}(t)$ denote the dynamic displacement and dynamic rotation at the kth node for the dynamic term. The expressions for the interpolation functions are given in the Appendix A.

The magnet B may rebound after it impacts the magnet C. To estimate the rebound displacement, consider a simplified model that the magnet B approaches magnet C and collision occurs between them. In this simplified model, the beam is neglected. The rebound velocity of magnet B after impact is $e\dot{w}_b|_{w_b = h - H_b/2}$, where $\dot{w}_b|_{w_b = h - H_b/2}$ denotes the impact velocity of magnet B at the instance just before the impact and e is the coefficient of restitution during the impact. By knowing the mass of the magnet, the kinetic energy of the magnet B can be calculated according to the impact velocity. During the rebound, a negative work is done by the magnetic force. By using the work–kinetic energy principle, the rebound displacement can be estimated. It should be noted that the above simple model overestimates the rebound displacement because we neglect the effects from the beam, which has an upward momentum during the impact and reduce the downward rebound displacement. For the case of small rebound displacement, one can assume that the contact point (corner point 1 shown in Figure 3b) remains no separation after impact. The other corner point 2 shown in Figure 3b is allowed to have motion. However, the degree of freedom of this point is constrained by the magnetic force between two magnets. In this model, an equivalent spring k_s between the magnets B and C is introduced to model this constraint, as shown in Figure 3b. For simplification without loss of generality, we set $t = 0$ at the instance of the impact. In the following, the analysis procedures of the beam after the impact will be derived in detail.

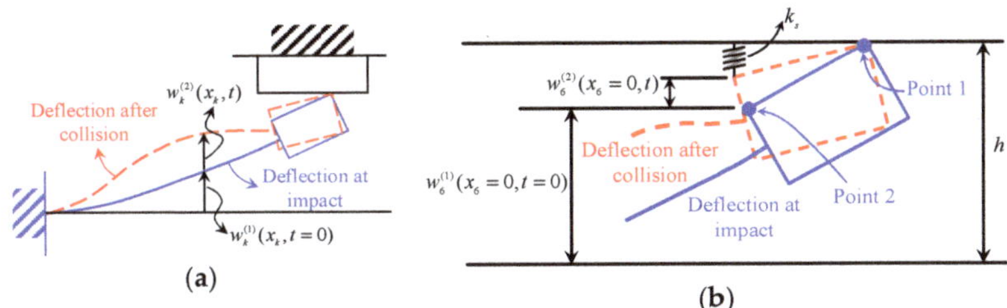

Figure 3. (**a**) Schematic of the beam deflection after impact; (**b**) The spring model of the contact.

In the equivalent spring model, the spring force $F_s(t)$ is given by

$$F_s(t) = k_s \delta(t) \tag{27}$$

where k_s denotes the equivalent spring constant and

$$\delta(t) = h - w_6^{(2)}(x_6 = 0, t) - w_6^{(1)}(x_6 = 0, t = 0) \tag{28}$$

As an estimation, k_s is written in the form

$$k_s = \frac{dF}{d\delta} \tag{29}$$

To derive the equation of motion, consider the elastic internal energy $V_e^{(2)}$, electrostatic internal energy $V_s^{(2)}$ and kinetic energy. The equations of motion can be derived by Lagrange mechanics:

$$\frac{d}{dt}\left(\frac{\partial L^{(2)}}{\partial \dot{q}_k^{(2)}}\right) - \frac{\partial L^{(2)}}{\partial q_k^{(2)}} = 0, \ k = 1, 2, \cdots, 6 \tag{30}$$

$$\frac{d}{dt}\left(\frac{\partial L^{(2)}}{\partial \dot{\theta}_k^{(2)}}\right) - \frac{\partial L^{(2)}}{\partial \theta_k^{(2)}} = 0, \ k = 1, 2, \cdots, 5 \tag{31}$$

where $L^{(2)} = T^{(2)} - V_e^{(2)} - V_e^{(2)}$. The kinetic energy and electrostatic internal energy for the second phase have the same forms as those mentioned in Equations (10)–(12) by changing superscript (1) by (2). The elastic internal energy after impact is

$$V_e^{(2)} = \frac{1}{2}\sum_{k=0}^{5}(EI)_k\int_0^{L_k}\left(\frac{\partial^2 w_k^{(2)}}{\partial x^2}\right)^2 dx_k - \frac{bd_{31}(t_m+t_p)V}{2s_{11}}\sum_{k=1}^{4}\int_0^{L_k}\frac{\partial^2 w_k^{(2)}}{\partial x^2}dx_k + \frac{1}{2}k_{eq}\left(h - q_6^{(2)} - q_6^{(1)}\right)^2 \tag{32}$$

Similarly, the circuit equation after impact can be derived in the same way. The equations of the motion and equation of circuit after impact can be written by

$$\mathbf{M}\ddot{\mathbf{q}}^{(2)} + \mathbf{C}\dot{\mathbf{q}}^{(2)} + \mathbf{K}\mathbf{q}^{(2)} + \alpha V^{(2)} = 0 \tag{33}$$

$$\eta\alpha^T\dot{\mathbf{q}}^{(2)} - C_p\dot{V}^{(2)} = \frac{V^{(2)}}{R} \tag{34}$$

where $\mathbf{M}$ is the mass matrix considering the masses of the beam and magnet, $\mathbf{C}$ is the damping matrix, $\mathbf{K}$ is the stiffness matrix considering both the elastic force of the beam and the equivalent spring, α is the electromechanical coupling matrix containing the piezoelectric constant, C_p is the capacitor of the PZT, V is the output voltage, R is the resistance, and

$$\mathbf{q}^{(2)} = \begin{bmatrix} q_1^{(2)} & \theta_1^{(2)} & \cdots & q_5^{(2)} & \theta_5^{(2)} & q_6^{(2)} \end{bmatrix}^T \tag{35}$$

The expressions for the matrices $\mathbf{M}$, $\mathbf{K}$ and α are given in the Appendix A. The proportional damping model is used in the present analysis, i.e.,

$$\mathbf{C} = \gamma_1\mathbf{M} + \gamma_2\mathbf{K} \tag{36}$$

where γ_1 and γ_2 are constants and can be determined by solving the following two equations:

$$\frac{\gamma_1}{\omega_1} + \omega_1\gamma_2 = 2\zeta_1 \tag{37}$$

$$\frac{\gamma_1}{\omega_2} + \omega_2\gamma_2 = 2\zeta_2 \tag{38}$$

where ω_k, ζ_k ($k = 1, 2$) are the natural frequency and damping ratio, respectively, of the first two vibration modes.

The boundary conditions for the dynamic displacement at $x_0 = 0$ and $x_6 = L_6$ after the impact are

$$q_0^{(2)} = 0,\ \theta_0^{(2)} = 0,\ \theta_7^{(2)} = 0 \tag{39}$$

At the instance that the impact occurs ($t = 0$), the initial conditions for the second phase are

$$q_k^{(2)}(t=0) = 0,\ \theta_k^{(2)}(t=0) = 0,\ \dot{q}_k^{(2)}(t=0) = \frac{1}{\dot{u}_b}W_k^{(1)}\left(\dot{w}_b\Big|_{w_b=h-\frac{H_b}{2}}\right),$$
$$\dot{\theta}_k^{(2)}(t=0) = \frac{1}{\dot{u}_b}\frac{dW_k^{(2)}}{dx_k}\left(\dot{w}_b\Big|_{w_b=h-\frac{H_b}{2}}\right),\ V^{(2)}(t=0) = V^{(1)}\Big|_{w_b=h-\frac{H_b}{2}} \tag{40}$$

where $\dot{w}_b\Big|_{w_b=h-\frac{H_b}{2}}$ denotes the velocity of the beam tip at the time just before the impact and can be determined from the solution of Equation (13) mentioned in phase 1.

By solving Equations (33) and (34) in conjunction with the boundary conditions Equation (39) and initial conditions Equation (40), the transient responses of displacements and voltage of the beam can be obtained.

The energy E_{PEH} harvested by the PEH can be calculated by

$$E_{\text{PEH}} = \int_{t_0}^{0} \frac{V^2}{R} dt + \int_{0}^{\Delta t} \frac{V^2}{R} dt \tag{41}$$

where Δt is the time interval for calculating the energy.

4. Fabrication of Prototype and Experimental Setup

Figure 4a illustrates the PEH prototype. The piezoelectric beam is clamped on a base. The magnets A and C are respectively mounted on two sliding bars, which can slide under the guidance of a pair of guiders. In the initial status, the beam magnet B is attracted by magnet A and is at rest. The upper sliding bar is placed at a position that the magnet C is at the position aligned to magnets A and B. By pushing the sliding bar to move right, the magnet A separates magnet B, and the beam moves upward and finally stopped by the magnet C. Then an impact occurs. In Figure 4b, the two bars are hidden for a better view for piezoelectric beam and magnets. The base, guiders and bars are made by acrylic to avoid interfering the magnetic fields.

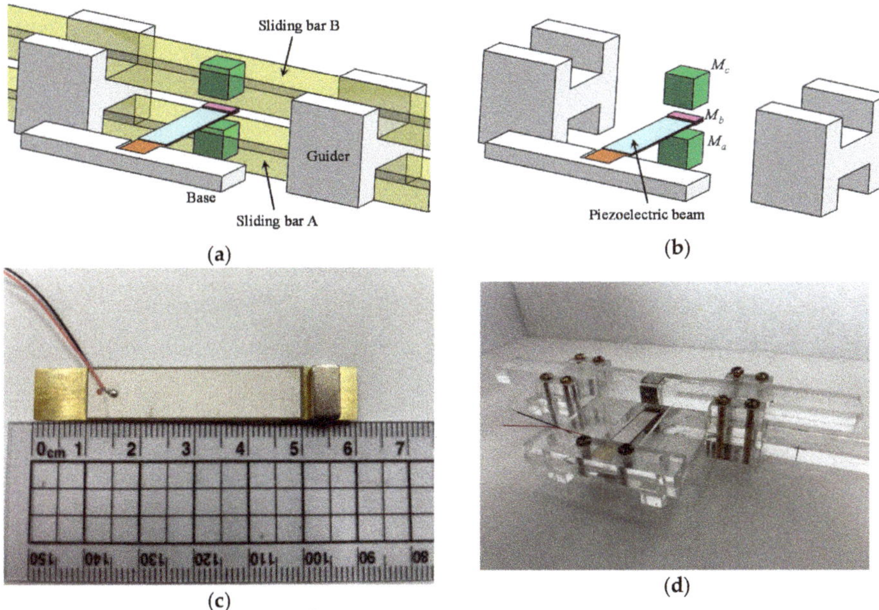

Figure 4. (**a**) The design of the PEH; (**b**) The design of the PEH with invisible slider bars; (**c**) Piezoelectric beam; (**d**) Prototype of the PEH.

The fabrication and assembly of the piezoelectric beam were provided by Eleceram Technology Co., Taiwan, as shown in Figure 4c. The geometric parameters of the beam and magnets are listed in Table 1. Two identical NdFeB magnets with a size of $10 \times 5 \times 0.8$ mm^3 were adhered at the beam tip to form the magnet B, which has an equivalent size of $10 \times 5 \times 1.6$ mm^3. Another two identical NdFeB magnets serve as magnets A and C, each of which has a size of $10 \times 10 \times 10$ mm^3. The geometric parameters of the magnets are also summarized in Table 1. Figure 4d shows the assembled prototype.

Table 1. The geometric parameters of the beam and magnets.

Parts	Geometric Parameters
Beam	$L_0 = 0.5$ mm, $L_1 = L_2 = L_3 = L_4 = 10$ mm, $L_5 = 1.5$ mm, $L_6 = 5$ mm, $t_p = 0.239$ mm, $t_m = 0.1$ mm, $b = 10$ mm
Magnet A	$L_a = H_a = W_a = 10$ mm
Magnet B	$L_b = 5$ mm, $H_b = 1.6$ mm, $W_b = 10$ mm
Magnet C	$L_c = H_c = W_c = 10$ mm

Figure 5 shows the setup for measuring the magnetic forces from one magnet to another for various distances between two magnets. In the experiment, a magnet was fixed on the fixture. A second magnet was placed on a jig that can freely slide along a guider. A force gauge (DigiTech DTG-10) withstood the second magnet at a gap between the two magnets and measured the magnetic force. Three magnet pairs A-A, B-B and A-B were considered in the magnetic force measurements for various gaps between the two magnets.

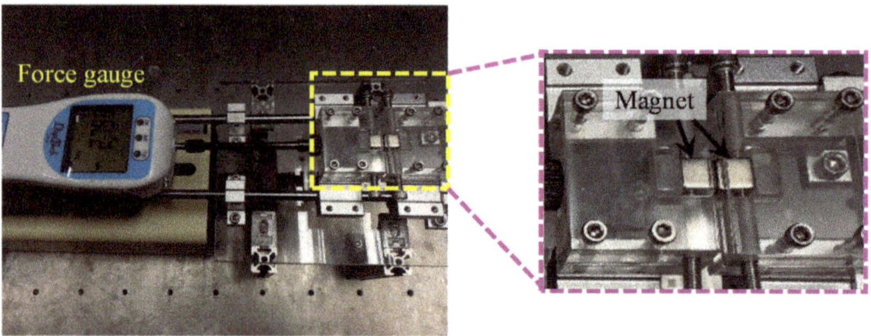

Figure 5. The experimental setup for the magnetic force measurement.

For the magnet pair A-A experiment, the magnetic force $F_{a-a,k}^{(\exp)}$ is measured for kth gap. Based on the 3-D magnetic force model described in Equation (2), the magnetic force $F_{a-a,k}^{(model)}$ of the magnet pair A-A can be written in the form:

$$F_{a-a,k} = \frac{M_a^2}{4\pi\mu_0}\left(\phi_{1,k} + \phi_{2,k} + \phi_{3,k} + \phi_{4,k}\right) \tag{42}$$

where the subscript k denotes the parameters calculated according to the kth gap. According to the least square method, the magnetic magnetization M_a of the magnet A can be determined solving the equation:

$$\frac{d}{dQ_{a-a}}\sum_{k=1}^{N}\left(F_{a-a,k}^{(\exp)} - F_{a-a,k}\right)^2 = 0 \tag{43}$$

where $Q_{a-a} = M_a^2$ and N denotes the number of gaps for the measurement. Following a similar procedure, the magnetization M_b can be also determined from the magnet pair B-B experiment.

To investigate the energy generated by the PEH, an oscilloscope (Tektronix DPO 4054B) was used to monitor the output voltage of the resistor connected to the PEH, as shown in Figure 6a. Because the motion of the sliding bar is moved by hand, the velocities of the magnet A change for different tests. The motion of the sliding bar (or the magnet A) was monitored by a high-speed camera (OLYMPUS i-SPEED 3) with a frame rate of 5000 Hz, as shown in Figure 6b. By analyzing the photos frame by frame, the positions of the sliding bar and the beam tip can be determined. The velocities of the beam magnet B can be calculated accordingly.

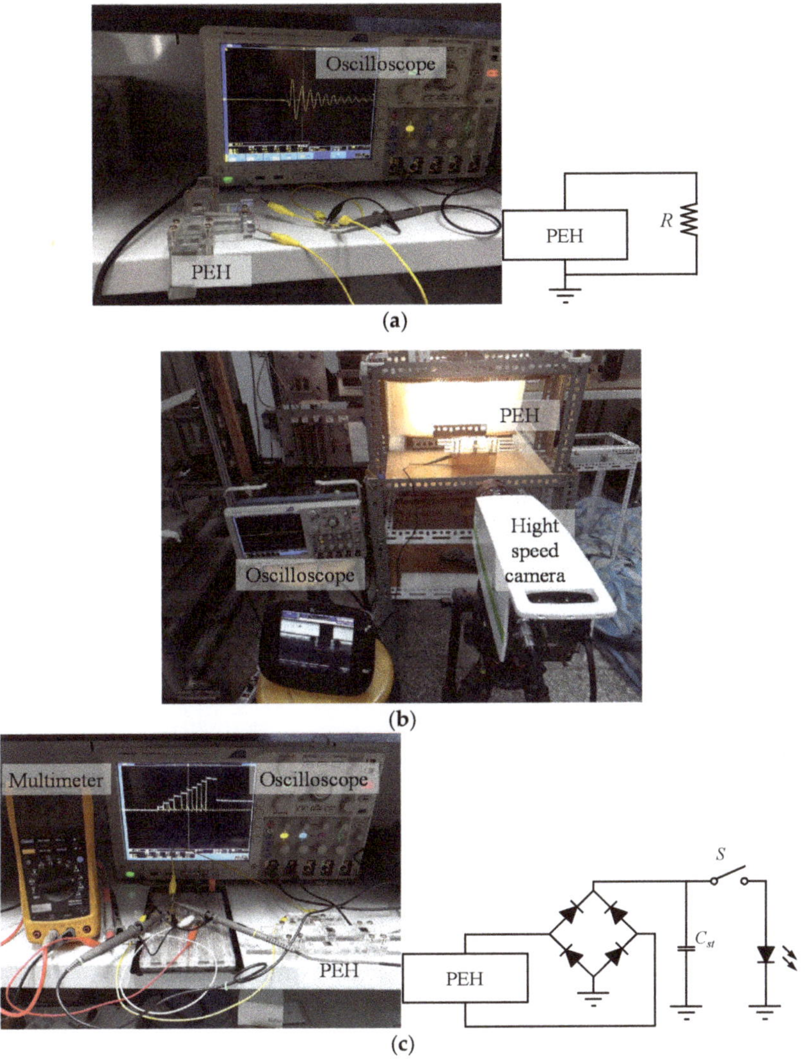

Figure 6. The experimental setup of the PEH: (**a**) voltage measurement of the PEH coupled with a resistance; (**b**) motion measurement of the PEH by the highspeed camera; (**c**) voltage measurement of the PEH coupled with a diode-bridge, capacitor, and LED.

Figure 6c shows experimental setup for light LED. Four diodes (1N4004) were connected to form a bridge. The AC voltages were rectified to DC signals and the electric charges were stored in a capacitor C_{st} of 10 μF. An LED was connected to the capacitor in parallel. In the experiments, the first impact was driven by magnet A and the collision between the magnets B and C. Put the magnet A back to the initial position and move the magnet C. Then the beam moved to magnet A and the second impact occurred. The process was repeated ten times and produced ten impacts. For each impact, the capacitor was charged and the voltage across the capacitor was boosted. After ten impacts, the switch S was switched to connect the LED loop. The capacitor discharged and a current flowed through and lighted up the LED. The oscilloscope was used to monitor the voltage across the capacitor during the charging and discharging in order to evaluate the energy stored in the capacitor. A multimeter (Fluke 189) was used to capture the peak current during the discharging in order to evaluate the maximum instantaneous power for lighting the LED.

5. Results and Discussions

In this section, the experimental and numerical results are presented to demonstrate the performance of the PEH. The material properties used in the dynamic model are listed in Table 2.

Table 2. The material properties used in the model.

Material	Property
Piezoelectric	$E_{pzt} = 1/s_{11} = 66\,\text{GPa}$, $\rho_{pzt} = 7900\,\text{kg/m}^3$, $d_{31} = 140 \times 10^{-12}\,\text{C/N}$, $\varepsilon_{33}/\varepsilon_0 = 2100$
Metal shim	$E_{metal} = 110\,\text{GPa}$, $\rho_{metal} = 8000\,\text{kg/m}^3$
Magnet	$\rho_{mag} = 7300\,\text{kg/m}^3$, $M_a = 0.9537\,\text{T}$, $M_b = 0.5147\,\text{T}$

5.1. Measurements of Magnetic Forces

The variation of the magnetic forces for various gaps are shown in Figure 7. Note that the magnetizations M_a, M_b, and M_c for the magnets A, B, and C are respectively unknown. By using the least square method described in Equation (43), the magnetizations M_a and M_b are determined to be 0.9537 T and 0.5147 T, respectively. In Figure 7, it is seen that the regression curves for magnet pairs A-A and B-B agree well with the measurement data. These two magnetizations are used to calculate the magnetic forces for the magnet pair A-B. In Figure 7, it is seen that the computation results for magnetic force between magnets A and B are validated by experiments. It indicates that Equation (2) can used to predict the magnetic force at high accuracy.

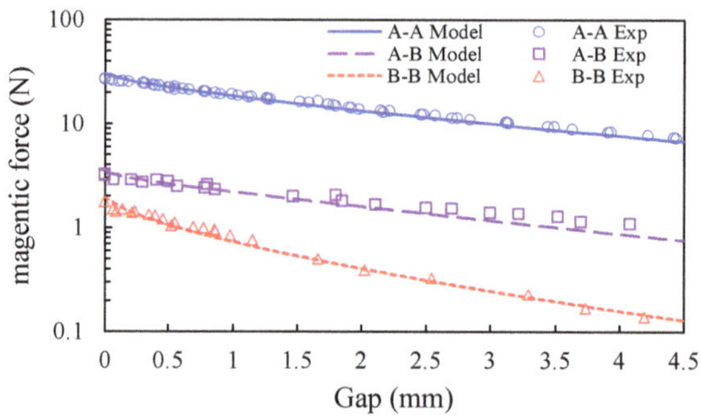

Figure 7. The comparisons of magnetic forces by measurements and model for magnet pairs A-A, B-B, and A-B for various gaps.

5.2. Effects of the Velocity of the Magnet A

The PEH was driven by finger pressing. The velocity v_a of the magnet A varies for different tests. To investigate the effects of the velocity v_a on the motion of the beam and output voltage, five tests to trigger the PEH have been tested. Figure 8 shows the measured beam tip deflections w_b for the first test. The deflection results show that the beam tip accelerated during its upward motion before impact. The measured positions u_y of the magnet A for the first test are also showed in Figure 8. It is seen that the position u_y of the magnet A exhibits a linear trend over time and the constant velocity assumption described in Equation (22) can be acceptable.

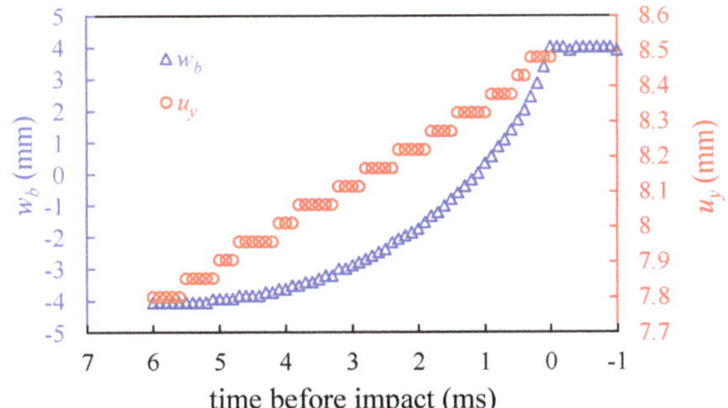

Figure 8. The measured deflection w_b of the beam tip and the position u_y of the magnet A before impact for the first test.

Table 3 lists the measured relative position u_0 between magnets A and B at the instance that the beam tip begins to move upward, and the measured velocities v_a of the magnet A for the five tests. The variation of relation position u_0 for different tests is small. Meanwhile, the velocity v_a varies from different tests because the motion of the magnet A was driven by hand. Among the five tests, the maximum, minimum and average values of v_a are 0.1337 m/s, 0.07763 m/s, and 0.09731 m/s, respectively. The measured impact velocity, which is defined as the beam tip velocity $\dot{w}_b|_{w_b=h-\frac{H_b}{2}}$ at the time just before impact, for the five tests are also listed in Table 3. It is observed that the relation between impact velocity and the moving magnet velocity v_a is insignificant. The average impact velocity for the five tests is 5.106 m/s. Assume that the coefficient of restitution during the impact is $e = 0.5$, then the rebound velocity of the magnet B is 2.553 m/s. By knowing the mass of 0.584 g for the magnet B, the kinetic energy after impact is 1.955 mJ. According to the magnetic force measurements, the magnetic attraction force is approximately 3 N when the magnet B is close to magnet C. By using the energy balance, the rebound displacement is 0.65 mm, which is quite small by comparing the whole moving distance. As mentioned in Section 3.2, the calculation of the rebound displacement is overestimated. It can be concluded that the small rebound assumption is acceptable. In Figure 8, the measurement data also show the beam tip deflection after impact. It is observed that the beam tip becomes stationary after impact, i.e., no rebound can be observed. Therefore, the model shown in Figure 3b can be acceptable.

Figure 9 shows the time history of the measured beam tip deflection for the five tests. Each test reveals almost identical beam tip deflection curve. It indicates that the beam tip motion is independent of the driving velocity v_a. The model calculation results agree well with the measured data, indicating that the model for the first phase can predict the motion of the beam tip at high accuracy.

Figure 10 shows the time history of magnetic forces applied on magnet B from magnets A and C during the phase before impact. As expected, the magnetic force from magnet C increases by time, while the magnetic force from magnet A is relatively small, especially for the time approaching impact. At the time of 1 ms before impact, the magnetic force from magnet C dominates the beam tip motion. The small force magnetic force from magnet A can be used to explain the insignificant contribution of v_a.

Figure 11 shows the open-circuit voltage for the five tests. The transient responses of the voltages before and after impact keep the same for various v_a. The voltage responses computed from the model are also shown in Figure 11. According to the voltage responses after impact shown in Figure 11, the motion mainly exhibits the fundamental mode although some minor higher order vibration modes are observed. The domination of the fundamental mode can be also seen in model calculations. It is seen that the model agrees with the measurements. By picking the peak and valley points of the transient voltage response, the natural frequency of the vibration can be determined. According to this method, the natural frequencies for the five tests in Figure 11 are 407.9, 403.3, 408.9, 413.6, and 410.9 Hz.

The average natural frequency for the five data is 408.9 Hz. The same method can be also applied in the time response obtained by the model and the result is 373.0 Hz.

Table 3. Measurement results of u_0, v_a, and impact velocity before impact for the five tests.

Test No.	1	2	3	4	5	Average
u_0 (mm)	7.80	7.79	7.85	7.87	7.75	7.81
v_a (m/s)	0.113	0.134	0.082	0.080	0.078	0.097
Impact velocity (m/s)	5.451	5.047	5.207	4.372	5.456	5.106

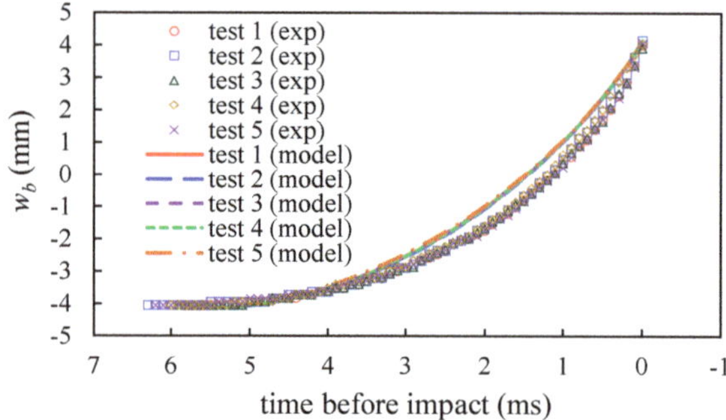

Figure 9. The comparisons of the deflection w_b of the beam tip before impact obtained by experiments and model.

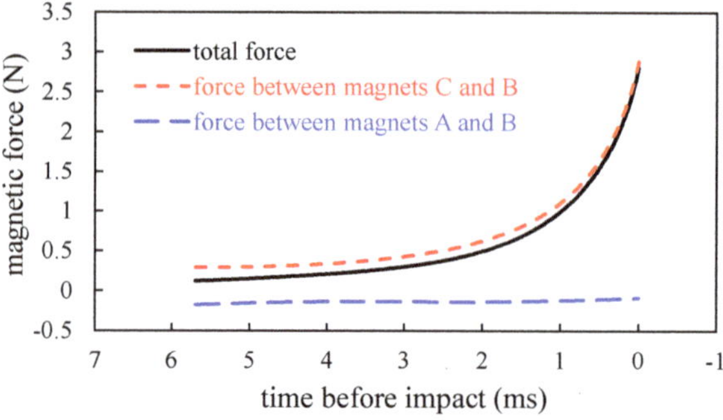

Figure 10. The computed magnetic forces by 3-D magnetic force model before impact.

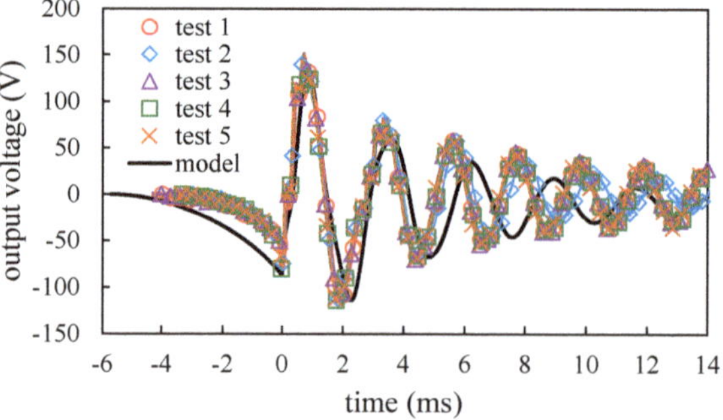

Figure 11. The comparisons of the open-circuit voltage responses of the PEH obtained by experiments and the dynamic model.

5.3. Energy Measurements of the PEH

Figure 12 shows the measured maximum $V_{\max}$ and energy E_{PEH} harvested by the PEH for various external resistances. For small external resistance, both $V_{\max}$ and E_{PEH} are small. The energy increases with the increase of R when R is less than 20 $k\Omega$. A maximum energy of 0.4045 mJ occurs at the optimum resistance $R_{\text{opt}} = 15\ k\Omega$. For a large R, $V_{\max}$ reaches a stable value and the average power becomes small. In Figure 12, the measured voltage and energy are also compared with the results computed by the model. The comparisons show that the results of the model agree well with the experimental data.

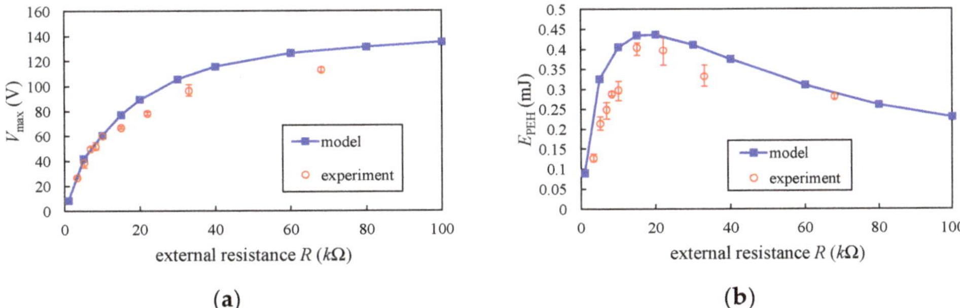

(a)　　　　　　　　**(b)**

Figure 12. (**a**) The maximum output voltage of the PEH for various external resistance R; (**b**) The energy harvested by the PEH for various external resistance R.

For the lighting up LED experiment, two tests were conducted and the voltages across the capacitor, as shown in Figure 6c, are plotted in Figure 13. For each test, the voltage was boosted to 6.2 V after ten impacts and the energy stored in the capacitor was 19.22 µJ. When the circuit was switched, the capacitor discharged then a current flowed through and lighted up the LED. The voltage drops during the discharging for the two tests were 3.44 V and 3.52 V. The maximum currents during the discharging for the two tests were 450 µA and 476 µA. The maximum instantaneous powers during the discharging for the two tests were 0.774 mW and 0.838 mW. The experiments showed that the energy stored in the capacitor can light up the LED.

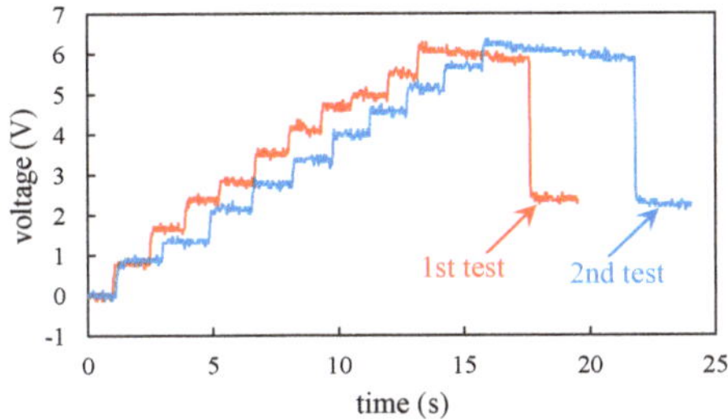

Figure 13. The measured voltage of the capacitor in the lighting up LED experiment.

6. Conclusions

In this study, the impact-driven PEH in the magnetic field is presented. The multi-DOF mathematical model was developed for solving the dynamic and vibration behaviors of the piezoelectric beam under the effects of magnetic fields. The prototype was also fabricated and the performance of the PEH was measured in detail. The conclusions of this study are summarized as follows:

(1) The 3-D magnetic force model were introduced to calculate the magnetic force between magnets. The magnetic force experimental setup was developed and the measured forces for various gaps between two magnets agree with the model.

(2) Based on the multi-DOF mathematical model, the deflections and voltages of the piezoelectric beam were investigated in detail. The model is divided by two phases. In the first phase, the motion of the piezoelectric beam is governed by the restoring force of the beam and the magnetic forces due to magnets. The second phase begins at the time of impact. The ending conditions of the first phase are imposed as the initial conditions in the analysis of the second phase. In the second phase, the transient vibration responses can be solved.

(3) To produce the impact, the magnet A was moved by hand and the consequent motions were triggered. The experimental results showed that the velocity v_a of magnet A varies from different tests. However, it was found that the variations of v_a have nearly no contribution on the beam motion in the first phase and the voltage responses in the second phase. This phenomenon was also observed in the model simulation.

(4) The voltage and energy outputs were measured for various external resistance R. The experiments showed that the voltage outputs increases with the increase of R. The energy output was observed to be low for both small and large R. The maximum energy output was found to be 0.4045 mJ at the optimum resistance $R_{opt} = 15\ k\Omega$. The voltage and energy outputs computed by the model for various resistances agree well with the measurements.

(5) In the lighting LED experiment, the voltage was charged and an energy of 19.22 µJ was stored in the capacitor by ten impacts. The experiments showed that the energy stored in the capacitor can light up the LED.

(6) The permanent magnets are brittle and easy to be damaged after impact. In the experiments, however, no cracks or damages have been observed in the magnets. It suggests that the impact velocity in the present PEH is not fast enough to damage the magnet. In addition, the PZT is also brittle and easy to be damaged, although it is not impacted directly by the magnet. A more detail stress analysis could be performed in the future for the damage evaluation of the brittle materials.

Author Contributions: Conceptualization, C.-D.C., Y.-H.W., and P.-W.S.; formal analysis, C.-D.C. and Y.-H.W.; methodology Y.-H.W. and P.-W.S.; validation, C.-D.C.; writing—original draft, C.-D.C. and Y.-H.W.; writing—review and editing, C.-D.C. All authors have read and agreed to the published version of the manuscript.

Funding: The authors are grateful to the Ministry of Science and Technology, Taiwan, for the financial support through grant MOST 107-2221-E-006-154-MY2.

Conflicts of Interest: The authors declare no conflict of interest.

Appendix A

The normalized deflection $W_k^{(1)}(x_k)$ appeared in Equation (1) is given by

$$
W_k^{(1)}(x_k) = \begin{cases}
\dfrac{1}{2(EI)_0}\left(\tfrac{1}{2}L_6 + \sum\limits_{i=0}^{5} L_i\right)x_0^2 - \dfrac{1}{6(EI)_0}x_0^3, & \text{for } k = 0 \\[2em]
\dfrac{1}{2(EI)_k}\left(\tfrac{1}{2}L_6 + \sum\limits_{i=k}^{5} L_i\right)x_k^2 - \dfrac{1}{6(EI)_k}x_k^3 + \left(\left.\dfrac{dW_{k-1}^{(1)}}{dx_{k-1}}\right|_{x_{k-1}=L_{k-1}}\right)x_k + \left.W_{k-1}^{(1)}\right|_{x_{k-1}=L_{k-1}}, & \text{for } k = 1,2,\cdots,5 \\[2em]
W_6^{(1)}(x_6) = \left(\left.\dfrac{dW_5^{(1)}}{dx_5}\right|_{x_5=L_5}\right)x_6 + \left.W_5^{(1)}\right|_{x_5=L_5}, & \text{for } k = 6
\end{cases}
$$

The interpolation functions appeared in Equation (26) is given by

$$N_{k1} = 1 - \frac{3x_k^2}{L_k^2} + \frac{2x_k^3}{L_k^3},\ N_{k2} = x_k - \frac{2x_k^2}{L_k} + \frac{x_k^3}{L_k^2},\ N_{k3} = \frac{3x_k^2}{L_k^2} - \frac{2x_k^3}{L_k^3},\ N_{k4} = -\frac{x_k^2}{L_k} + \frac{x_k^3}{L_k^2},\ k = 0, 2, \ldots, 6$$

The element $M_{i\text{-}j}$ $(i, j = 1, 2, \ldots, 11)$ at ith row and jth column of the mass matrix **M** appeared in Equation (33) is given by

$$M_{1\text{-}1} = \tfrac{13}{35}\Big[L_0(\rho A)_0 + L_1(\rho A)_1\Big],\ M_{1\text{-}2} = M_{2\text{-}1} = -\tfrac{11}{210}\Big[L_0^2(\rho A)_0 - L_1^2(\rho A)_1\Big],$$

$$M_{1\text{-}3} = M_{3\text{-}1} = \tfrac{9}{70}L_1(\rho A)_1,\ M_{1\text{-}4} = M_{4\text{-}1} = -\tfrac{13}{420}L_1^2(\rho A)_1,$$

$$M_{2\text{-}2} = \tfrac{1}{105}\Big[L_0^3(\rho A)_0 + L_1^3(\rho A)_1\Big],\ M_{2\text{-}3} = M_{3\text{-}2} = \tfrac{13}{420}L_1^2(\rho A)_1,$$

$$M_{2\text{-}4} = M_{4\text{-}2} = -\tfrac{1}{140}L_1^3(\rho A)_1,\ M_{3\text{-}3} = \tfrac{13}{35}\Big[L_1(\rho A)_1 + L_2(\rho A)_2\Big],$$

$$M_{3\text{-}4} = M_{4\text{-}3} = -\tfrac{11}{210}\Big[L_1^2(\rho A)_1 - L_2^2(\rho A)_2\Big],\ M_{3\text{-}5} = M_{5\text{-}3} = \tfrac{9}{70}L_2(\rho A)_2,$$

$$M_{3\text{-}6} = M_{6\text{-}3} = -\tfrac{13}{420}L_2^3(\rho A)_2,\ M_{4\text{-}4} = \tfrac{1}{105}\Big[L_1^3(\rho A)_1 + L_2^3(\rho A)_2\Big],$$

$$M_{4\text{-}5} = M_{5\text{-}4} = \tfrac{13}{420}L_2^2(\rho A)_2,\ M_{4\text{-}6} = M_{6\text{-}4} = -\tfrac{1}{140}L_2^3(\rho A)_2,$$

$$M_{5\text{-}5} = \tfrac{13}{35}\Big[L_2(\rho A)_2 + L_3(\rho A)_3\Big],\ M_{5\text{-}6} = M_{6\text{-}5} = -\tfrac{11}{210}\Big[L_2^2(\rho A)_2 - L_3^2(\rho A)_3\Big],$$

$$M_{5\text{-}7} = M_{7\text{-}5} = \tfrac{9}{70}L_3(\rho A)_3,\ M_{5\text{-}8} = M_{8\text{-}5} = -\tfrac{13}{420}L_3^2(\rho A)_3,$$

$$M_{6\text{-}6} = \tfrac{1}{105}\Big[L_2^2(\rho A)_2 + L_3^2(\rho A)_3\Big],\ M_{6\text{-}7} = M_{7\text{-}6} = \tfrac{13}{420}L_3^2(\rho A)_3,$$

$$M_{6\text{-}8} = M_{8\text{-}6} = -\tfrac{1}{140}L_3^2(\rho A)_3,\ M_{7\text{-}7} = \tfrac{13}{35}\Big[L_3(\rho A)_3 + L_4(\rho A)_4\Big],$$

$$M_{7\text{-}8} = M_{8\text{-}7} = -\tfrac{11}{210}\Big[L_3^2(\rho A)_3 - L_4^2(\rho A)_4\Big],\ M_{7\text{-}9} = M_{9\text{-}7} = \tfrac{9}{70}L_4(\rho A)_4,$$

$$M_{7\text{-}10} = M_{10\text{-}7} = -\tfrac{13}{420}L_4^2(\rho A)_4,\ M_{8\text{-}8} = \tfrac{1}{105}\Big[L_3^3(\rho A)_3 + L_4^3(\rho A)_4\Big],$$

$$M_{8\text{-}9} = M_{9\text{-}8} = \tfrac{13}{420}L_4^2(\rho A)_4,\ M_{8\text{-}10} = M_{10\text{-}8} = -\tfrac{1}{140}L_4^3(\rho A)_4,$$

$$M_{9\text{-}9} = \tfrac{13}{35}\Big[L_4(\rho A)_4 + L_5(\rho A)_5\Big],\ M_{9\text{-}10} = M_{10\text{-}9} = -\tfrac{11}{210}\Big[L_4^2(\rho A)_4 - L_5^2(\rho A)_5\Big],$$

$$M_{9\text{-}11} = M_{11\text{-}9} = \frac{L_5(13L_5 + 54L_6)(\rho A)_5}{420L_6},\ M_{10\text{-}10} = \tfrac{1}{105}\Big[L_4^3(\rho A)_4 + L_5^3(\rho A)_5\Big],$$

$$M_{10\text{-}11} = M_{11\text{-}10} = \frac{L_5^2(3L_5 + 13L_6)(\rho A)_5}{420L_6},$$

$$M_{11\text{-}11} = \frac{1}{105L_6^2}\Big[\big(L_5^3 + 11L_5^2L_6 + 39L_5L_6^2\big)(\rho A)_5 + 35L_6^3(\rho A)_6\Big]$$

The element $K_{i\text{-}j}$ $(i, j = 1, 2, \ldots, 11)$ at ith row and jth column of the stiffness matrix **K** appeared in Equation (33) is given by

It is noted that the elements not mentioned above for the mass matrix and stiffness matrix are zero.

$$K_{1-1} = 12\left[\frac{(EI)_0}{L_0^3} + \frac{(EI)_1}{L_1^3}\right], \; K_{1-2} = K_{2-1} = 6\left[-\frac{(EI)_0}{L_0^2} + \frac{(EI)_1}{L_1^2}\right], \; K_{1-3} = K_{3-1} = -\frac{12(EI)_1}{L_1^3},$$

$$K_{1-4} = K_{4-1} = \frac{6(EI)_1}{L_1^2}, \; K_{2-2} = 4\left[\frac{(EI)_0}{L_0} + \frac{(EI)_1}{L_1}\right], \; K_{2-3} = K_{3-2} = -\frac{6(EI)_1}{L_1^2},$$

$$K_{2-4} = K_{4-2} = \frac{2(EI)_1}{L_1}, \; K_{3-3} = 12\left[\frac{(EI)_1}{L_1^3} + \frac{(EI)_2}{L_2^3}\right], \; K_{3-4} = K_{4-3} = 6\left[-\frac{(EI)_1}{L_1^2} + \frac{(EI)_2}{L_2^2}\right],$$

$$K_{3-5} = K_{5-3} = -\frac{12(EI)_2}{L_2^3}, \; K_{3-6} = K_{6-3} = \frac{6(EI)_2}{L_2^2}, \; K_{4-4} = 4\left[\frac{(EI)_1}{L_1} + \frac{(EI)_2}{L_2}\right],$$

$$K_{4-5} = K_{5-4} = -\frac{6(EI)_2}{L_2^2}, \; K_{4-6} = K_{6-4} = \frac{2(EI)_2}{L_2}, \; K_{5-5} = 12\left[\frac{(EI)_2}{L_2^3} + \frac{(EI)_3}{L_3^3}\right],$$

$$K_{5-6} = K_{6-5} = 6\left[-\frac{(EI)_2}{L_2^2} + \frac{(EI)_3}{L_3^2}\right], \; K_{5-7} = K_{7-5} = -\frac{12(EI)_3}{L_3^3}, \; K_{5-8} = K_{8-5} = \frac{6(EI)_3}{L_3^2},$$

$$K_{6-6} = 4\left[\frac{(EI)_2}{L_2} + \frac{(EI)_3}{L_3}\right], \; K_{6-7} = K_{7-6} = -\frac{6(EI)_3}{L_3^2}, \; K_{6-8} = K_{8-6} = \frac{2(EI)_3}{L_3},$$

$$K_{7-7} = 12\left[\frac{(EI)_3}{L_3^3} + \frac{(EI)_4}{L_4^3}\right], \; K_{7-8} = K_{8-7} = 6\left[-\frac{(EI)_3}{L_3^2} + \frac{(EI)_4}{L_4^2}\right],$$

$$K_{7-9} = K_{9-7} = -\frac{12(EI)_4}{L_4^3}, \; K_{7-10} = K_{10-7} = \frac{6(EI)_4}{L_4^2}, \; K_{8-8} = 4\left[\frac{(EI)_3}{L_3} + \frac{(EI)_4}{L_4}\right],$$

$$K_{8-9} = K_{9-8} = -\frac{6(EI)_4}{L_4^2}, \; K_{8-10} = K_{10-8} = \frac{2(EI)_4}{L_4}, \; K_{9-9} = 12\left[\frac{(EI)_4}{L_4^3} + \frac{(EI)_5}{L_5^3}\right],$$

$$K_{9-10} = K_{10-9} = 6\left[-\frac{(EI)_4}{L_4^2} + \frac{(EI)_5}{L_5^2}\right], \; K_{9-11} = K_{11-9} = -\frac{6(L_5+2L_6)(EI)_5}{L_5^3 L_6},$$

$$K_{10-10} = 4\left[\frac{(EI)_4}{L_4} + \frac{(EI)_5}{L_5}\right], \; K_{10-11} = K_{11-10} = -\frac{2(L_5+3L_6)(EI)_5}{L_5^2 L_6},$$

$$K_{11-11} = k_{eq} + \frac{4\left(L_5^2+3L_5L_6+3L_6^2\right)(EI)_5}{L_5^2 L_6^2}$$

The electromechanical coupling matrix $\boldsymbol{\alpha}$ is given by

$$\boldsymbol{\alpha} = \frac{bd_{31}\left(t_m + t_p\right)}{s_{11}}\begin{bmatrix} 0 & 1 & 0 & 0 & 0 & 0 & 0 & 0 & 0 & -1 & 0 \end{bmatrix}^T$$

References

1. Roundy, S.; Wright, P.K.; Rabaey, J. A study of low level vibrations as a power source for wireless sensor nodes. *Comput. Commun.* **2003**, *26*, 1131–1144. [CrossRef]
2. Anton, S.R.; Sodano, H.A. A review of power harvesting using piezoelectric materials (2003–2006). *Smart Mater. Struct.* **2007**, *16*, R1–R21. [CrossRef]
3. Sarker, M.R.; Julai, S.; Sabri, M.F.M.; Said, S.M.; Islam, M.M.; Tahir, M. Review of piezoelectric energy harvesting system and application of optimization techniques to enhance the performance of the harvesting system. *Sens. Actuators A* **2019**, *300*, 111634. [CrossRef]
4. Covaci, C.; Gontean, A. Piezoelectric Energy Harvesting Solutions: A Review. *Sensors* **2020**, *20*, 3512. [CrossRef]
5. Zhu, D.; Tudor, M.J.; Beeby, S.P. Strategies for increasing the operating frequency range of vibration energy harvesters: A review. *Meas. Sci. Technol.* **2010**, *21*, 022001. [CrossRef]
6. Pasquale, G.D.; Kim, S.G.; Pasquale, D.D. GoldFinger: Wireless human–machine interface with dedicated software and biomechanical energy harvesting system. *IEEE/ASME Trans. Mechatron.* **2016**, *21*, 565–575. [CrossRef]

7. Elvin, N.G.; Elvin, A.A. Vibrational energy harvesting from human gait. *IEEE/ASME Trans. Mechatron.* **2013**, *18*, 637–644. [CrossRef]

8. Umeda, M.; Nakamura, K.; Ueha, S. Analysis of transformation of mechanical impact energy to electrical energy using a piezoelectric vibrator. *Jpn. J. Appl. Phys.* **1996**, *35*, 3267–3273. [CrossRef]

9. Gu, L.; Livermore, C. Impact-driven frequency up-converting coupled vibration energy harvesting device for low frequency operation. *Smart Mater. Struct.* **2011**, *20*, 045004. [CrossRef]

10. Halim, M.A.; Park, J.Y. Theoretical modeling and analysis of mechanical impact driven and frequency up-converted piezoelectric energy harvester for low-frequency and wide-bandwidth operation. *Sens. Actuators A Phys.* **2014**, *208*, 56–65. [CrossRef]

11. Zhang, J.; Qin, L. A tunable frequency up-conversion wideband piezoelectric vibration energy harvester for low-frequency variable environment using a novel impact- and rope-driven hybrid mechanism. *Appl. Energy* **2019**, *240*, 26–34. [CrossRef]

12. Huang, M.; Hou, C.; Li, Y.; Liu, H.; Wang, F.; Chen, T.; Yang, Z.; Tang, G.; Sun, L. A low-frequency MEMS piezoelectric energy harvesting system based on frequency-up conversion mechanism. *Micromachines* **2019**, *10*, 639. [CrossRef]

13. Febbo, M.; Machado, S.P.; Osinaga, S.M. A novel up-converting mechanism based on double impact for non-linear piezoelectric energy harvesting. *J. Phys. D Appl. Phys.* **2020**, *53*, 475501. [CrossRef]

14. Wickenheiser, A.; Garcia, E. Broadband vibration-based energy harvesting improvement through frequency up-conversion by magnetic excitation. *Smart Mater. Struct.* **2010**, *19*, 065020. [CrossRef]

15. Pillatsch, P.; Yeatman, E.; Holmes, A. Magnetic plucking of piezoelectric beams for frequency up-converting energy harvesters. *Smart Mater. Struct.* **2014**, *23*, 25009–25020. [CrossRef]

16. Xue, T.; Roundy, S. On magnetic plucking configurations for frequency up-converting mechanical energy harvesters. *Sens. Actuators A Phys.* **2017**, *253*, 101–111. [CrossRef]

17. Pozzi, M.; Zhu, M. Plucked piezoelectric bimorphs for knee-joint energy harvesting: Modelling and experimental validation. *Smart Mater. Struct.* **2011**, *20*, 055007. [CrossRef]

18. Kuang, Y.; Ruan, T.; Chew, Z.J.; Zhu, M. Energy harvesting during human walking to power a wireless sensor node. *Sens. Actuators A Phys.* **2017**, *254*, 69–77. [CrossRef]

19. Wei, S.; Hu, H.; He, S. Modeling and experimental investigation of an impact-driven piezoelectric energy harvester from human motion. *Smart Mater. Struct.* **2013**, *22*, 105020. [CrossRef]

20. Erturk, A.; Hoffmann, J.; Inman, D.J. A piezomagnetoelastic structure for broadband vibration energy harvesting. *Appl. Phys. Lett.* **2009**, *94*, 254102. [CrossRef]

21. Ramlan, R.; Brennan, M.J.; Mace, B.R.; Kovacic, I. Potential benefits of a non-linear stiffness in an energy harvesting device. *Nonlinear Dyn.* **2010**, *59*, 545–558. [CrossRef]

22. Erturk, A.; Inman, D.J. Broadband piezoelectric power generation on high-energy orbits of the bistable Duffing oscillator with electromechanical coupling. *J. Sound Vib.* **2011**, *330*, 2339–2353. [CrossRef]

23. Tang, L.; Yang, Y. A nonlinear piezoelectric energy harvester with magnetic oscillator. *Appl. Phys. Lett.* **2012**, *101*, 094102. [CrossRef]

24. Wang, C.; Zhang, Q.; Wang, W. Low-frequency wideband vibration energy harvesting by using frequency up-conversion and quin-stable nonlinearity. *J. Sound Vib.* **2017**, *399*, 169–181. [CrossRef]

25. Xu, Z.; Yang, H.; Zhang, H.; Ci, H.; Zhou, M.; Wang, W.; Meng, A. Design and analysis of a magnetically coupled multi-frequency hybrid energy harvester. *Sensors* **2019**, *19*, 3203. [CrossRef] [PubMed]

26. Yang, C.L.; Chen, K.W.; Chen, C.D. Model and Characterization of a Press-Button Type Piezoelectric Energy Harvester. *IEEE/ASME Trans. Mechatron.* **2019**, *24*, 132–143. [CrossRef]

27. Akoun, G.; Yonnet, J.P. 3D analytical calculation of the forces exerted between two cuboidal magnets. *IEEE Trans. Magn.* **1984**, *20*, 1962–1964. [CrossRef]

28. Agashe, J.S.; Arnold, D.P. A study of scaling and geometry effects on the forces between cuboidal and cylindrical magnets using analytical force solutions. *J. Phys. D Appl. Phys.* **2008**, *42*, 105001. [CrossRef]

29. Upadrashta, D.; Yang, Y. Finite element modeling of nonlinear piezoelectric energy harvesters with magnetic interaction. *Smart Mater. Struct.* **2015**, *24*, 045042. [CrossRef]

30. Fu, H.; Yeatman, E.M. A methodology for low-speed broadband rotational energy harvesting using piezoelectric transduction and frequency up-conversion. *Energy* **2017**, *125*, 152–161. [CrossRef]
31. He, L.; Wang, Z.; Wu, X.; Zhang, Z.; Zhao, D.; Tian, X. Analysis and experiment of magnetic excitation cantilever-type piezoelectric energy harvesters for rotational motion. *Smart Mater. Struct.* **2020**, *29*, 055043. [CrossRef]

Publisher's Note: MDPI stays neutral with regard to jurisdictional claims in published maps and institutional affiliations.

Article

Impact-Driven Energy Harvesting: Piezoelectric Versus Triboelectric Energy Harvesters

Panu Thainiramit, Phonexai Yingyong and Don Isarakorn *

Department of Instrumentation and Control Engineering, Faculty of Engineering,
King Mongkut's Institute of Technology Ladkrabang, Bangkok 10520, Thailand; panu.th@kmitl.ac.th (P.T.);
62601019@kmitl.ac.th (P.Y.)
* Correspondence: don.is@kmitl.ac.th

Received: 3 September 2020; Accepted: 4 October 2020; Published: 15 October 2020

Abstract: This work investigated the mechanical and electrical behaviors of piezoelectric and triboelectric energy harvesters (PEHs and TEHs, respectively) as potential devices for harvesting impact-driven energy. PEH and TEH test benches were designed and developed, aiming at harvesting low-frequency mechanical vibration generated by human activities, for example, a floor-tile energy harvester actuated by human footsteps. The electrical performance and behavior of these energy harvesters were evaluated and compared in terms of absolute energy and power densities that they provided and in terms of these energy and power densities normalized to unit material cost. Several aspects related to the design and development of PEHs and TEHs as the energy harvesting devices were investigated, covering the following topics: construction and mechanism of the energy harvesters; electrical characteristics of the fabricated piezoelectric and triboelectric materials; and characterization of the energy harvesters. At a 4 mm gap width between the cover plate and the stopper (the mechanical actuation components of both energy harvesters) and a cover plate pressing frequency of 2 Hz, PEH generated 27.64 mW, 1.90 mA, and 14.39 V across an optimal resistive load of 7.50 kΩ, while TEH generated 1.52 mW, 8.54 μA, and 177.91 V across an optimal resistive load of 21 MΩ. The power and energy densities of PEH (4.57 mW/cm^3 and 475.13 μJ/cm^3) were higher than those of TEH (0.50 mW/cm^3, and 21.55 μJ/cm^3). However, when the material cost is taken into account, TEH provided higher power and energy densities per unit cost. Hence, it has good potential for upscaling, and is considered well worth the investment. The advantages and disadvantages of PEH and TEH are also highlighted as main design factors.

Keywords: piezoelectric energy harvesting; triboelectric energy harvesting; low-frequency vibration energy harvesting; direct-force generator

---

1. Introduction

Investigation of energy harvesting from the ambient environment is increasing rapidly as a result of today's efficient energy consumption requirement and a need for low environmental impact. Energy sources to be harvested are usually waste energy from activities such as machine vibration [1], transportation [2], human motion [3], wind flow [4], and ocean wave [5]. This source of direct force acting on the generators is the most promising candidate for harvesting energy for self-powered electronic and sensing systems [6]. As they are waste energy, they do not affect the environment any more negatively than when they are left as waste and are not getting harvested. Harvesting waste energy does not disrupt the local ecosystem and does not cause any pollution or global warming [7]. For this work, the kinetic energy source was mechanical vibration from human activities. Although the electrical output of waste energy harvesting is less than other alternative energy sources, it is sufficient for powering small electronic devices, which are rapidly being introduced for many applications in people's daily routine [8].

There are many energy harvesting techniques, such as piezoelectric, triboelectric, pyroelectric, and electromagnetic transductions [8,9]. Piezoelectric energy harvesters (PEHs) and triboelectric energy harvesters (TEHs) have gained much attention in the last decade, as demonstrated by a large number of studies on the conversion of mechanical vibration energy to electrical energy [9,10]. Several of those were about the characterization and optimization of piezoelectric materials for energy harvesting, sensing, and actuating. For example, there has been a study on piezoelectric material on various substrates for a specific purpose, such as achieving small size, a higher degree of flexibility, being more stretchable, possessing a desired degree of roughness, and low cost [11]. Phosy Panthongsy et al. [12] proposed two different piezoelectric test benches. The piezoelectric material was used in the form of a cantilever, of which one end was attached to a holder on the base, and the other end was attached to a freely moving proof mass. The first test bench generated its electrical output by having the proof mass magnetize an iron bar on the cover plate to produce stress on the cantilever. The cantilever then freely oscillated after the cover plate was pressed and released. That test bench had a complex structure and was designed to prevent the cantilever from over-bending. On the other hand, the second type of test bench had a simple structure and a higher electrical performance than the first type because electrical energy was generated each time the cover plate was pressed and each time it was released. In another study, Don Isarakorn et al. [13] investigated the behavior of double-stage energy harvesting floor tile. They varied the excitation acceleration and gap width of the cover plate, observed the results, and found that the accelerations of the moving cover plate and gap width were directly proportional to the electrical output. Although that harvester was not designed for harvesting energy at their resonance frequency, it had a potential to harvest energy from low and variable frequency mechanical vibration sources, for instance, human motion [14].

A triboelectric energy harvester converts mechanical energy to electrical energy from friction or temporary contact between two different triboelectric materials. Its basic energy harvesting mechanism is based on combined contact electrification and electrostatic induction [15]. TEH is a promising technology that has attracted much attention and developed rapidly [16–20]. Four basic energy harvesting modes of TEHs are vertical contact-separation (CS) mode, lateral-sliding (LS) mode, single-electrode (SE) mode, and freestanding triboelectric-layer (FT) mode [21]. Examples of studies on characterization and optimization of the electrical performance of triboelectric and electrode materials, device fabrication, and environment condition control are the following studies [22–27]. In addition, Cun Xin Lu et al. [28] investigated the temperature effect on the triboelectric electrical output of a single-electrode mode and reported that, at a low temperature, the TEH was able to generate high electrical output, and vice versa. The relative humidity is also a critical factor that affects triboelectric charge density. Therefore, a device properly installed in a controlled environment could be highly efficient [29]. Moreover, enlarging the size of a TEH and parallelly connecting many of them in the same device has the potential to increase the electrical output [30]. In addition, hybridizing TEH with another energy harvesting device that operates by a different harvesting mechanism has been investigated and found to increase energy production efficiency [31–33].

Even though there have been numerous research studies on the electrical output performance of PEHs and TEHs, including applications with an individual generator and combined generators, none of them have compared the voltages, powers, and energies provided by them as well as their produced energy densities and costs. Therefore, in this study, we evaluated and compared the electrical performance and behavior of TEH and PEH in terms of their electrical output density per material cost. This development of energy harvester design based on low-frequency vibration for harvesting mechanical energy from human motion or activities had the potential to be used in practical applications and is well worth the investment.

The rest of the paper is structured as follows. Section 2 describes the construction and working principles of both PEH and TEH test benches for harvesting energy from low-frequency mechanical vibration sources. Section 3 describes the details of the fabrication material of both PEHs and TEHs, as well as the experimental methods and setups. Section 4 presents the experimental results of both

PEHs and TEHs, including the electrical output characteristics, the electrical performance comparison, and the potential of practical applications. Section 5 discusses the experimental results related to gap width, electrical performance, power density, energy density, unit cost consideration, and summary of the advantages and disadvantages. Lastly, Section 6 concludes the paper.

2. Construction and Mechanism of the Test Bench

The structures of the test benches—two different energy harvesters (PEH and TEH)—were designed and developed. Their mechanisms, the measurement schematic, and their output voltage characteristics are described in Sections 2.1 and 2.2, respectively.

2.1. Piezoelectric Energy Harvester Test Bench

The purpose of the piezoelectric energy harvester test bench was to harvest waste energy from mechanical vibration at a low frequency into usable electrical energy. The cover plate of the test bench absorbed the excitation impact from a pneumatic actuator. A pressure valve regulator was used to control the compressed air pressure into the pneumatic actuator. Springs were used to separate the cover plate from the stopper of the base and to store the mechanical energy after the cover plate was pressed, then the mass tip of the cantilever bounced and oscillated freely in the air, as shown in Figure 1a, and the images of actual PEH test bench are shown in Figure 1b. The essential element for energy conversion was the piezoelectric cantilever. One end of it was side-mounted to a holder on the cover plate, and the other end was attached to a proof mass. This cantilever generates electricity in two stages. The first stage was when the cover plate was pressed. The strain within the beam caused by the pressing of the cover plate allows the cantilever beam to oscillate freely afterward. After the cover plate was pressed by an actuator, the stress in the cover plate spring pushes it back upward. The mass tip of the piezoelectric cantilever sprung upward, causing the beam to bend and freely oscillate until the next actuating period. This occurrence caused another round of electricity generation, as shown in Figure 2.

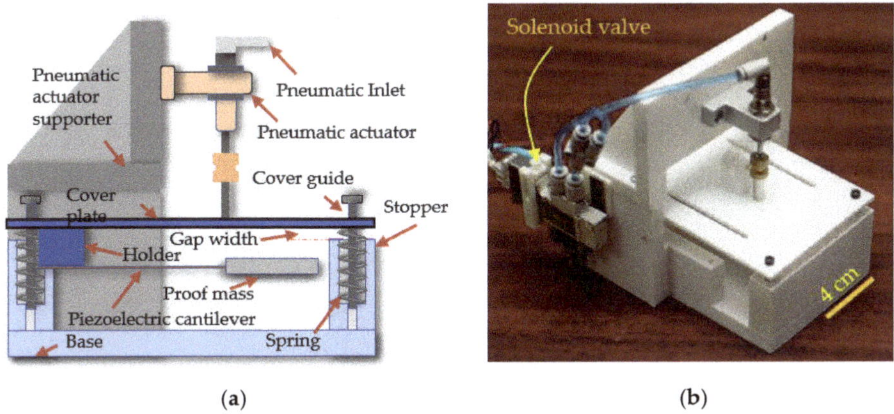

(a)

(b)

Figure 1. Piezoelectric energy harvester (PEH) test bench: (**a**) cross-section of PEH; (**b**) photograph of PEH test bench.

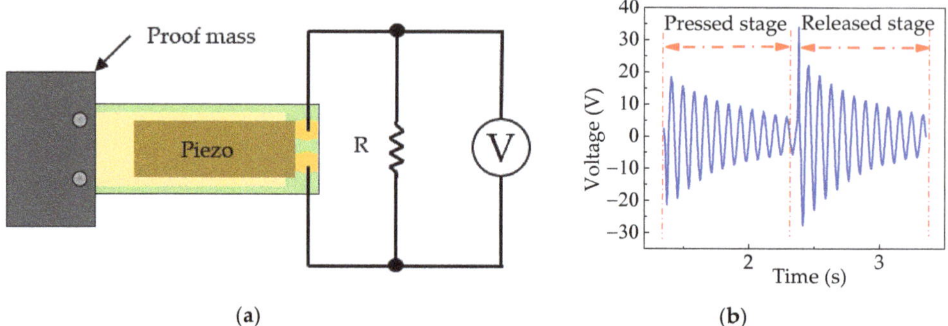

Figure 2. (**a**) Measurement setup of PEH, and (**b**) typical output voltage of PEH test bench.

2.2. Triboelectric Energy Harvester Test Bench

The primary electricity generation mechanism of the triboelectric energy harvester was charge accumulation by friction or by temporary contact between two different triboelectric materials. The design and operation of the TEH test bench design were not very different from those of the PEH test bench. The top electrode mounted under the cover plate absorbed direct force from the pneumatic actuator. The pneumatic pressure and the pressing frequency applied to the pneumatic actuator were controlled by a pressure valve regulator and a function generator, respectively. The base supported the cover guide, the springs, and the pneumatic actuator supporter, as shown in Figure 3a, and the images of the actual TEH test bench is shown in Figure 3b. In this research, the TEH test bench was a contact mode single-electrode triboelectric energy harvester. The electrical output is generated by pressing the cover plate downward; subsequently, the top electrode contacted the triboelectric material, and the contact electrification created tribo-charges on the surface of the triboelectric material attached to the bottom electrode and on the top electrode, which was also a triboelectric material; then, the charges were transferred from the bottom to the top electrode through an external circuit as a result of electrostatic transduction [15].

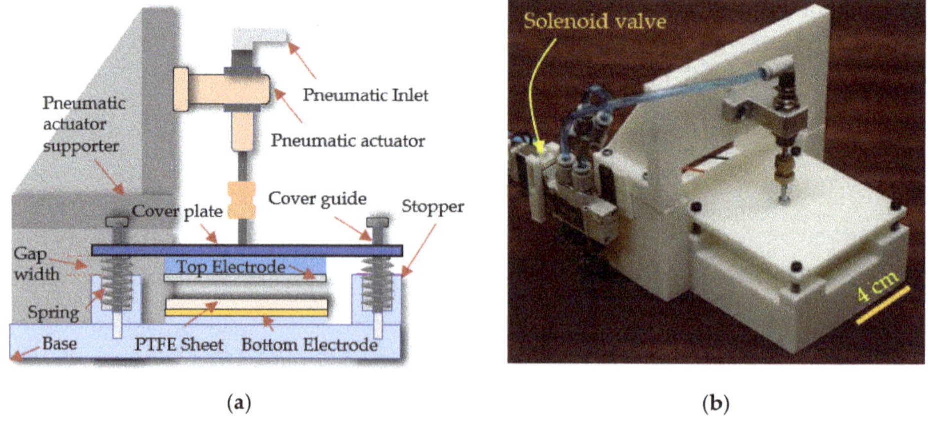

Figure 3. Triboelectric energy harvester (TEH) test bench: (**a**) Cross-section of TEH test bench; (**b**) Photograph of TEH test bench. PTFE, polytetrafluoroethylene.

Figure 4a shows the measurement setup, including the TEH source connected to the maximum power tracking (MTP) circuit, for measuring the electrical characteristics across the resistive loads [34].

The TEH had a very high output impedance, so the output resistive load had a very high resistance as well. When the resistive load was higher than a measurement probe of the oscilloscope, the TEH output power could not be determined accurately. This MPT technique was thus applied to measure the output power. The output voltage generation occurred in two stages: a pressed-stage and a released-stage, as shown in Figure 4b.

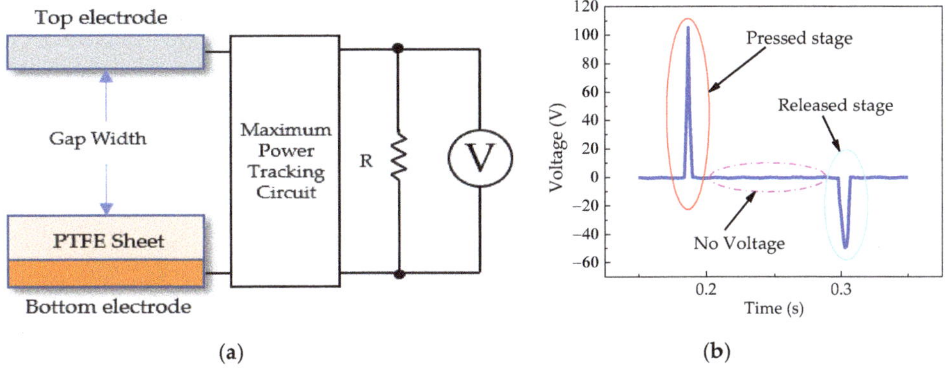

Figure 4. (**a**) Measurement setup of TEH, and (**b**) typical output voltage of TEH.

3. Materials and Methods

The types, sizes, costs, and critical properties of the fabricated PEH and TEH materials used in this work are reported below as well as the experimental methods and setup.

3.1. Fabrication Materials

A piezoelectric material, we used a Midé S230-J1FR-1808XB piezoelectric module ($7.1 \times 2.54 \times 0.076$ cm^3) in the piezoelectric cantilever configuration. One end of it was attached to a holder at the cover plate, and the other end was mounted with a proof mass (stainless-steel, $3 \times 2.6 \times 0.6$ cm^3, total mass 40 g), and total PEH volume was 6.05 cm^3. The cost of this bimorph cantilever was US$274 per unit, discounted if bought in a large quantity. For example, for 100 units, the price per unit was only US$82 [35]. The physical and mechanical properties of this piezoelectric cantilever are shown in Table 1 [36].

Table 1. Physical and electrical properties of the piezoelectric material.

Property (Unit)	Value
Density ρ (Kg/m^3)	7800
Mechanical Q (Q_m)	60
Elastic (Young's modulus) $Y^E{}_3$ (N/m^2)	5×10^{10}
Relative dielectric constant $K^T{}_3$ (@ 1 kHz)	2100
Piezoelectric coefficient d_{33} (pC/N)	500
Piezoelectric coefficient d_{31} (pC/N)	-210
Piezoelectric voltage constant g$_{33}$ (Vm/N)	23×10^{-3}
Piezoelectric voltage constant g$_{31}$ (V-m/N)	-10.4×10^{-3}
Coupling coefficient K_{33}	0.74
Coupling coefficient K_{31}	0.37
Polarizing field (E_p) V/m	$>1.7 \times 10^6$
Initial depolarizing field E_c (V/m)	$\sim4 \times 10^5$
Coercive field E_c (V/m)	$\sim1.0 \times 10^6$

For the TEH test bench, Wang et al. [37] listed the materials in the triboelectric series. A material will reach more negative charge when touching a material to the bottom of the series. The farther away two materials were selected from each other on the series, the greater charge will be transferred. Aluminium appears near the top of the series, whereas polytetrafluoroethylene (PTFE) is near the bottom, so Al and PTFE were chosen for this project. A PTFE sheet ($5 \times 5 \times 0.1$ cm^3) is attached to the bottom electrode (primary electrode), and the assembly is mounted on the base of the test bench. The bottom electrode ($5 \times 5 \times 0.01$ cm^3) is a copper foil (MT 8113C copper foil tape conductive adhesive). The top electrode ($5 \times 5 \times 0.012$ cm^3) is an aluminum foil (3M 425 Aluminum foil tape) mounted on a supporter on the cover plate, and the total volume of TEH was 3.05 cm^3. This top electrode functioned both as a triboelectric material and a reference electrode. The average price of a PTFE sheet, including the top electrode and bottom electrode, was US$1.44. In a large quantity, the price could be discounted to US$1.17 per square centimeter. The physical and mechanical properties of the PTFE are shown in Table 2 [38].

Table 2. Physical properties of the polytetrafluoroethylene (PTFE) sheet.

Property (Unit)	Value
Density (g/cm^3)	2.3–2.45
Water absorption (%)	>0.01
Tensile strength (kg/cm^2)	140–350
Flexural strength (kg/cm^2)	16.4
Rockwell hardness	D55
Izod impact strength (kg cm/ cm with notch)	2.5–2.7
Friction coefficients	0.10–0.04
Coefficient of linear thermal expansion (x 10^{-5}/°C)	7.0–10.0
Thermal conductivity (kcal/m. Hr. °C)	6.0
Heat distortion temperature (°C)	120
Heat resistance (°C)	−70–260
Dielectric breakdown strengths (kV/mm)	19
Coefficient of volume resistance (Ohm-cm)	10^{18}

3.2. Experimental Set-Up and Methods

This section aims to present an overview of the experimental setup, including the mechanical energy input, which was a critical part of our direct mechanical force applied to an energy harvesting device.

The mechanical energy input that we used was a constant air pressure of 600 kPa from a pneumatic pump, regulated by a pressure valve regulator. The compressed air pressure was the pneumatic actuator (CJPB6-15 from SMC Corporation) that provided a direct mechanical force 17.20 N on the cover plate of both types of test benches, and a spring, with the spring constant of 263 N/m, was used for separating the cover plate from the stopper. The acceleration of the moving cover plate was measured with an accelerometer (EI-CALC). The frequency of the application of direct force was controlled by a function generator (GW INSTEK AFG-2225), which could be set to be any desired frequency that was in harmony with the low-frequency actuation of the mechanical parts of both types of test benches. An oscilloscope (Tektronix TDS-3032B) was used to measure the open-circuit voltage and the output voltage across resistive loads through an MPT circuit, as shown in Figure 4a. Both PEHs and TEHs have experimented at room temperature.

The equations related to this experiment are explained below. For calculating the output power (P) for optimal load identification, the equation was Equation (1). In Equation (1), V_L was the output voltage across the resistive loads R_L. For calculating the total energy harvested $E(t_n)$, the equation was Equation (2), where t_n was the total time while the triboelectric was working $V_{L(t_M)}$ was the voltage across the resistive load at a time t_m and Δt was the sampling time interval. We used these equations in the same way that they were used in [12,13]:

$$P = \frac{V_L{}^2}{R_L} \tag{1}$$

and

$$E(t_n) = \sum_{m=1}^{n} \frac{V_L{}^2(t_m)}{R_L} \Delta t, \text{ for } n > 0 \tag{2}$$

We divided the power and energy obtained from Equations (1) and (2) with the volume and cost of material of each of the two energy harvesters to determine and compare their power and energy densities as well as cost-effectiveness.

The procedural steps in the experiments of both PEH and TEH are as follows.

(1) Designing and fabricating test benches that used the same pneumatic actuator to apply a proper direct force to their cover plate and doing the same thing for the spring that pushed the cover plate away from the stopper. Images of actual PEH and TEH test benches are shown in Figures 1b and 3b, respectively.

(2) Using resistive loads to connect the top electrode to the bottom electrode to identify their optimal load and peak power by measuring the peak output voltage across resistive load while varying the gap width between the cover plate and the stopper from 1 mm to 4 mm.

(3) Measuring and recording the output voltages of PEH and TEH for a 4 mm gap width while varying the cover plate pressing frequency from 0.5 Hz to 5 Hz in an increment of 0.5. The measured voltages were used in energy computation.

(4) Calculating the electrical performances of the PEH and TEH normalized by material volume and cost into output voltage, power density, and energy density per unit cost, which would be immensely useful for designing and developing a practical energy harvesting device.

4. Experimental Results

The PEH and TEH converted mechanical energy into electrical energy by deformation of its cantilever and by making temporary contact between triboelectric materials, respectively. It was difficult to compare most of the conventional parameters related to energy harvester, so we focused on their practical utility; we compared their power and energy densities. We applied an optimal input of direct force to the cover plate of each type of test bench and measured the electrical output. For proper comparison, we used the same input for exciting the cover plates of both types of the test bench, as well as the same pneumatic pressure and gap width between the cover plate and the stopper; the initial acceleration of the moving cover plate was 0.93 g. At gap widths of 1, 2, 3, and 4 mm, the acceleration when the cover plate impacted on the stopper was 3.05 g, 3.77 g, 4.06 g, and 4.28 g, respectively. The g is the gravitational acceleration, where 1 g implies 9.81 m/s^2. The electrical performances of the PEH and TEH are presented in Sections 4.1 and 4.2, respectively. The experimental results of the two techniques in various scenarios are compared and discussed in Section 4.3.

4.1. Electrical Output Characteristics of PEH Test Bench

The electrical performance of the PEH test bench was determined by connecting resistive loads between the top electrode and the bottom electrode mounted on the piezoelectric cantilever. Figure 5 displays the electrical output characteristics across variable resistive loads. A constant pneumatic pressure was applied to the cover plate under the condition of a 4 mm gap width between the cover plate and the stopper. The output current was directly proportional to the output voltage across the resistive loads. The peak output voltage, peak output current, and peak output power across 7.50 kΩ optimal loads were 14.39 V, 1.92 mA, and 27.64 mW, respectively. Figure 6 shows the output power under the conditions of 1–4 mm gap width. The peak output powers across the optimal load were found to be 4.12 mW, 8.07 mW, 11.10 mW, and 27.64 mW, respectively. Therefore, it was clear that the output power was directly proportional to the gap width between the cover plate and the stopper.

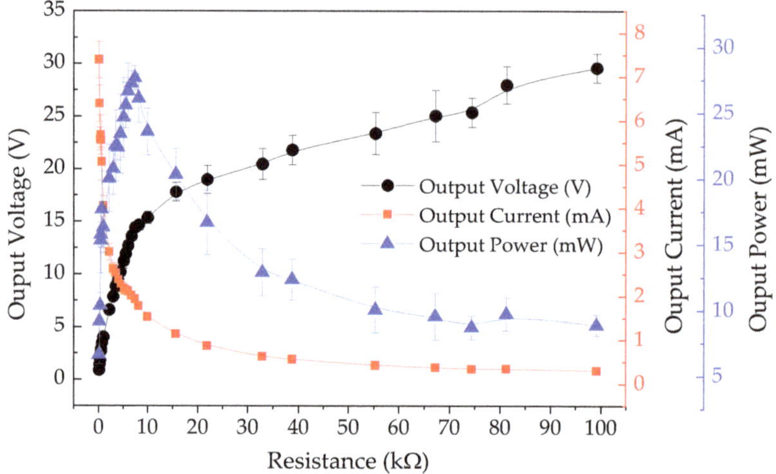

Figure 5. Electrical output characteristics of PEH across resistive loads, at 4 mm gap width.

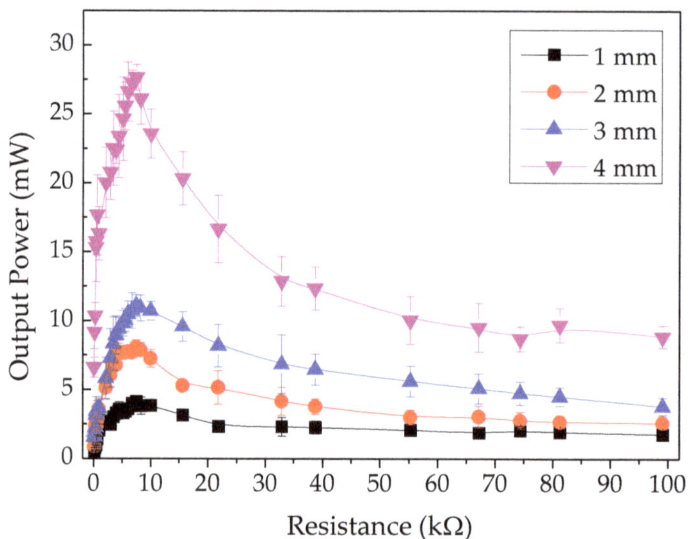

Figure 6. Output power at various gap widths from 1 mm to 5 mm.

Figure 7 shows the output voltage and energy across the optimal load at 10 s after the cover plate was actuated and at four different gap widths from 1 mm to 4 mm under the pressing frequency of 2 Hz. The output voltage and energy tended to slightly increase when the gap width was varied from 1 mm to 4 mm. At gap widths of 1, 2, 3, and 4 mm, the average peak voltage and total energy are 6 V and 1.03 mJ, 7 V and 1.29 mJ, 8 V and 2.37 mJ, and 9 V and 2.88 mJ, respectively.

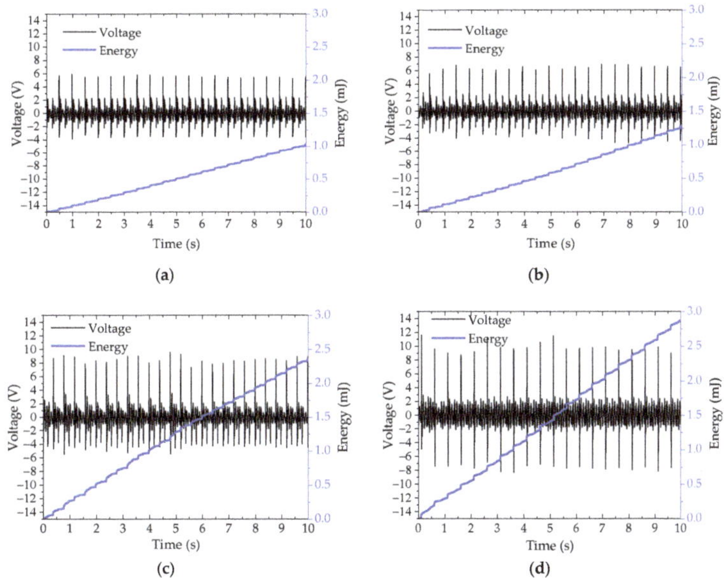

Figure 7. Output voltage and energy curve of PEH across the optimal load at 2 Hz with varying gap widths of (**a**) 1 mm, (**b**) 2 mm, (**c**) 3 mm, and (**d**) 4 mm.

4.2. Electrical Output Characteristics of TEH Test Bench

The TEH test bench was designed to harvest energy from low frequency impacts, so the electrical performance of this device was investigated by applying a constant pneumatic pressure to produce a direct force on the cover plate. The electrical output parameters at 4 mm gap width are shown in Figure 8. The output currents were directly proportional to the output voltages across resistive loads, the same as those observed in PEH. TEH had an optimal load of 21 MΩ. Its peak output voltage, peak output current, and peak output power were 177.91 V, 8.54 μA, and 1.52 mW, respectively. Its electrical performance varied with the gap width. The peak output power varied with the velocity of the top electrode moving down to press on the contact on the PTFE sheet and the gap width. It slightly increased with gap width: 0.44 mW for 1 mm gap width, 0.88 mW for 2 mm, 1.20 mW for 3 mm, and 1.52 mW for 4 mm gap width, as shown in Figure 9. The output voltage and energy at four different gap widths at 2 Hz and at 10 s after actuation are shown in Figure 10. The average peak voltage and energy tended to increase slightly with gap width: 100 V and 23.96 μJ, 130 V and 39.84 μJ, 170 V and 58.51 μJ, and 175 V and 65.73 μJ for 1, 2, 3, and 4 mm gap width, respectively.

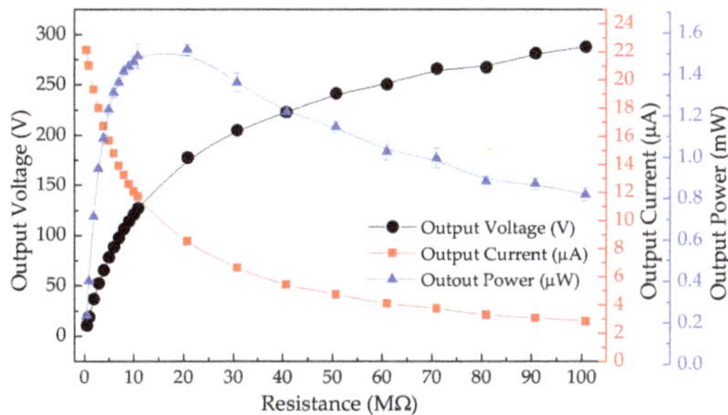

Figure 8. Electrical output parameters of TEHs across resistive loads at 4 mm gap width.

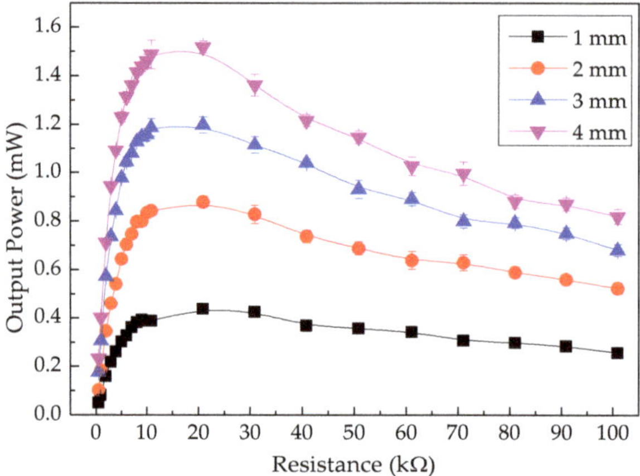

Figure 9. Output power of TEH test bench at various gap widths from 1 mm to 4 mm.

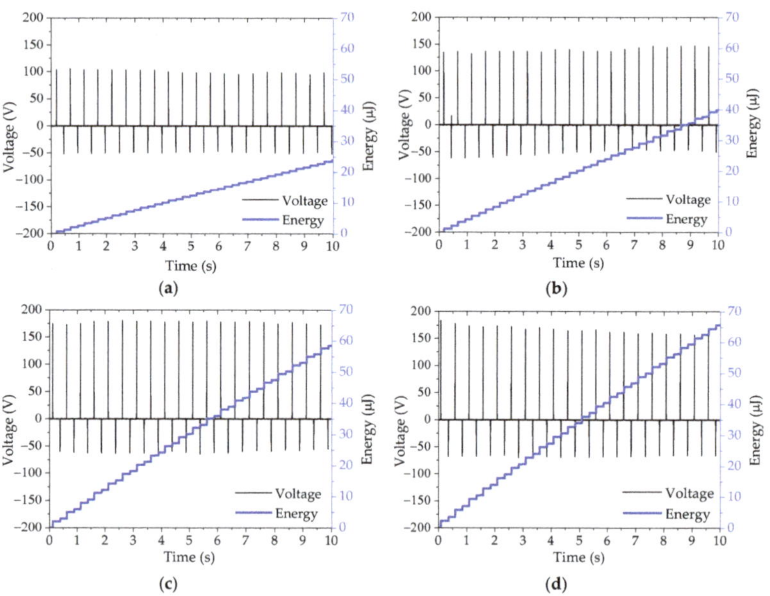

Figure 10. Output voltage and energy curve of TEH across the optimal load under 2 Hz with varying gap width; (**a–d**) are of 1 mm to 4 mm gap width in an increment of 1 mm.

4.3. Electrical Output Characteristics of PEH and TEH Test Benches

PEHs and TEHs were readily accessible technology for energy harvesting. It was still not clear which one was better in terms of energy harvesting based on direct force excitation. The electrical output parameters determined were output voltage, power density, and energy density.

4.3.1. Output Voltage Comparison

Output voltage is a critical parameter for comparing PEH and TEH. The column graph in Figure 11 illustrates the peak output voltage across the optimal resistive load at different gap widths. It can be seen that the output voltage of both types of test bench was directly proportional to the gap width, and the output voltage at the gap widths from 1 mm to 3 mm of TEH were more than 17 times higher than the voltage density of PEH. At 4 mm gap width, the voltage density of TEH was 14 times higher than that of PEH. As an application may require a high operational voltage, TEH would be better than PEH for this kind of application.

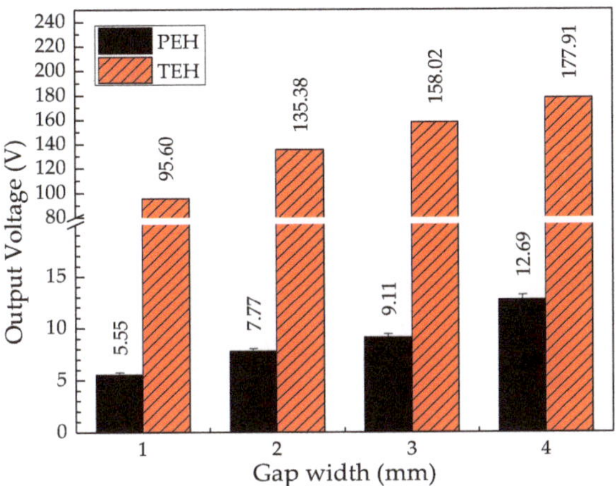

Figure 11. Output voltage of PEH and TEH.

4.3.2. Power Density Comparison between PEH and TEH

Figure 12 shows a column bar graph of the power density of PEH and TEH. The power density of PEH, for any gap widths, was higher than that of TEH. The power density of both devices was higher when the gap width was larger (between 1 and 4 mm). The power density of PEH at a gap width from 1 mm to 3 mm was more than four times higher than the power density of TEH at any gap width. At a 4 mm gap width, the power density of PEH was 9 times higher than that of TEH: 4.57 mW/cm^3 for PEH and 0.50 mW/cm^3 for TEH. PEH and TEH output powers are shown in Figures 6 and 9. Both of them exhibited good potential for powering a small electronic device, of which some examples are shown in Table 3.

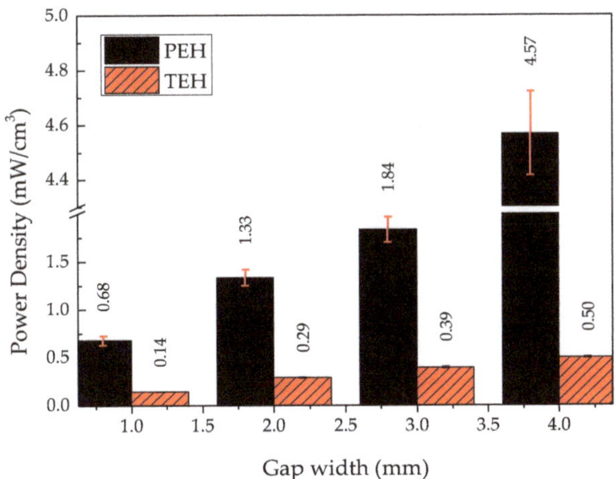

Figure 12. Power densities of PEH and TEH.

Table 3. Power consumption of common small electronic devices.

Device	Power Consumption	PEH	TEH
Watches [39]	3–10 µW	applicable	applicable
Smoke detector [40]	4.95 µW	applicable	applicable
Peacemakers [39]	25–80 µW	applicable	applicable
Capacitive strain gauge [41]	600 µW	applicable	applicable
Hearing aids [42]	<1.4 mW	applicable	applicable
CO_2 sensor [43]	<3.5 mW	applicable	not applicable
Digital clocks [39]	13 mW	applicable	not applicable
Light-emitting diode [39]	25 mW–100 mW	applicable	not applicable

4.3.3. Energy Density Comparison between PEH and TEH

The energy densities of both types of energy harvesting devices were investigated under the conditions of direct force application on the cover plate, of variable pressing frequency from 0.5 Hz to 5 Hz, and of an optimal load for PEH and TEH, which was 7.5 kΩ and 21 MΩ, respectively.

Figure 13 shows the energy densities provided by PEH and TEH at a 4 mm gap width and under a variable pressing frequency between 0.5 and 5 Hz. It can be observed that PEH provided a higher energy density than TEH, under every pressing frequency from 0.5 to 5 Hz, in increments of 0.5 Hz.

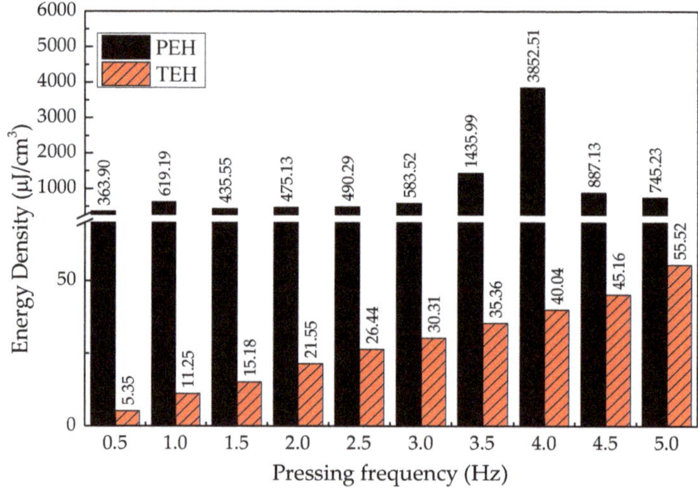

Figure 13. Energy densities of PEH and TEH under the variable pressing frequency of 0.5–5 Hz and at a 4 mm gap width.

The energy density of PEH slightly increased with input excitation frequency from 0.5 Hz to 3 Hz and reached the highest output energy density of 3.85 mJ/cm³ at the input pressing frequency of 4 Hz, but dropped dramatically at the excitation frequencies of 4.5 Hz and 5 Hz.

The resonance frequency of the cantilever in the PEH test bench was about 12 Hz, but we intended to harvest electrical energy from a source, providing a frequency of mechanical vibration that was lower than this resonance frequency—from the frequency of human steps, walking leisurely, at around 1–2 Hz. The varying energy density of PEH with pressing frequency might be the result of the harmonics of the direct-force excitation frequency. The outcome of this phenomenon can be observed in the dramatic decrease of peak output voltage as the moving cover plate stopped when it hit the stopper or the cover guide, as shown in Figure 7. The energy density of TEH increased linearly and reached the highest value of 55.52 µJ/cm³ at the highest pressing frequency of 5 Hz.

4.3.4. Power Density and Energy Density per Unit Cost

The cost of fabrication material is a critical parameter for the practical application of a device. In terms of price, TEH was much better than PEH, shown in Figure 14, but in terms of power and energy densities, PEH was more able than TEH. The power density per unit cost of TEH at a gap width from 1 mm to 3 mm was about 40 times higher than that of PEH, and the power density per unit cost of TEH at a 4 mm gap width was 20 times higher than that of PEH.

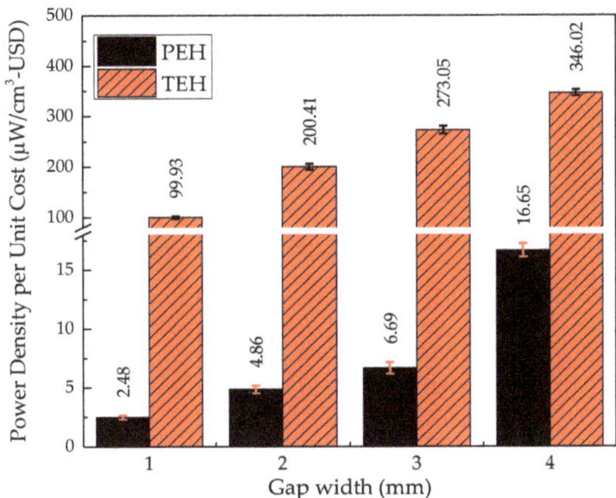

Figure 14. Power density per unit cost of PEH and TEH at a 4 mm gap width.

Overall, the energy density per unit cost of TEH was much higher than that of PEH because triboelectric material for fabrication of TEH was much less expensive than piezoelectric material for fabrication of PEH. The energy density per unit cost of TEH increased linearly with the pressing frequency—38.68 μJ/cm^3-USD at a 5 Hz pressing frequency. It was over 14 times higher than the energy density per unit cost of PEH. The energy density per unit cost of PEH reached the highest value of 14.04 μJ/cm^3-USD at the 4 Hz pressing frequency, but it is still much lower than that of TEH, as shown in Figure 15.

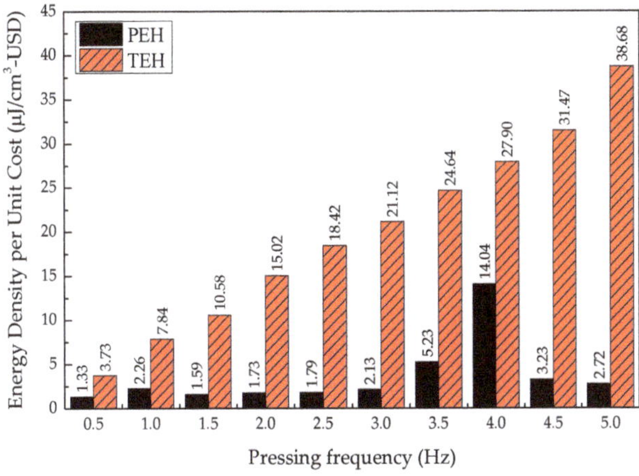

Figure 15. Energy density per unit cost of PEH and TEH at a 4 mm gap width.

5. Discussion

The electrical performances of a developed piezoelectric energy harvester (PEH) and a triboelectric energy harvester (TEH) actuated by direct mechanical force at a low frequency were investigated. In this section, we discuss three topics regarding the electrical performance outcomes: the relationship between the gap width separating the cover plate and the stopper and the electrical outputs of PEH and TEH, power and energy densities per unit cost, and aspects to be considered in real implementation.

5.1. Gap Width and Electrical Output

It was found that the electrical output of either PEH or TEH was directly proportional to the gap width between the cover plate and the stopper. They tended to generate a higher electrical output (voltage, current, power, and energy) at a larger gap width. In this study, some pressing frequencies highly excited the piezoelectric cantilever, inducing an electrical output of PEH (Figure 5), but some other frequencies depressed it because of their out-of-phase nature to the resonance-frequency [44]. Even though the generated piezoelectric power (Figure 6) was much higher than the generated triboelectric power (Figure 9), PEH power output depended more strongly on the resonance frequency of its structure than TEH did [45], limiting the practical design of a PEH structure compared with that of a TEH structure. The electrical power of TEH increased linearly with gap width and excitation (or pressing the cover plate) frequency (Figure 9). The output energy of TEH was more stable than that of PEH, because TEH does not depend on the resonance frequency of its structure, as presented in Figure 10. This stability was observed in the experimental outcomes as a linear dependence of electrical output with excitation frequency.

The tested gap widths were limited to 4 mm because, when it was larger than that, the cantilever of PEH would overbend, and the tip mass would hit the cover plate and the base, leading to severe damage to the cantilever. In contrast, the TEH structure benefited from a wider gap width, within a limit. A wider gap width of no more than ten times the thickness of tribo-material would generate a higher power from this energy harvester [46]. The final velocity with which the two triboelectric materials touched each other, as well as the surface roughness and configuration of the two tribo-electric materials, were the main design parameters for TEH energy harvester [21,47].

5.2. Power and Energy Densities per Unit Cost

To usefully compare the performances of PEH and TEH, their power and energy densities should be investigated relative to their volume. Their practical applications absolutely depend on the economy of the initial investment capital. In this study, the power and energy densities were investigated at their respective maximum power, corresponding to an optimum resistance. The optimum resistance of these two energy harvesters was very different because of the different characteristics (e.g., operating frequency, number of generators, and type of electrical connection) of the two electricity-generating materials [48]. Figure 12 shows that the power density of PEH increased exponentially with gap width. On the other hand, the power density of TEH increased linearly with gap width; this related phenomenon was also reported by Niu et al. [49]. They investigated the electrical output from TEH, which is related to the gap width regarding the charging behaviour of the TEH capacitance model. Regarding cost, as shown in Figure 14, TEH provided a much higher power density for a very low cost compared with PEH. Moreover, it was easier to fabricate and could use a wide-range of triboelectric materials.

Another important condition to consider in this kind of study was the time period for electrical output data collection. Figure 13 shows the outcomes of the highest-peak energy density provided by PEH. As can be observed, the maximum energy densities of PEH achieved by different pressing frequencies were drastically different. In contrast, TEH's maximum energy density was proportional to the pressing frequency. There was no resonance frequency effect. Figure 15, moreover, shows that the maximum energy density per unit cost increased linearly. Furthermore, TEH not only provided a much higher energy density per unit cost than PEH, but also was not sensitive to variation in ambient input parameters—that is, TEH was not sensitive to resonance excitation from human activities.

In addition, the output performance for piezoelectric and triboelectric energy harvesters is compared and summarized in Table 4; this table lists references, size of the harvester, voltage, power, power density, operating frequency, and optimal load. However, this table shows the related correspondence that we found regarding our research. These parameters (e.g., power density and optimal load) appeared to follow the same trend in our research, but the operating frequency in this table was extremely different due to our research being suitable for the floor-tile energy harvester.

Table 4. Output performances of PEH and TEH.

Reference	Harvester Size	Voltage (V)	Power	Power Density	Frequency (Hz)	Optimal Load
1. Piezoelectric energy harvester						
Panthongsy et al. [1]	2.80 cm^3	7.57	0.58 mW	2.07 mW/cm^3	20.83	99 kΩ
Yang et al. [50]	241 mm^3	22.50	2.53 mW	10.50 mW/cm^3	44.00	200 kΩ
Dai et al. [51]	2.26 cm^3	24.47	1.06 mW	0.47 mW/cm^3	51.00	564.7 kΩ
Sriramdas et al. [52]	108 mm^3	3.96	8.1 μW	0.08 mW/cm^3	30.80	1 MΩ
Ma et al. [53]	270 mm^3	1.46	3.18 μW	0.012 mW/cm^3	93.00	700 kΩ
Dhakar et al. [54]	102.08 mm^3	6.32	40 μW	0.39 mW/cm^3	36.00	1 MΩ
Lee et al. [55]	0.023 mm^3	0.84	1.38 μW	61.3 mW/cm^3	255.9.	510 kΩ
Song et al. [56]	0.11 mm^3	0.03	0.023 μW	0.209 mW/cm^3	48.00	40 kΩ
Zou et al. [57]	400 mm^3	12.29	387 μW	0.968 mW/cm^3	9.90	390 kΩ
Morimoto et al. [58]	0.26 mm^3	0.51	5.3 μW	20.46 mW/cm^3	126.00	50 kΩ
2. Triboelectric energy harvester						
Zhang et al. [5]	18.75 cm^2	36	162 μW	86.4 mW/cm^2	-	8 MΩ
Jurado et al. [6]	64 cm^2	3.83	307.88 μW	19.24 μW/cm^2	150	10 MΩ
Shamsuddin et al. [17]	118.16 cm^2	40.00	17 μW	0.14 μW/cm^2	-	60 MΩ
Yang et al. [18]	36 cm^2	749.40	9.36 mW	260 μW/cm^2	3.2	60 MΩ
Uddin et al. [19]	2.28 cm^2	16.20	36 μW	15.8 μW/cm^2	3	1 MΩ
Xia et al. [26]	6 cm^2	48.00	2.88 mW	480 μW/cm^2	-	800 kΩ
Kim et al. [27]	36 cm^2	13	210 μW	48 μW/cm^2	-	20 MΩ
Mule et al. [29]	4 cm^2	774.59	10 mW	2.54 mW/cm^2	-	60 MΩ

5.3. Aspects to Be Considered in Realistic Implementation of the Energy Harvesters

When it comes to real applications, for example, the floor-tile energy harvester, optimal design parameters of TEH would be more easily changed to handle unexpected problems because the structural design and arrangement of parts of the test bench did not affect its resonance frequency. Our study, on the other hand, addressed a different issue. We focused on an upscale application of converting the force from human walking steps in a crowded area into electricity, that is, a floor-tile energy harvester. As listed in Table 5, the advantages and disadvantages of PEH and TEH in this scenario can function as the main design factors of a more modern version of these energy harvesters. It should be noted that the detail in Table 5 gives the guideline for the selection of transducers for a floor tile energy harvester, while other designs may rank the trade-offs differently.

Table 5. Advantages and disadvantages of PEH and TEH.

	Piezoelectric Energy Harvester	Triboelectric Energy Harvester
Advantages	- Low internal resistance - Two stages of output voltage generation - High current, power, and energy density - Low output voltage (able to power small electronic devices)	- Simple construction - Easy fabrication - Smaller displacement than that required by the piezoelectric energy harvester - Low cost - Two stages of output voltage generation - Energy density is proportional to pressing frequency
Disadvantages	- Complex construction - High cost of material - Low output voltage - Pressing frequency may affect energy density - Large gap displacement may over-bend the cantilever	- High internal resistance - Low current, power, and energy densities - Environmental condition may affect its electrical output characteristics - High output voltage (in term of power management circuit design)

6. Conclusions

This study developed two types of energy harvesters based on direct force excitation. Piezoelectric and triboelectric energy harvesters (PEHs and TEHs) were newly designed to be embedded in floor tiles. Test benches of a piezoelectric energy harvester (PEH) and a triboelectric energy harvester (TEH) were developed with the same kind of mechanical actuator and excitation parameters. From the outcomes of electrical performance experiments, PEH was shown to provide higher power and energy densities than those of TEH, and those densities were dependent on the induced displacement gap between the cover plate (mechanical actuator) and the stopper as well as the actuating frequency. PEH was better in this regard than TEH because the piezoelectric material mounted on the cantilever of PEH could be bent to a high degree, and hence was able to provide high power. However, even though PEH provided higher power and energy densities in absolute terms, TEH provided higher power and energy densities per unit cost. TEH may be a better alternative for the future because the costs of triboelectric materials were much lower than the costs of piezoelectric materials. Future work can be on investigating and designing multiple triboelectric material strips in a generator to boost output power or an investigation of a hybrid design [59,60] to take full advantage of the strengths of both types of electricity-generating materials.

Author Contributions: Conceptualization, D.I.; methodology, D.I.; validation, P.T., P.Y., and D.I.; writing—original draft preparation, P.Y.; writing—review and editing, D.I. and P.T.; visualization, P.Y.; supervision, D.I.; project administration, D.I.; funding acquisition, D.I. All authors have read and agreed to the published version of the manuscript.

Funding: This work was supported by King Mongkut's Institute of Technology Ladkrabang Research Fund [KREF146302].

Conflicts of Interest: The authors declare no conflict of interest.

References

1. Panthongsy, P.; Isarakorn, D.; Hamamoto, K. A test bench for characterization of piezoelectric frequency up-converting energy harvesters. In Proceedings of the 2018 15th International Conference on Electrical Engineering/Electronics, Computer, Telecom-Munications and Information Technology (ECTI-CON), Chiang Rai, Thailand, 18–21 July 2018.
2. Ruan, J.J.; Lackhart, R.A.; Janphuang, P.; Quintero, A.V.; Briand, D.; Rooij, N. An automatic test bench for complete characterization of vibration-energy harvesters. *IEEE Trans. Instrum. Meas.* **2013**, *62*, 2966–2973. [CrossRef]
3. Yang, Z.; Zhou, S.; Zu, J.; Inman, D. High-performance piezoelectric energy harvesters and their application. *Joule* **2018**, *2*, 1–56. [CrossRef]
4. Rase, M.S.; Halim, M.A.; Cho, H.O.; Park, J.Y. A wrist-band coupled, human skin based triboelectric generator for harvesting biomechanical energy. In Proceedings of the 2015 Transducers—2015 18th International Conference on Solid-State Sensors, Actuators and Microsystems (TRANSDUCERS), Anchorage, AK, USA, 21–25 June 2015.
5. Zhang, X.; Guo, Y.; Wang, Y.; Zhang, H.; Brugger, J. A transparent silk-fibroin-based triboelectric microgenerator for airflow energy harvesting. In Proceedings of the 2017 IEEE 12th International Conference on Nano/Micro Engineered and Molecular Systems (NEMS), Los Angeles, CA, USA, 9–12 April 2017.
6. Jurado, U.T.; Pu, S.H.; White, N.M. A contact-separation mode triboelectric nanogenerator for ocean wave impact energy harvesting. In Proceedings of the 2017 IEEE SENSORS, Glasgow, Scotland, UK, 29 October–1 November 2017.
7. Kılkış, Ş.; Krajačić, G.; Duić, N.; Montorsi, L.; Wang, Q.; Rosen, M.A.; Al-Nimr, M.A. Research frontiers in sustainable development of energy, water and environment systems in a time of climate crisis. *Energy Convers. Manag.* **2019**, *199*, 1–21. [CrossRef]
8. Unsal, O.F.; Bedeloglu, A.C. Recent trends in flexible nanogenerators: A review. *Mater. Sci. Res. India* **2018**, *15*, 114–130. [CrossRef]
9. Rathore, S.; Sharma, S.; Swain, B.P.; Ghadai, R.K. A critical review on triboelectric nanogenerator. *IOP Conf. Ser. Mater. Sci. Eng.* **2018**, *337*, 1–16. [CrossRef]
10. Askari, H.; Khajepour, A.; Khamesee, M.B.; Saadatnia, Z.; Wang, Z.L. Piezoelectric and triboelectric nanogenerators: Trends and impacts. *Nano Today* **2018**, *22*, 10–13. [CrossRef]
11. Promsawat, N.; Wichaiwong, W.; Pimpawat, P.; Changsarn, K.; Phimol, K.; Promsawat, M.; Janphuang, P. A study of flexible piezoelectric generators by sputtering ZnO thin film on PET substrate. *Integr. Ferroelectr.* **2019**, *195*, 220–229. [CrossRef]
12. Panthongsy, P.; Isarakorn, D.; Hamamoto, K.; Janphuang, P. Performance and behavior analysis of piezoelectric energy harvesting floor tiles. In Proceedings of the 2019 5th International Conference on Engineering, Applied Sciences and Technology (ICEAST), Luang Prabang, Laos, 2–5 July 2019.
13. Isarakorn, D.; Jayasvasti, S.; Panthongsy, P.; Janphuang, P.; Hamamoto, K. Design and evaluation of double-stage energy harvesting floor tile. *Sustainability* **2019**, *11*, 5582. [CrossRef]
14. Pachi, A.; Ji, T. Frequency and velocity of people walking. *Struct. Eng.* **2005**, *83*, 36–40.
15. Salauddin, M.; Park, J.Y. A low frequency vibration driven, miniaturized and hybridized electromagnetic and triboelectric energy harvester using dual Halbach array. In Proceedings of the 2017 19th International Conference on Solid-State Sensors, Actuators and Microsystems (TRANSDUCERS), Kaohsiung, Taiwan, 1 February 2017.
16. Niu, S.; Wang, Z.L. Theoretical systems of triboelectric nanogenerators. *Nano Energy* **2014**, *14*, 161–192. [CrossRef]
17. Shamsuddin; Khan, S.A.; Rahimoon, A.Q.; Abro, A.; Ali, M.; Hussain, I.; Ahmed, F. Biomechanical Energy Harvesting by Single Electrode-based Triboelectric Nanogenerator. In Proceedings of the 2019 2nd International Conference on Computing, Mathematics and Engineering Technologies (iCoMET), Sukkur, Pakistan, 30–31 January 2019.
18. Yang, B.; Zeng, W.; Peng, Z.H.; Liu, S.R.; Chen, K.; Tao, X.M. A fully verified theoretical analysis of contact-mode triboelectric nanogenerators as a wearable power source. *Adv. Energy Mater.* **2016**, *6*, 1–8. [CrossRef]

19. Uddin, A.I.; Chung, G.S. A self-powered active hydrogen sensor using triboelectric effect. In Proceedings of the 2016 IEEE SENSORS, Orlando, FL, USA, 30 October–3 November 2016.

20. Huang, H.; Li, X.; Liu, S.; Hu, S.; Sun, Y. TriboMotion: A self-powered triboelectric motion sensor in wearable Internet of Things for human activity recognition and energy harvesting. *IEEE Internet Things J.* **2018**, *5*, 4441–4453. [CrossRef]

21. Wu, C.; Wang, A.C.; Ding, W.; Guo, H.; Wang, Z.L. Triboelectric nanogenerator: A foundation of the energy for the new era. *Adv. Energy Mater.* **2018**, *9*, 1–25. [CrossRef]

22. Bertacchini, A.; Larcher, L.; Lasagni, M.; Pavan, P. Ultra-low-cost triboelectric energy harvesting solutions for embedded sensor systems. In Proceedings of the 15th IEEE International Conference on Nanotechnology, Rome, Italy, 27–30 July 2015.

23. Zhou, T.; Zhang, L.; Xue, F.; Tang, W.; Zhang, C.; Wang, Z.L. Multilayered electret films based triboelectric nanogenerator. *Nano Res.* **2016**, *9*, 1442–1451. [CrossRef]

24. Wen, R.; Guo, J.; Yu, A.; Zhai, J.; Wang, L.Z. Humidity-resistive triboelectric nanogenerator fabricated using metal organic framework composite. *Adv. Funct. Mater.* **2019**, *9*, 1–9. [CrossRef]

25. Dharmasena, R.D.I.G.; Deane, J.H.B.; Silva, S.R.P. Nature of power generation and output optimization criteria for triboelectric nanogenerators. *Adv. Energy Mater.* **2018**, *8*, 1–11. [CrossRef]

26. Xia, K.; Zhu, Z.; Zhang, H.; Du, C.; Wang, R.; Xu, Z. High output compound triboelectric nanogenerator based on paper for self-powered height sensing system. *IEEE Trans. Nanotechnol.* **2018**, *17*, 1217–1223. [CrossRef]

27. Kim, K.; Song, G.; Park, C.; Yun, K.S. Multifunctional woven structure operating as triboelectric energy harvester, capacitive tactile sensor array, and piezoresistive strain sensor array. *Sensors* **2017**, *17*, 2582. [CrossRef]

28. Lu, C.X.; Han, C.B.; Gu, G.Q.; Chen, J.; Yang, Z.W.; Jiang, T.; He, C.; Wang, Z.L. Temperature effect on performance of triboelectric nanogenerator. *Adv. Eng. Mater.* **2017**, *19*, 1–8. [CrossRef]

29. Mule, A.R.; Dudem, B.; Graham, S.A.; Yu, J.S. Humidity sustained wearable pouch-type triboelectric nanogenerator for harvesting mechanical energy from human activities. *Adv. Funct. Mater.* **2019**, *29*, 1–11. [CrossRef]

30. Sriphan, S.; Charoonsuk, T.; Maluangnont, T.; Vittayakorn, N. High-performance hybridized composited-based piezoelectric and triboelectric nanogenerators based on BaTiO3/PDMS composite film modified with Ti0.8O2 nanosheets and silver nanopowders cofillers. *ACS Appl. Energy Mater.* **2019**, *2*, 3840–3850. [CrossRef]

31. Jung, W.S.; Kang, M.G.; Moon, H.G.; Baek, S.H.; Yoon, S.J.; Wang, Z.L.; Kim, S.W.; Kang, C.Y. High output piezo/triboelectric hybrid generator. *Sci. Rep.* **2015**, *5*, 1–6. [CrossRef] [PubMed]

32. Wang, X.; Yang, B.; Liu, J.; He, Q.; Guo, H.; Yang, C.; Chen, X. Flexible triboelectric and piezoelectric coupling nanogenerator based on electrospinning P(VDF-TRFE) nanowires. In Proceedings of the 2015 28th IEEE International Conference on Micro Electromechanical Systems (MEMS), Estoril, Portugal, 18–22 January 2015.

33. Zhang, H.; Galayko, D.; Basset, P. A conditioning system for high-voltage electrostatic/triboelectric energy harvesters using bennet doubler and self-actuated hysteresis switch. In Proceedings of the Transducers 2019-Eurosensors XXXIII, Berlin, Germany, 23–27 June 2019.

34. Isarakorn, D.; Jayasvasti, S.; Thainiramit, P.; Yingyong, P. Power Measurement Circuit for Energy Harvesting Devices. Thai Patent Application No. 2001005793, 2 September 2020.

35. Mide Technology. Piezoelectric Bending Transducer (S230-J1FR-1808XB). Available online: https://piezo.com/-products/piezoelectric-bending-transducer-s230-j1fr-1808xb (accessed on 27 August 2020).

36. Material Properties. Piezoelectric Materials-PZT-5H. Available online: https://support.piezo.com/article/62-material-properties (accessed on 27 August 2020).

37. Wang, Z.L.; Lin, L.; Chen, J.; Niu, S.; Zi, Y. *Triboelectric Nanogenerators*; Springer International Publishing: Basel, Switzerland, 2016; pp. 1–19.

38. Indyplastic. Engineering Plastic Sheets–PTFE. Available online: https://www.indyplasticsheet.com/th/pages/4645 (accessed on 27 August 2020).

39. Carames, T.M.; Lamas, P.F. Towards the internet of smart clothing: A review on IoT wearables and garments for creating intelligent connected E-textiles. *Electronic* **2018**, *7*, 405. [CrossRef]

40. Luis, J.A.; Galan, J.A.G.; Espigado, J.A. Low power wireless smoke alarm system in home fires. *Sensors* **2015**, *15*, 20717–20729. [CrossRef]

41. Zeiser, R.; Fellner, T.; Wilde, J. Capacitive strain gauge on flexible polymer substrates for wireless, intelligent systems. *J. Sens. Sens. Syst.* **2014**, *3*, 77–86. [CrossRef]

42. Hemminger, L.T.; Walters, E.G. Reducing hearing aid power consumption using truncated-matrix multipliers. *Glob. J. Res. Eng.* **2013**, *13*.
43. Gibson, D.R.; MacGregor, C. Self-powered non-dispersive infra-red CO_2 gas sensor. *J. Phys. Conf. Ser.* **2011**, *307*. [CrossRef]
44. Torah, R.N.; Beeby, S.P.; Tudor, M.J.; O'Donnell, T.; Roy, S. Development of a Cantilever Beam Generator Employing Vibration Energy Harvesting. In Proceedings of the 6th International Workshop on Micro and Nanotechnology for Power Generation and Energy Conversion Applications (PowerMEMS 2006), Berkeley, CA, USA, 29 November—1 December 2006; pp. 181–184.
45. Zhu, D.; Tudor, M.J.; Beeby, S.P. Strategies for increasing the operating frequency range of vibration energy harvesters: A review. *Meas. Sci. Technol.* **2010**, *21*, 2. [CrossRef]
46. Wang, Z.L.; Lin, L.; Chen, J.; Niu, S.; Zi, Y. Triboelectric nanogenerator: *Freestanding* triboelectric-layer mode. In *Triboelectric Nanogenerators*; Green Energy and Technology; Springer International Publishing: Cham, Switzerland, 2016; pp. 109–153. ISBN 978-3-319-40038-9.
47. Yingyong, P.; Thainiramit, P.; Vittayakorn, N.; Isarakorn, D. Performance and behavior analysis of single-electrode triboelectric nanogenerator for energy harvesting floor tiles. In Proceedings of the 17th International Conference on Electrical Engineering/Electronics, Computer, Telecommunications, and Information Technology (ECTI-CON), Phuket, Thailand, 24–27 June 2020; pp. 514–517.
48. Kazmierski, T.J.; Beeby, S. *2011. Energy Harvesting Systems: Principles, Modeling and Applications*; Springer Science & Business Media: Berlin, Germany, 2010.
49. Niu, S.; Liu, Y.; Zhou, Y.S.; Wang, S.; Lin, L.; Wang, Z.L. Optimization of triboelectric nanogenerator charging systems for efficient energy harvesting and storage. *IEEE Trans. Electron. Devices* **2015**, *62*, 641–647.
50. Yang, Z.; Wang, Y.Q.; Zuo, L.; Zu, J. Introducing arc-shaped piezoelectric elements into energy harvesters. *Energy Convers. Manag.* **2017**, *148*, 260–266. [CrossRef]
51. Dai, X.; Wen, Y.; Li, P.; Yang, J.; Zhang, G. Modeling, characterization and fabrication of vibration energy harvester using Terfenol-D/PZT/Terfenol-D composite transducer. *Sens. Actuators Phys.* **2009**, *156*, 350–358. [CrossRef]
52. Sriramdas, R.; Chiplunkar, S.; Cuduvally, R.M.; Pratap, R. Performance enhancement of piezoelectric energy harvesters using multilayer and multistep beam configurations. *IEEE Sens. J.* **2015**, *15*, 3338–3348. [CrossRef]
53. Ma, M.; Xia, S.; Li, Z.; Xu, Z.; Yao, X. Enhanced energy harvesting performance of the piezoelectric unimorph with perpendicular electrodes. *Appl. Phys. Lett.* **2014**, *105*, 043905-1–043905-4. [CrossRef]
54. Dhakar, L.; Liu, H.; Tay, F.E.H.; Lee, C. A new energy harvester design for high power output at low frequencies. *Sens. Actuators Phys.* **2013**, *199*, 344–352. [CrossRef]
55. Lee, B.S.; Lin, S.C.; Wu, W.J.; Wang, X.Y.; Chang, P.Z.; Lee, C.K. Piezoelectric MEMS generators fabricated with an aerosol deposition PZT thin film. *J. Micromech. Microeng.* **2009**, *19*, 065014-1–065014-8. [CrossRef]
56. Song, H.-C.; Kumar, P.; Maurya, D.; Kang, M.-G.; Reynolds, W.T.; Jeong, D.-Y.; Kang, C.-Y.; Priya, S. Ultra-low resonant piezoelectric MEMS energy harvester with high power density. *J. Microelectromec. Syst.* **2017**, *26*, 1226–1234. [CrossRef]
57. Zou, H.-X.; Zhang, W.-M.; Li, W.-B.; Hu, K.-M.; Wei, K.-X.; Peng, Z.-K.; Meng, G. A broadband compressive-mode vibration energy harvester enhanced by magnetic force intervention approach. *Appl. Phys. Lett.* **2017**, *110*, 163–904. [CrossRef]
58. Morimoto, K.; Kanno, I.; Wasa, K.; Kotera, H. High-efficiency piezoelectric energy harvesters of c-axis-oriented epitaxial PZT films transferred onto stainless steel cantilevers. *Sens. Actuators Phys.* **2010**, *163*, 428–432. [CrossRef]
59. Ji, S.H.; Lee, W.; Yun, J.S. All-in-One Piezo-Triboelectric Energy Harvester Module Based on Piezoceramic Nanofibers for Wearable Devices. *ACS Appl. Mater. Interfaces* **2020**, *12*, 18609–18616. [CrossRef]
60. Rodrigues, C.A.; Gomes, A.; Ghosh, A.; Pereira, A.; Ventura, J. Power-generating footwear based on a triboelectric-electromagnetic-piezoelectric hybrid nanogenerator. *Nano Energy* **2019**, *62*, 660–666. [CrossRef]

Publisher's Note: MDPI stays neutral with regard to jurisdictional claims in published maps and institutional affiliations.

Letter

Performance of An Electromagnetic Energy Harvester with Linear and Nonlinear Springs under Real Vibrations

Tra Nguyen Phan, Sebastian Bader * and Bengt Oelmann

Department of Electronics Design, Mid Sweden University, 85170 Sundsvall, Sweden; tra.phan@miun.se (T.N.P.); bengt.oelmann@miun.se (B.O.)
* Correspondence: sebastian.bader@miun.se

Received: 25 August 2020; Accepted: 21 September 2020; Published: 23 September 2020

Abstract: The introduction of nonlinearities into energy harvesting in order to improve the performance of linear harvesters has attracted a lot of research attention recently. The potential benefits of nonlinear harvesters have been evaluated under sinusoidal or random excitation. In this paper, the performances of electromagnetic energy harvesters with linear and nonlinear springs are investigated under real vibration data. Compared to previous studies, the parameters of linear and nonlinear harvesters used in this paper are more realistic and fair for comparison since they are extracted from existing devices and restricted to similar sizes and configurations. The simulation results showed that the nonlinear harvester did not generate higher power levels than its linear counterpart regardless of the excitation category. Additionally, the effects of nonlinearities were only available under a high level of acceleration. The paper also points out some design concerns when harvesters are subjected to real vibrations.

Keywords: energy harvester; electromagnetic; real vibration; nonlinearities

---

1. Introduction

Recently, scavenging energy from the ambient environment has become a more attractive research topic due to its wide range of applications. Electrical energy can be extracted from a wide variety of sources such as solar, chemical, thermal, radio frequency, and vibration. Among those, vibration energy harvesting (VEH) is a promising alternative due to the availability of vibration sources in many application environments [1,2]. VEH converts the ambient mechanical vibration into electrical energy. Based on the transduction mechanism, the VEH can be further classified into three main categories. They include electrostatic energy harvesters, electromagnetic energy harvesters, and piezoelectric energy harvesters. Each one has its own advantages and drawbacks. Thanks to the robustness and low-cost design, the electromagnetic energy harvester has attracted considerable attention from researchers [3,4].

The majority of previous research has focused on linear resonant energy harvesters [5–7]. For this type of design, the output power of the harvester can reach the optimal value when the resonant frequency of the oscillator matches the dominant frequency of the ambient vibration. Thus, such a linear device requires high precision during its manufacturing process. It also places critical performance limitations, especially when the excitation frequency in applications changes over time. The reason is that the output power drops significantly when the external excitation frequency deviates from resonance conditions.

Prior works have proposed some solutions to broaden the frequency spectrum. Ooi et al. [8] utilized a novel dual-resonator method consisting of two separate resonator systems to improve the frequency response range. Cammarano et al. [9] examined the ability to tune a resonant energy harvester by coupling it to variable generalized electrical loads. Another widely popular approach is to utilize nonlinear effects in mechanical oscillators. The nonlinearities can be broadly classified into being monostable, bistable, or multistable, depending on the number of stable equilibrium states. The most common method to design such nonlinear systems is the introduction of a nonlinear restoring force through mechanical structures or permanent magnets. Mann et al. [10] proposed a nonlinear energy harvester using a permanent magnet sandwiched between two other permanent magnets. This monostable nonlinearity is obtained from the effect of magnetic levitation. Under harmonic base excitation, the frequency response of the nonlinear system expands over a wider bandwidth. The monostable harvester utilizing magnetic levitation effect was further modified by using a magnetic rolling pendulum [11] to enhance performance in both the primary and subharmonic resonance regions. Cottone et al. [12] introduced a bistable electromagnetic energy harvester, which employs a clamped-clamped buckled beam working as a nonlinear spring to achieve a large bandwidth response. This bistable configuration was shown to produce higher power as compared with monostable regimes under an optimal acceleration level. Lan and Qin [13] added a small magnet at the middle of two fixed magnets in a bistable energy harvester to reduce the barrier and improve the performance under random excitations. In order to achieve multistability, cantilever beams with tip magnet [14–16] or magnetic levitation [17] can be utilized. Due to shallower potential wells, tri- and quadstable systems can easily achieve interwell oscillations at lower frequency ranges and weaker base excitations compared to bistable systems. Recently, Nammari et al. [18] presented an enhanced design that combines both mechanical and magnetic springs to introduce additional stiffness nonlinearities. Non-dimensional analyses demonstrated that the proposed design results in more harvested power than the linear version. Another approach proposed by Wang et al. [19] utilized preloading and mechanical stoppers to introduce a piecewise linear stiffness in vibration systems. The multiple nonlinear effects were proven to have significant influences on the system response.

The potential benefits of introducing nonlinearity to VEH designs have been evaluated in previous studies. However, most of the works have been done with the assumption that the input excitations are sinusoidal, colored noise, or Gaussian white noise [20–22]. Only a limited number of studies have focused on the harvester performance under real-world ambient vibrations [23–25]. Beeby et al. [23] presented the comparison of output power from linear and nonlinear harvesters under vibration data taken from measurements of a diesel ferry engine, heat and power pump, car engine, and white noise vibration. The parameters in their paper were chosen quite freely and, thus, are not linked to a concrete harvester implementation. Green et al. [24] assessed the effectiveness of current nonlinear harvesters subjected to human motion and bridge vibrations only. It was concluded in their paper that the potential benefits of nonlinear energy harvester solutions are sensitive to the nature of ambient vibration sources. Rantz and Roundy [25] considered a broad range of real vibrations and provided a comparative analysis of the theoretical maximum output power that linear and nonlinear harvester architectures can reach under these inputs. These optimal values may not be obtained in the real devices with design restrictions.

The present study extracts the parameters from actual linear and monostable nonlinear electromagnetic energy harvester implementations and numerically analyzes their performance under a wide range of real vibration excitations. The monostable energy harvester with Duffing-type nonlinearities is of particular interest here. The goal is to evaluate the performance of the monostable Duffing-type harvester compared to its linear counterpart when subjected to inputs of different characteristics, i.e., the number of dominant frequencies, the stationary, or the noise effects. The vibration signals were collected with several types of acquisition kits and can be downloaded from The NiPS Laboratory "Real Vibration" database [26]. Two electromagnetic energy harvester designs, including linear and nonlinear, as reported by Mallick et al. [27] were under investigation for comparison. These two designs were restricted in the same sizes and similar configurations for a fair comparison. The rest of the paper is organized as follows. Section 2 describes the device configurations and modeling of the electromagnetic energy harvesters. The classification and properties of selected real vibration signals are presented in Section 3. The simulation results and discussions are shown in Section 4. Finally, Section 5 concludes the paper.

2. Electromagnetic Energy Harvester

This section presents the electromagnetic energy harvester configuration used for examination. The total energy harvester system is modeled and simulated in Matlab/Simulink environment. The system responses under sinusoidal signal with frequency sweep and amplitude sweep are also included.

2.1. Device Configuration

The device configuration used for the investigation in this paper was proposed by Mallick et al. [27]. It consists of four main parts: spring, magnets, copper coil, and frame. The magnets are attached to the center top of the spring structure while the coil is assembled on the glass slide of the frame, which is separated from the spring by the spacers as shown in Figure 1. When the magnets move up and down under external vibration, the relative displacement between the magnets and coil changes. As a result, voltage is induced into the coil according to Faraday's law of induction. Depending on the spring structure, the resonator can be linear or nonlinear. Figure 2 displays two spring designs used for comparison. The clamped-free configuration shown in Figure 2a results in only the linear term, while the fixed-fixed spring arms configuration shown in Figure 2b causes nonlinear stretching. For a fair comparison, both linear and nonlinear harvesters have the same spring size, similarly oriented magnets, and the same proof mass sizes.

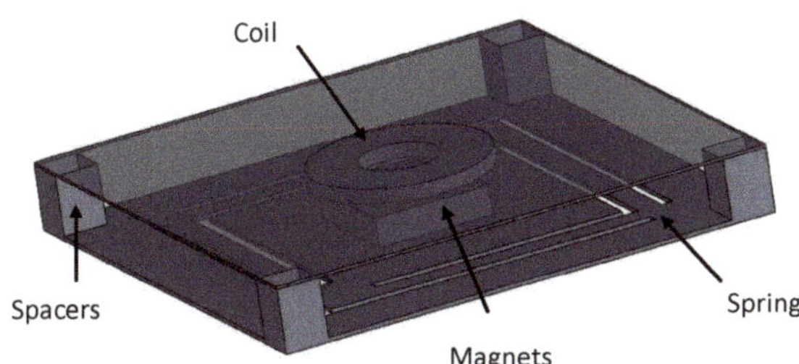

Figure 1. Electromagnetic energy harvester device proposed by Mallick et al . [27].

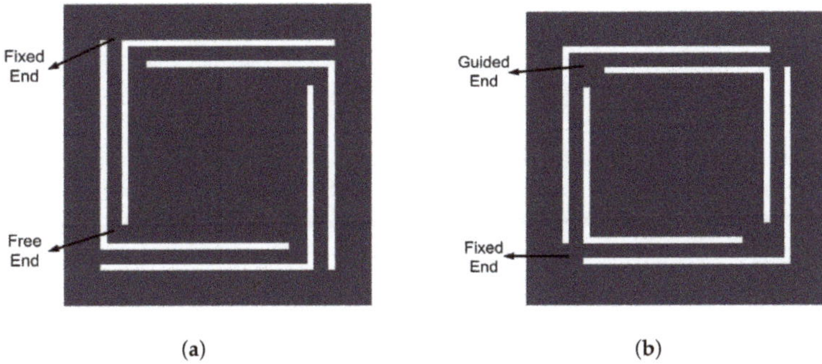

Figure 2. Different spring structures: (**a**) Linear spring. (**b**) Nonlinear spring. [27].

2.2. Model

The energy harvester can be modeled as a spring-mass-damper system with base excitation. The governing differential equation of the electromechanical system is given by

$$m\ddot{x} + c\dot{x} + F(x) + \gamma I = -m\ddot{z} \tag{1}$$

where m is the inertial mass, x is the relative displacement between the mass and the frame, c is the mechanical damping ratio, $F(x)$ is the generalized spring force, γ is the electromagnetic coupling coefficient, I is the induced current, and z is the input vibration. For a linear harvester, the storing force is proportional to displacement

$$F(x) = kx \tag{2}$$

where k is the linear stiffness coefficient. In the case of the nonlinear harvester, it was shown in the study of Mallick et al. [27] that the storing force can be modeled as the nonlinear spring force similar to the hardening-spring Duffing oscillator

$$F(x) = kx + k_n x^3 \tag{3}$$

where k_n is the nonlinear stiffness coefficient. Then, Equation (1) can be rewritten as

$$m\ddot{x} + 2m\rho\omega_0\dot{x} + kx + k_n x^3 + \gamma I = -m\ddot{z} \tag{4}$$

where ρ is the mechanical damping coefficient, and ω_0 is the resonant frequency. For the linear system, the term with the nonlinear stiffness will be ignored. The induced current can be modeled in the following electrical circuit:

$$L\dot{I} + RI - \gamma\dot{x} = 0 \tag{5}$$

where L is the electromagnetic inductance, and R is the total resistance combining the coil resistance R_c and the load resistance R_L. Neglecting the inductance of the coil, which is commonly accepted for low frequencies, the following equation can be derived from Equations (4) and (5)

$$m\ddot{x} + 2m\rho\omega_0\dot{x} + \frac{\gamma^2}{R}\dot{x} + kx + k_n x^3 = -m\ddot{z} \tag{6}$$

The voltage across a load resistance R_L and the corresponding load power generated in the system are given by

$$V_L(t) = \gamma\dot{x}\left(\frac{R_L}{R_L + R_c}\right) \tag{7}$$

and

$$P_L(t) = \frac{V_L(t)^2}{R_L} \tag{8}$$

Figure 3 shows the overall model implemented in the Matlab/Simulink environment. The parameters used in the model were obtained from Mallick et al. [27] and are listed in Table 1. All simulations were run with the ode45 solver. The variable-step was chosen with a maximum step size of 1×10^{-6} and a relative tolerance of 10^{-9}.

Table 1. Parameters of the nonlinear electromagnetic energy harvester proposed by Mallick et al. [27].

Parameters	Value	Units
Linear stiffness k	185.66	N·m^{-1}
Nonlinear stiffness k_n	3.56×10^9	N·m^{-3}
Electromagnetic coupling coefficient γ	0.035	Wb·m^{-1}
Inertial mass m	16×10^{-5}	Kg
Mechanical damping coefficient ρ	0.015	

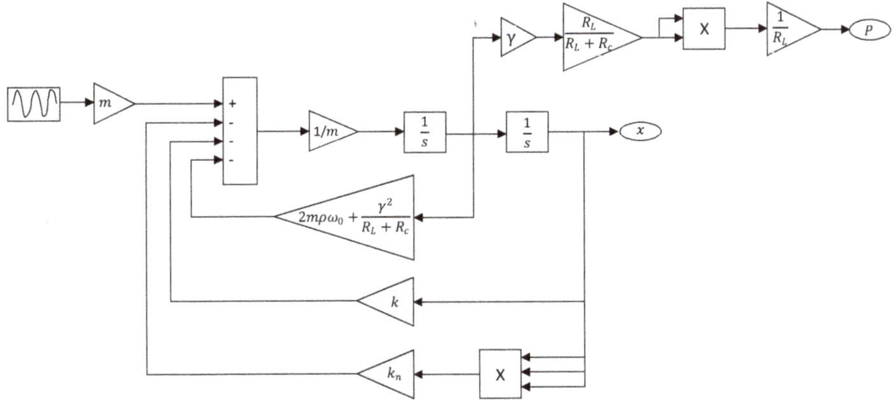

Figure 3. Simulink model.

Figure 4 shows the output power of linear and nonlinear transducers under linear frequency up and down sweeps with different levels of acceleration amplitude. The frequency of the sinusoidal input signal was swept from 140 to 240 Hz during 120 s for each sweeping direction. Then, the average output power for each 1.2 s time interval was calculated as the output power at the corresponding frequency. Under a low acceleration level, linear and nonlinear systems had similar responses. However, when acceleration increased, the frequency response of the nonlinear transducer divided into two different branches for up and down sweep, respectively. The up-sweep curve has a wide bandwidth while the down-sweep curve has a narrow bandwidth. This behavior is similar to that of the Duffing oscillator. Bifurcation occurs over the region where multiple stable roots occur in the steady-state equation. Depending on the sweep rate and initial conditions, the frequency response can either follow the high- or the low-energy branch. It can be seen that the nonlinear harvester can broaden the bandwidth compared to the linear response at the expense of the output power in the down-sweep case.

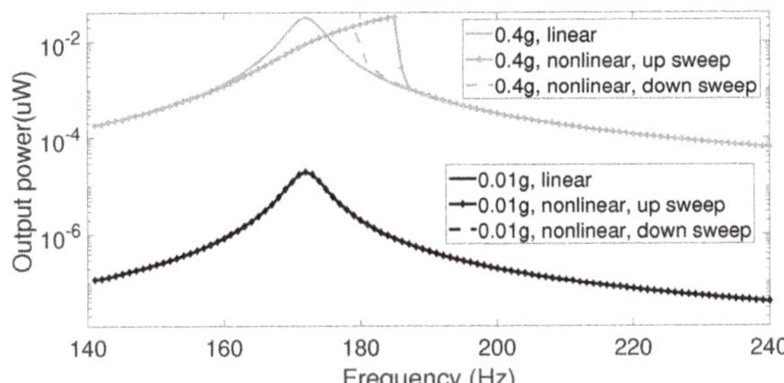

Figure 4. Frequency–response curves for the linear and nonlinear transducer under frequency up and down sweep.

Similar phenomena can be observed when the acceleration amplitude is swept, as shown in Figure 5. In this case, the output power is simulated under amplitude up and down sweep with different values of frequency. At a certain frequency, the hysteresis in the nonlinear harvester can be detected.

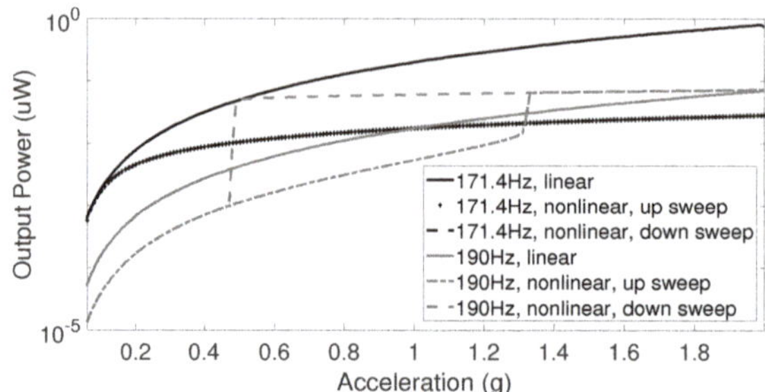

Figure 5. Amplitude–response curves for the linear and nonlinear transducer under amplitude up and down sweep.

3. Real Vibration Database

The NiPS Laboratory [26] provides a website of a digital database for real vibration signals. The vibrations are collected from everyday life activities of cars, trains, airplanes, and even human beings. The database is recorded with various devices. There are three axes of vibration data for each signal in the database. Rantz and Roundy [28] have presented several approaches to categorize these signals. One of those is based on their spectrograms. To facilitate the classification procedure, Rantz and Roundy [28] have filtered the spectrogram by considering only frequency content with the values of A^2/ω greater than $1/2$ the maximum value of A^2/ω in each FFT frame, where A is the input acceleration amplitude and ω is the input frequency. Then, classification is based on the dominant frequency number of the filtered spectrogram. In this paper, some representative signals are selected to examine the performance of the linear and nonlinear harvesters. These signals cover typical types of vibration data such as signals with one dominant frequency, signals with two dominant frequencies, and stochastic signals. For each signal, the ordinary and the filtered spectrograms are displayed for classification. The spectrograms and properties are shown in the following.

3.1. One-Dominant-Frequency Signals

Figure 6 shows the spectrogram for the signal with the title "airplane passenger table" in the Y direction. The acquisition kit used to collect the signal is an iPhone with a sampling frequency of 100 Hz. As can be seen, over the length of the signal, there was one dominant frequency at 27 Hz. The acceleration amplitude of the data at the dominant frequency varied over time, and we also saw some noise. The signal had a low acceleration level with the maximum acceleration amplitude value of just 10 mg.

Another one-dominant-frequency signal displayed in Figure 7 is the one with the title "car in highway" in the X direction. This dominant frequency was not constant as in the previous signal but varied during the first 80 s. After the start-up time, the signal was stable and the dominant frequency reached 19 Hz. In terms of acceleration degree, the car-in-highway signal had a higher level than that of the airplane passenger table signal.

3.2. Two-Dominant-Frequency Signals

The air-pump signal in the Z direction and the aquarium signal in the X direction both consist of two dominant frequencies as shown in Figures 8 and 9. The acceleration amplitudes of these dominant frequencies also changed over time. The main differences between these two signals are that the two dominant frequencies in the second signal were closer to each other as compared to that in the first signal, and the acceleration intensity of the first signal was much higher than that of the second one. For the air-pump signal, two dominant frequencies were apparent, which occurred at 35 and 44 Hz, and the maximum acceleration amplitude reached 400 mg. Otherwise, two dominant frequencies of the aquarium signal were observed at 44 and 46 Hz, and the maximum amplitude had a value of 25 mg only.

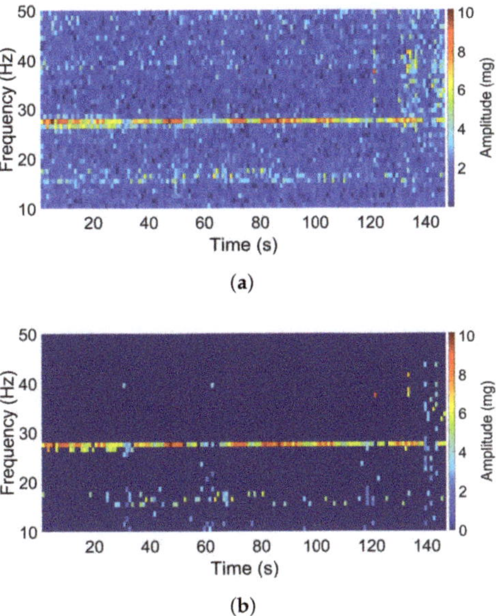

(a)

(b)

Figure 6. Spectrogram of the airplane passenger table signal (Y direction): (**a**) Original. (**b**) Filtered.

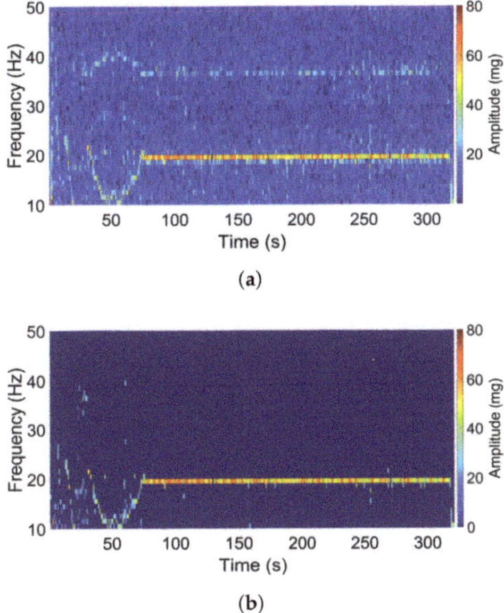

(a)

(b)

Figure 7. Spectrogram of the car-in-highway signal (X direction): (**a**) Original. (**b**) Filtered.

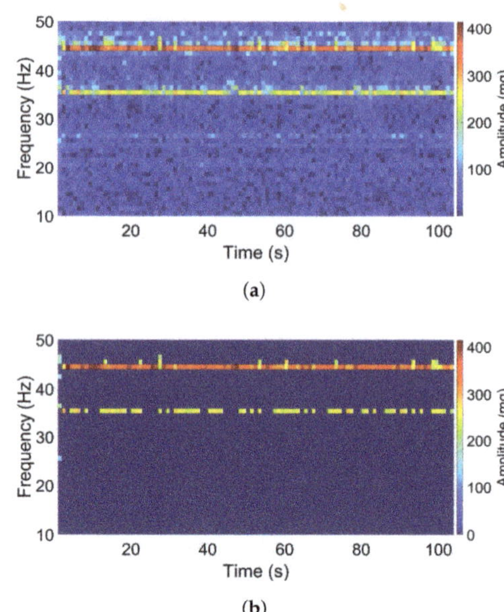

(a)

(b)

Figure 8. Spectrogram of the air-pump signal (Z direction): (**a**) Original. (**b**) Filtered.

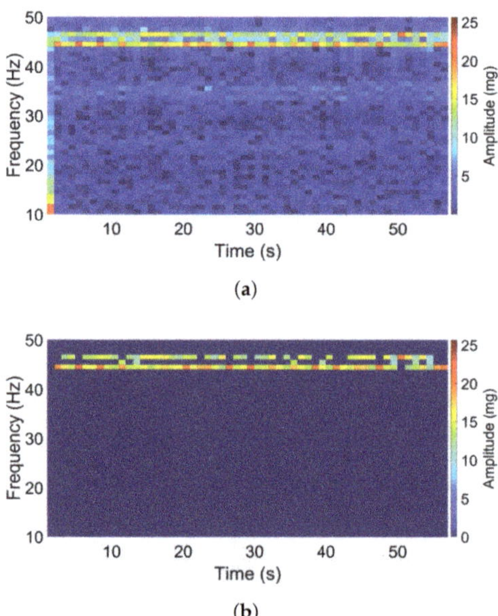

(a)

(b)

Figure 9. Spectrogram of the aquarium signal (Z direction): (**a**) Original. (**b**) Filtered.

3.3. Stochastic Signals

Figures 10 and 11 show signals with the stochastic property. It is hard to see any dominant frequency from the original spectrograms. Even with the assistance of the filtered spectrograms, it is difficult to classify these signals in terms of the dominant frequency number. The vibration content with significant levels covers a broad band of frequencies. The signal titled "Acoustic guitar" was recorded by an EVAL-ADXL345Z acquisition kit with a sampling rate of 100 Hz. Its spectrogram shows that the frequency content varied over the whole bandwidth of 50 Hz. For the signal from the bike, a slam stick [29] was used to collect the data at a sampling rate of 3134 Hz. It is clear from its spectrogram that the band of significant frequencies appeared at the lower part of the bandwidth. It was also noted that different degrees of acceleration can be seen from these two samples.

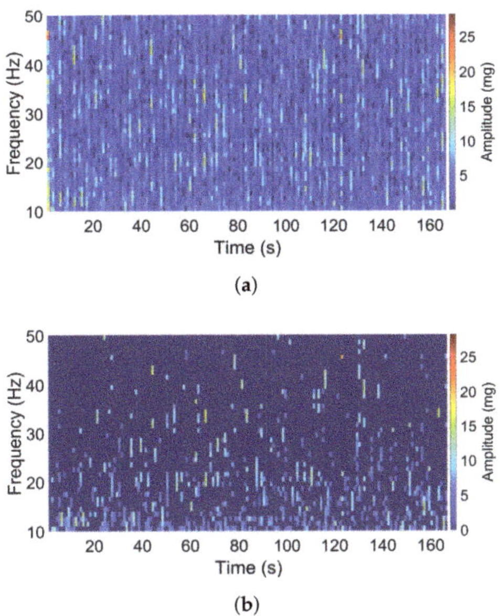

Figure 10. Spectrogram of the acoustic guitar signal (Z direction): (**a**) Original. (**b**) Filtered.

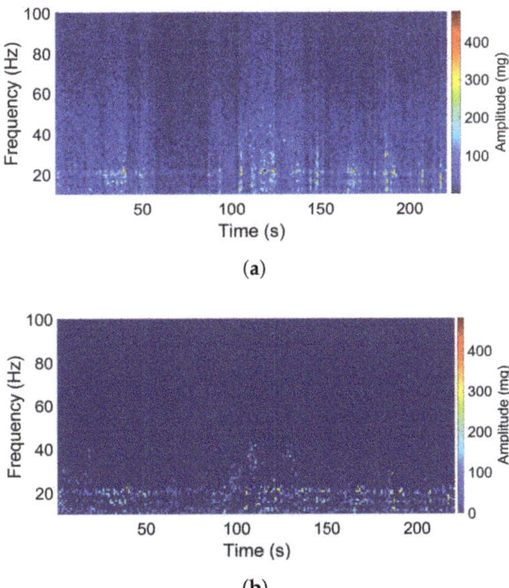

Figure 11. Spectrogram of the bike signal (Z direction): (**a**) Original. (**b**) Filtered.

4. Results and Discussion

In this section, different scenarios are set up to investigate the performance of the linear and nonlinear harvesters under the real vibration signals described in the previous section. The simulation results for each situation are presented and discussed. Throughout the simulations, the frequency range of the real vibrations was modified by changing their sampling frequency in order to match the harvester configuration. The following procedure was used to define the center frequency of the vibration data. The center frequency of dominant signals was chosen to be the frequency of the most significant amplitude. For those with stochastic properties, the center frequency was considered as the frequency component in the middle of the relevant frequency range. For example, the significant frequency content for the bike signal was in the range of 0–41 Hz; thus, a frequency of 20.5 Hz, which is in the middle of this frequency range, should be chosen as the center frequency. The output power generated from the simulations is the average load power over the length of the signal.

Table 2 expresses the output power of the linear and nonlinear harvesters when the center frequency of the input vibration data is adjusted to match the linear resonant frequency. It can be seen that in this case, the linear harvester and the nonlinear harvester had similar output powers under a low level of acceleration, regardless of the vibration data category. It is also observed that when the input acceleration amplitudes increased, the linear harvester produced more output power than the nonlinear harvester.

Table 2. Simulation results of output power of the linear and nonlinear harvesters at resonant frequency.

Vibration Data/Direction	Maximum Acceleration Amplitude (mg)	Center Freq. (Hz)	# Dominant Frequencies	Linear Output Power (W)	Nonlinear Output Power (W)
Airplane passenger table/Z	10	27	1	7.49×10^{-12}	7.49×10^{-12}
Car in highway/X	80	19	1	3.92×10^{-10}	3.82×10^{-10}
Aquarium/Z	25	44	2	3.26×10^{-11}	3.26×10^{-11}
Air pump/Z	400	44	2	1.53×10^{-8}	6.6×10^{-9}
Bike/Z	450	20.5	NA	2.05×10^{-9}	1.51×10^{-9}
Acoustic guitar/Z	28	25	NA	4.37×10^{-12}	4.37×10^{-12}

While under a small excitation amplitude, linear and nonlinear harvesters behaved similarly and generated the peak output power at the resonant frequency; this was not true under high acceleration amplitude. When higher acceleration was applied, the frequency response of the nonlinear systems bent to the right and reached the maximum point at the frequency differing from the linear resonant frequency. The optimal point depends on the acceleration level and the initial state of the system. Since real vibrations have varied frequency and amplitude over time, it is difficult to determine the frequency at which the optimal output power of the nonlinear harvester can be reached. Thus, it is of interest to investigate the output power of the linear and nonlinear harvesters when the center frequency of the vibration data with high acceleration is swept through a frequency range around the linear resonant frequency. The simulation results for this scenario are presented in Figures 12–14. It is shown that the center frequency at which the nonlinear system produces the maximum output power was higher than the linear resonant frequency. It is also noted that this peak output power was still lower than the optimal value from the linear harvester in all cases. This can be explained by the fact that the frequency response of the nonlinear system was not kept on the high-energy branch all the time. Due to the complexity of the real vibration, the system may follow either a high- or low-energy branch, in which the output power is significantly reduced in the case of the lower branch.

When subjected to signals with low acceleration levels such as the airplane passenger seat signal, aquarium signal, and acoustic guitar signal, there was no difference between output response of the linear and nonlinear harvesters. Therefore, it may be beneficial to examine the maximum output power from these two harvesters when the vibration amplitude is scaled by a factor. The first step is to multiply the vibration data by an element and then to sweep the center frequency of these data over a range similar to previous simulations. The optimal output powers of the linear and nonlinear harvesters are collected for each scale factor, and the final results are plotted in Figures 15–17.

From these figures, we can see that the higher the scale factor is, the more different the peak output power values of the linear and nonlinear systems are. When the scale factor increases, the linear harvester generates more output power than the nonlinear one. For vibrations that have dominant frequencies, these observations are more obvious compared to those with the stochastic property. This is because the stochastic signals have frequency contents that cover a broader range than the signals with dominant frequencies.

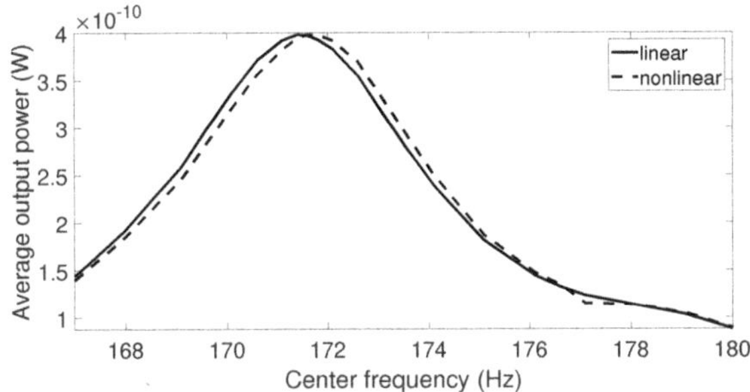

Figure 12. The average output power response for the linear and nonlinear prototypes under car-in-highway excitation when the center frequency of the excitation is swept.

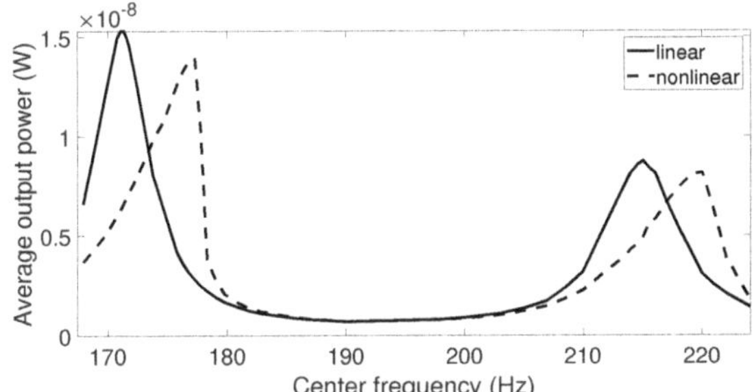

Figure 13. The average output power response for the linear and nonlinear prototypes under air-pump excitation when the center frequency of the excitation is swept.

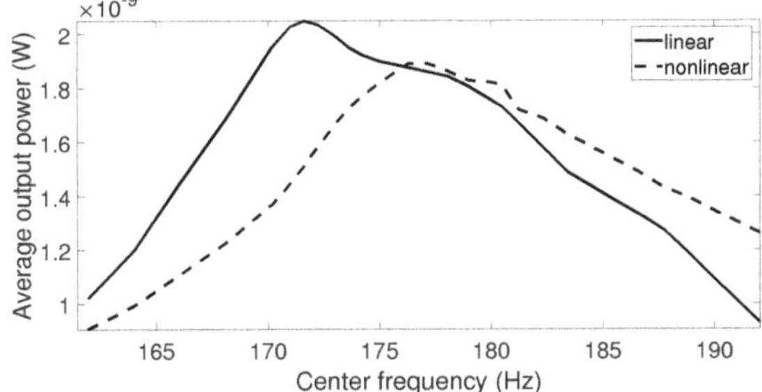

Figure 14. The average output power response for the linear and nonlinear prototypes under bike excitation when the center frequency of the excitation is swept.

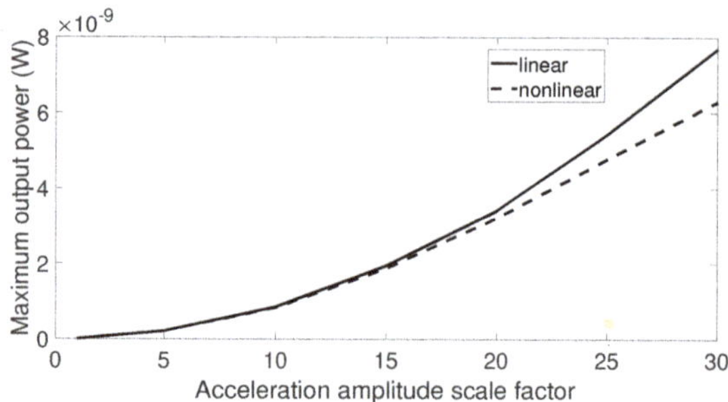

Figure 15. Maximum output power response for the linear and nonlinear prototypes under center frequency sweeping airplane passenger seat excitation when the acceleration amplitude is scaled by a factor.

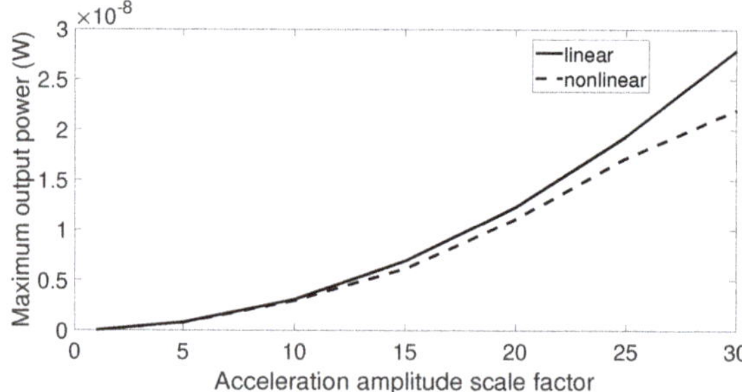

Figure 16. Maximum output power response for the linear and nonlinear prototypes under center frequency sweeping aquarium excitation when the acceleration amplitude is scaled by a factor.

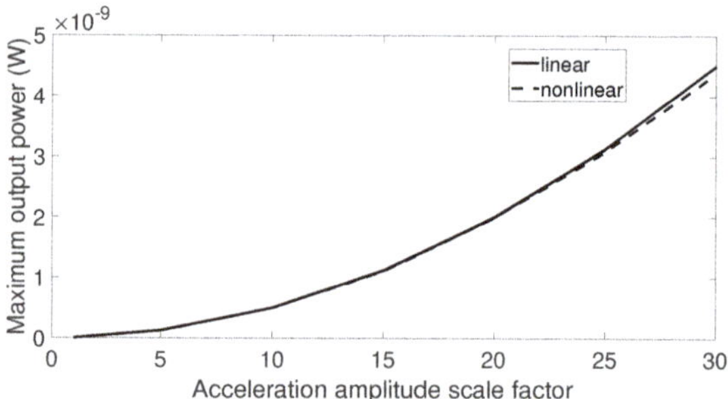

Figure 17. Maximum output power response for the linear and nonlinear prototypes under center frequency sweeping acoustic guitar excitation when the acceleration amplitude is scaled by a factor.

It can also be noted that the excitations from the current investigated database have a relatively low amplitude, which is in the range of 0–500 mg. Since nonlinear harvesters behave much like linear harvesters regardless of the excitation types under low excitation, the effect of the nonlinearity cannot be seen in such conditions. Thus, nonlinear harvesters require the input vibrations with high acceleration amplitude in order to unlock the nonlinearity and obtain the wider bandwidth.

Moreover, it can be seen that, for the current database, the hardening nonlinear harvester did not produce higher output power compared to the linear counterpart. This agrees in general with the results by Beeby et al. [23], where only one case was reported in which the nonlinear harvester outperformed the linear one. As shown in their paper, under car-engine excitation, the nonlinear harvester can produce 137 mJ of energy compared to the value of 130 mJ generated by the linear harvester. The excitation from the car engine has one dominant frequency, which varied over the range of nearly 50 Hz during 45 min of investigation. Most signals from the current database have a shorter time recording in the range of several hundred seconds. In most application scenarios, frequency variations as large as the one of a car traveling and varying speeds are not so common.

Finally, it is worth to consider the coexistence of the energy branches for the nonlinear harvesters under high acceleration. Since the properties of real vibrations are affected by small variations and do not follow a clear up- or down-sweep, it is difficult to maintain high-energy branch output at all time. Therefore, there may be much less power collected from the nonlinear harvesters than expected if the system falls in the low-energy orbit. Several strategies have been proposed to alter the system orbit. These methods utilize either mechanical impacts [30,31] or electrical perturbation [32–34] to achieve the desired response. However, these proposed approaches may require precise control, energy investment, or a complex system to be performed, which has so far not been demonstrated to be energetically beneficial. Thus, a much simpler and low-power method to capture high-energy orbit is still under further investigation.

5. Conclusions and Future Study

This paper aimed to investigate the performance of electromagnetic energy harvesters with linear and nonlinear springs under real vibration data. The parameters for the harvesters were extracted from actual devices, with restrictions in the size and configurations for a fair comparison. Some typical signals from the downloaded database were selected, classified, and described. Different modifications have been made to these signals in order to create various scenarios under which the output power from the linear and nonlinear harvesters were collected and examined.

The addition of nonlinearities was first introduced to broaden the bandwidth of energy harvesters [10,35]. However, Daqaq et al. [36] demonstrated that under Gaussian white noise excitation, stiffness-type nonlinearities did not provide any benefits in terms of output power compared to the linear harvesters. Our paper shows that the conclusion regarding the devices excited by Gaussian white noise is still applicable to the specific studied harvesters subjected to real ambient vibrations. The power from linear and nonlinear harvesters is evaluated under different levels of excitation as well as under different properties of the dominant frequency. In none of the investigated cases did the specific investigated nonlinear harvester provide a higher average output power than its linear counterpart. The presented results are limited to a certain type of nonlinear harvester under the assumption that electromagnetic induction is neglected. Thus, the conclusion confirmed under these conditions may need to be investigated further for other types of nonlinear harvesters under more complicated scenarios.

Furthermore, some properties of the external vibrations and the frequency response of the systems were discussed in this paper. The ambient vibrations investigated in the collection of real-world signals have low dominant frequencies and low acceleration amplitude. These factors need to be taken into account when it comes to the design of relevant harvesters.

Author Contributions: Conceptualization, T.N.P., S.B. and B.O.; Data curation, T.N.P.; Formal analysis, T.N.P.; Investigation, T.N.P.; Methodology, S.B. and B.O.; Software, T.N.P.; Supervision, S.B. and B.O.; Validation, T.N.P.; Writing—original draft, T.N.P.; Writing—review & editing, S.B. and B.O. All authors have read and agreed to the published version of the manuscript.

Acknowledgments: This work is financially supported by the Knowledge Foundation under grant NIIT 20180170.

Conflicts of Interest: The authors declare no conflict of interest.

References

1. Amirtharajah, R.; Chandrakasan, A.P. Self-powered signal processing using vibration-based power generation. *IEEE J. Solid-State Circuits* **1998**, *33*, 687–695. [CrossRef]
2. Arroyo, E.; Badel, A.; Formosa, F. Energy harvesting from ambient vibrations: Electromagnetic device and synchronous extraction circuit. *J. Intell. Mater. Syst. Struct.* **2013**, *24*, 2023–2035. [CrossRef]

3. Beeby, S.P.; Torah, R.N.; Tudor, M.J.; Glynne-Jones, P.; O'Donnell, T.; Saha, C.R.; Roy, S. A micro electromagnetic generator for vibration energy harvesting. *J. Micromech. Microeng.* **2007**, *17*, 1257–1265. [CrossRef]

4. Halim, M.A.; Cho, H.; Park, J.Y. Design and experiment of a human-limb driven, frequency up-converted electromagnetic energy harvester. *Energy Convers. Manag.* **2015**, *106*, 393–404. [CrossRef]

5. El-hami, M.; Glynne-Jones, P.; White, N.M.; Hill, M.; Beeby, S.; James, E.; Brown, A.D.; Ross, J.N. Design and fabrication of a new vibration-based electromechanical generator. *Sens. Actuator A-Phys.* **2001**, *92*, 335–342. [CrossRef]

6. Roundy, S.; Wright, P.K. A piezoelectric vibration based generator for wireless electronics. *Smart Mater. Struct.* **2004**, *13*, 1131–1142. [CrossRef]

7. Gu, L.; Livermore, C. Passive self-tuning energy harvester for extracting energy from rotational motion. *Appl. Phys. Lett.* **2010**, *97*, 081904. [CrossRef]

8. Ooi, B.L.; Gilbert, J.M. Design of wideband vibration-based electromagnetic generator by means of dual-resonator. *Sens. Actuator A-Phys.* **2014**, *213*, 9–18. [CrossRef]

9. Cammarano, A.; Burrow, S.G.; Barton, D.; Carrella, A.; Clare, L. Tuning a resonant energy harvester using a generalized electrical load. *Smart Mater. Struct.* **2010**, *19*, 055003. [CrossRef]

10. Mann, B.P.; Sims, N.D. Energy harvesting from the nonlinear oscillations of magnetic levitation. *J. Sound Vib.* **2009**, *319*, 515–530. [CrossRef]

11. Kuang, Y.; Hide, R.; Zhu, M. Broadband energy harvesting by nonlinear magnetic rolling pendulum with subharmonic resonance. *Appl. Energy* **2019**, *255*, 113822. [CrossRef]

12. Cottone, F.; Basset, P.; Vocca, H.; Gammaitoni, L. Electromagnetic Buckled Beam Oscillator for Enhanced Vibration Energy Harvesting. In Proceedings of the 2012 IEEE International Conference on Green Computing and Communications, Besancon, France, 20–23 November 2012; pp. 624–627.

13. Lan, C.; Qin, W. Enhancing ability of harvesting energy from random vibration by decreasing the potential barrier of bistable harvester. *Mech. Syst. Signal Process.* **2017**, *85*, 71–81. [CrossRef]

14. Zhou, S.; Cao, J.; Inman, D. J.; Lin, J.; Liu, S.; Wang, Z. Broadband tristable energy harvester: Modeling and experiment verification. *Appl. Energy* **2014**, *133*, 33–39. [CrossRef]

15. Mei, X.; Zhou, S.; Yang, Z.; Kaizuka, T.; Nakano, K. A tri-stable energy harvester in rotational motion: modeling, theoretical analyses and experiments. *J. Sound Vib.* **2020**, *469*, 115142. [CrossRef]

16. Zhou, Z.; Qin, W.; Zhu, P. A broadband quad-stable energy harvester and its advantages over bi-stable harvester: simulation and experiment verification. *Mech. Syst. Signal Process.* **2017**, *84*, 158–168. [CrossRef]

17. Gao, M.; Wang, Y.; Wang, Y.; Wang, P. Experimental investigation of non-linear multi-stable electromagnetic-induction energy harvesting mechanism by magnetic levitation oscillation. *Appl. Energy* **2018**, *220*, 856–875. [CrossRef]

18. Nammari, A.; Bardaweel, H. Design enhancement and non-dimensional analysis of magnetically-levitatednonlinear vibration energy harvesters. *J. Intell. Mater. Syst. Struct.* **2017**, *28*, 2810–2822. [CrossRef]

19. Wang, X.; Chen, C.; Wang, N.; San, H.; Yu, Y.; Halvorsen, E.; Chen, X. A frequency and bandwidth tunable piezoelectric vibration energy harvester using multiple nonlinear techniques. *Appl. Energ.* **2017**, *190*, 368–375. [CrossRef]

20. Cottone, F.; Vocca, H.; Gammaitoni, L. Nonlinear energy harvesting. *Phys. Rev. Lett.* **2009**, *102*, 080601. [CrossRef]

21. Marinkovic, B.; Koser, H. Demonstration of wide bandwidth energy harvesting from vibrations. *Smart Mater. Struct.* **2012**, *21*, 065006. [CrossRef]

22. Nguyen, S.D.; Halvorsen, E.; Jensen, G.U. Wideband MEMS energy harvester driven by colored noise. *J. Microelectromech. Syst.* **2013**, *22*, 892–900. [CrossRef]

23. Beeby, S.P.; Wang, L.R.; Zhu, D.B.; Weddell, A.S.; Merrett, G.V.; Stark, B.; Szarka, G.; Al-Hashimi, B.M. A comparison of power output from linear and nonlinear kinetic energy harvesters using real vibration data. *Smart Mater. Struct.* **2013**, *22*, 075022. [CrossRef]

24. Green, P.L.; Papatheou, E.; Sims, N.D. Energy harvesting from human motion and bridge vibrations: An evaluation of current nonlinear energy harvesting solutions. *J. Intell. Mater. Syst. Struct.* **2013**, *24*, 1494–1505. [CrossRef]

25. Rantz, R.; Roundy, S. Characterization of Real-world Vibration Sources and Application to Nonlinear Vibration Energy Harvesters. *Energy Harvest. Syst.* **2017**, *4*, 67–76. [CrossRef]

26. Neri, I.; Travasso, F.; Mincigrucci, R.; Vocca, H.; Orfei, F.; Gammaitoni, L. A Real Vibration Database for Kinetic Energy Harvesting Application. *J. Intell. Mater. Syst. Struct.* **2012**, *23*, 2095–2101. [CrossRef]

27. Mallick, D.; Amann, A.; Roy, S. A nonlinear stretching based electromagnetic energy harvester on FR4 for wideband operation. *Smart Mater. Struct.* **2015**, *24*, 015013. [CrossRef]

28. Rantz, R.; Roundy, S. Characterization of Real-World Vibration Sources with a View toward Optimal Energy Harvesting Architectures. In *Industrial and Commercial Applications of Smart Structures Technologies*; SPIE: Bellingham, WA, USA, 2016; pp. 98010P.

29. Vibration Data from REAL VIBRATIONS Database, NiPS Laboratory. Available online: https://realvibrations.nipslab.org/slamstick (accessed on 21 August 2020).

30. Erturk, A.; Inman, D. J. Broadband piezoelectric power generation on high-energy orbits of the bistable Duffing oscillator with electromechanical coupling. *J. Sound Vib.* **2011**, *330*, 2339–2353. [CrossRef]

31. Zhou, S. X.; Cao, J. Y.; Inman, D. J.; Liu, S. S.; Wang, W.; Lin, J. Impact-induced high-energy orbits of nonlinear energy harvesters. *Appl. Phys. Lett.* **2015**, *106*, 093901. [CrossRef]

32. Mallick, D.; Amann, A.; Roy, S. Surfing the high energy output branch of nonlinear energy harvesters. *Phys. Rev. Lett.* **2016**, *117*, 197701. [CrossRef]

33. Lan, C. B.; Tang, L. H.; Qin, W. Y. Obtaining high-energy responses of nonlinear piezoelectric energy harvester by voltage impulse perturbations. *Eur. Phys. J. Appl. Phys.* **2017**, *79*, 20902. [CrossRef]

34. Yan, L.; Lallart, M.; Karami, A. Low-cost orbit jump in nonlinear energy harvesters through energy-efficient stiffness modulation. *Sens. Actuators A* **2019**, *285*, 676–684. [CrossRef]

35. Ferrari, M.; Ferrari, V.; Guizzetti, M.; Ando, B.; Baglio, S.; Trigona, C. Improved energy harvesting from wideband vibrations by nonlinear piezoelectric converters. *Proc. Chem.* **2009**, *1*, 1203–1206. [CrossRef]

36. Daqaq, M. F. Response of uni-modal Duffing-type harvesters to random forced excitations. *J. Sound Vib.* **2010**, *329*, 3621–3631. [CrossRef]

MDPI
St. Alban-Anlage 66
4052 Basel
Switzerland
Tel. +41 61 683 77 34
Fax +41 61 302 89 18
www.mdpi.com

Sensors Editorial Office
E-mail: sensors@mdpi.com
www.mdpi.com/journal/sensors